Property of
(626) 675-5154

Understanding
Uncertainty

WILEY SERIES IN PROBABILITY AND STATISTICS

Established by WALTER A. SHEWHART and SAMUEL S. WILKS

Editors: *David J. Balding, Noel A. C. Cressie, Garrett M. Fitzmaurice, Harvey Goldstein, Iain M. Johnstone, Geert Molenberghs, David W. Scott, Adrian F. M. Smith, Ruey S. Tsay, Sanford Weisberg*
Editors Emeriti: *Vic Barnett, J. Stuart Hunter, Joseph B. Kadane, Jozef L. Teugels*

A complete list of the titles in this series appears at the end of this volume.

UNDERSTANDING UNCERTAINTY

Revised Edition

Dennis V. Lindley
Minehead, Somerset, England

Copyright © 2014 by John Wiley & Sons, Inc. All rights reserved.

Published by John Wiley & Sons, Inc., Hoboken, New Jersey.
Published simultaneously in Canada.

No part of this publication may be reproduced, stored in a retrieval system, or transmitted in any form or by any means, electronic, mechanical, photocopying, recording, scanning, or otherwise, except as permitted under Section 107 or 108 of the 1976 United States Copyright Act, without either the prior written permission of the Publisher, or authorization through payment of the appropriate per-copy fee to the Copyright Clearance Center, Inc., 222 Rosewood Drive, Danvers, MA 01923, (978) 750-8400, fax (978) 750-4470, or on the web at www.copyright.com. Requests to the Publisher for permission should be addressed to the Permissions Department, John Wiley & Sons, Inc., 111 River Street, Hoboken, NJ 07030, (201) 748-6011, fax (201) 748-6008, or online at http://www.wiley.com/go/permission.

Limit of Liability/Disclaimer of Warranty: While the publisher and author have used their best efforts in preparing this book, they make no representations or warranties with respect to the accuracy or completeness of the contents of this book and specifically disclaim any implied warranties of merchantability or fitness for a particular purpose. No warranty may be created or extended by sales representatives or written sales materials. The advice and strategies contained herein may not be suitable for your situation. You should consult with a professional where appropriate. Neither the publisher nor author shall be liable for any loss of profit or any other commercial damages, including but not limited to special, incidental, consequential, or other damages.

For general information on our other products and services or for technical support, please contact our Customer Care Department within the United States at (800) 762-2974, outside the United States at (317) 572-3993 or fax (317) 572-4002.

Wiley also publishes its books in a variety of electronic formats. Some content that appears in print may not be available in electronic formats. For more information about Wiley products, visit our web site at www.wiley.com.

Library of Congress Cataloging-in-Publication Data:

Lindley, D. V. (Dennis Victor), 1923-
  Understanding uncertainty / Dennis V. Lindley, Minehead, Somerset, England.– Revised edition.
    pages cm
  Includes bibliographical references and index.
  ISBN 978-1-118-65012-7 (cloth)
  1. Probabilities. 2. Uncertainty–Mathematics. 3. Decision making–Mathematics.
  4. Mathematical statistics. I. Title.
  QA273.L534 2013
  519.2–dc23

2013013977

Printed in the United States of America

10 9 8 7 6 5 4 3 2 1

Contents

PREFACE XI

PROLOGUE XIII

Chapter 1
UNCERTAINTY   1

1.1. Introduction, 1
1.2. Examples, 2
1.3. Suppression of Uncertainty, 11
1.4. The Removal of Uncertainty, 13
1.5. The Uses of Uncertainty, 15
1.6. The Calculus of Uncertainty, 17
1.7. Beliefs, 18
1.8. Decision Analysis, 20

Chapter 2
STYLISTIC QUESTIONS   23

2.1. Reason, 23
2.2. Unreason, 26
2.3. Facts, 28
2.4. Emotion, 29
2.5. Normative and Descriptive Approaches, 31
2.6. Simplicity, 33
2.7. Mathematics, 35
2.8. Writing, 37
2.9. Mathematics Tutorial, 38

Chapter 3
PROBABILITY   45

3.1.   Measurement, 45
3.2.   Randomness, 48
3.3.   A Standard for Probability, 50
3.4.   Probability, 52
3.5.   Coherence, 54
3.6.   Belief, 56
3.7.   Complementary Event, 58
3.8.   Odds, 60
3.9.   Knowledge Base, 63
3.10. Examples, 66
3.11. Retrospect, 68

Chapter 4
TWO EVENTS   69

4.1.   Two Events, 69
4.2.   Conditional Probability, 72
4.3.   Independence, 75
4.4.   Association, 77
4.5.   Examples, 79
4.6.   Supposition and Fact, 81
4.7.   Seeing and Doing, 82

Chapter 5
THE RULES OF PROBABILITY   85

5.1.   Combinations of Events, 85
5.2.   Addition Rule, 87
5.3.   Multiplication Rule, 89
5.4.   The Basic Rules, 92
5.5.   Examples, 95
5.6.   Extension of the Conversation, 98

5.7. Dutch Books, 101
5.8. Scoring Rules, 103
5.9. Logic Again, 105
5.10. Decision Analysis, 106
5.11. The Prisoners' Dilemma, 107
5.12. The Calculus and Reality, 110
5.13. Closure, 112

Chapter 6
BAYES RULE 113

6.1. Transposed Conditionals, 113
6.2. Learning, 116
6.3. Bayes Rule, 118
6.4. Medical Diagnosis, 119
6.5. Odds Form of Bayes Rule, 123
6.6. Forensic Evidence, 125
6.7. Likelihood Ratio, 127
6.8. Cromwell's Rule, 129
6.9. A Tale of Two Urns, 131
6.10. Ravens, 135
6.11. Diagnosis and Related Matters, 138
6.12. Information, 140

Chapter 7
MEASURING UNCERTAINTY 143

7.1. Classical Form, 143
7.2. Frequency Data, 145
7.3. Exchangeability, 147
7.4. Bernoulli Series, 151
7.5. De Finetti's Result, 152
7.6. Large Numbers, 154
7.7. Belief and Frequency, 157
7.8. Chance, 161

Chapter 8
THREE EVENTS 165

8.1. The Rules of Probability, 165
8.2. Simpson's Paradox, 168
8.3. Source of the Paradox, 170
8.4. Experimentation, 171
8.5. Randomization, 173
8.6. Exchangeability, 176
8.7. Spurious Association, 181
8.8. Independence, 183
8.9. Conclusions, 186

Chapter 9
VARIATION 189

9.1. Variation and Uncertainty, 189
9.2. Binomial Distribution, 191
9.3. Expectation, 195
9.4. Poisson Distribution, 197
9.5. Spread, 201
9.6. Variability as an Experimental Tool, 204
9.7. Probability and Chance, 206
9.8. Pictorial Representation, 208
9.9. Probability Densities, 212
9.10. The Normal Distribution, 213
9.11. Variation as a Natural Phenomenon, 217
9.12. Ellsberg's Paradox, 219

Chapter 10
DECISION ANALYSIS 225

10.1. Beliefs and Actions, 225
10.2. Comparison of Consequences, 227
10.3. Medical Example, 231
10.4. Maximization of Expected Utility, 234

10.5. More on Utility, 236
10.6. Some Complications, 238
10.7. Reason and Emotion, 240
10.8. Numeracy, 242
10.9. Expected Utility, 245
10.10. Decision Trees, 246
10.11. The Art and Science of Decision Analysis, 249
10.12. Further Complications, 252
10.13. Combination of Features, 256
10.14. Legal Applications, 260

Chapter 11
SCIENCE 265

11.1. Scientific Method, 265
11.2. Science and Education, 266
11.3. Data Uncertainty, 268
11.4. Theories, 271
11.5. Uncertainty of a Theory, 276
11.6. The Bayesian Development, 278
11.7. Modification of Theories, 281
11.8. Models, 284
11.9. Hypothesis Testing, 287
11.10. Significance Tests, 291
11.11. Repetition, 293
11.12. Summary, 296

Chapter 12
EXAMPLES 299

12.1. Introduction, 299
12.2. Cards, 300
12.3. The Three Doors, 301
12.4. The Problem of Two Daughters, 305
12.5. Two More Daughters and Cardano, 309
12.6. The Two Envelopes, 313

12.7. Y2K, 316
12.8. UFOs, 317
12.9. Conglomerability, 321
12.10. Efron's Dice, 323

Chapter 13
PROBABILITY ASSESSMENT   327

13.1. Nonrepeatable Events, 327
13.2. Two Events, 329
13.3. Coherence, 333
13.4. Probabilistic Reasoning, 336
13.5. Trickle Down, 337
13.6. Summary, 341

Chapter 14
STATISTICS   343

14.1. Bayesian Statistics, 343
14.2. A Bayesian Example, 346
14.3. Frequency Statistics, 350
14.4. Significance Tests, 355
14.5. Betting, 360
14.6. Finance, 365

EPILOGUE   375

SUBJECT INDEX   383

INDEX OF EXAMPLES   391

INDEX OF NOTATIONS   393

Preface

There are some things that you, the reader of this preface, know to be true, and others that you know to be false; yet, despite this extensive knowledge that you have, there remain many things whose truth or falsity is not known to you. We say that you are *uncertain* about them. You are uncertain, to varying degrees, about everything in the future; much of the past is hidden from you; and there is a lot of the present about which you do not have full information. Uncertainty is everywhere and you cannot escape from it.

Truth and falsity are the subjects of logic, which has a long history going back at least to classical Greece. The object of this book is to tell you about work that has been done in the twentieth century about uncertainty. We now know that uncertainty has to obey three rules and that, once they are understood, uncertainty can be handled with almost as much confidence as ordinary logic. Our aim is to tell you about these rules, to explain to you why they are inevitable, and to help you use them in simple cases. The object is not to make you an expert in uncertainty but merely to equip you with enough skill, so that you can appreciate an uncertain situation sufficiently well to see whether another person, lawyer, politician, scientist, or journalist is talking sense, posing the right questions, and obtaining sound answers. We want you to face up to uncertainty, not hide it away under false concepts, but to understand it and, moreover, to use the recent discoveries so that you can act in the face of uncertainty more sensibly than would have been possible without the skill. This is a book for the layman, for you, for everyone, because all of us are surrounded by uncertainty.

However, there is a difficulty, the rules really need to be written in the language of mathematics and most people have a distaste for mathematics. It would have been possible for the book to have been written entirely in English, or equally in Chinese, but the result would have been cumbersome and, believe me, even harder to understand. The presentation cries out for the use of another language; that of mathematics. For mathematics is essentially another language, rather a queer one, that is unfamiliar to us. However, you do not, for this book, need to understand this language completely; only a small part of it will be required. It is somewhat like an English speaker needing about six characters from Chinese out of the many thousands that the language uses. This book uses part of the language of mathematics, and this part is explained carefully with, I hope, enough motivation for you to be convinced of its advantages. There is almost no technical use of mathematics, and what there is can be appreciated as easily as ordinary arithmetic.

There is one feature of our uncertain world that may either distress or excite you, I hope the latter, in that it does not always behave like common sense might suggest. The most striking example is Simpson's paradox, in Chapter 8, where a medical treatment appears to be bad for both the men and the women, but good for all of us. We will apply the ideas about uncertainty to the law, to science, to economics, and to politics with sometimes surprising results.

The prologue tells something about how this book came to be written. The final version owes a great deal to José Bernardo, Ian Evett, and Tony O'Hagan who read a draft and made many constructive proposals, almost all of which have been eagerly incorporated. In addition, Jay Kadane read the draft with a keen, critical eye, made valuable suggestions, and persuaded me not to ride too vigorously into fields where I had more passion than sense. The final version is much improved as a result of their kind efforts.

PREFACE TO THE REVISED EDITION

The principal change from the original edition is the inclusion of an additional Chapter 14, describing the impact the ideas of this book have on statistics, betting, and finance. The treatment of one problem (§§12.4 and 12.5) has been enlarged because of developments between the two editions. Efron's dice have been discussed because some readers have queried an important assumption. Minor changes have been made in the interests of clarity, several kindly suggested by Mervyn Stone. I would like to thank my daughter, Rowan, for help with the logistics, without which this new edition would not have been possible; and Steve Quigley at Wiley for persuading me to undertake the revision.

Prologue

Almost all my professional life has been spent in academe as a statistician. In my first appointment in Cambridge, I was required to lecture for six hours each week during half of the year and personally to supervise some students. Admittedly, the preparation of new lecture courses took a lot of time, one occupying the whole of the 4 month summer vacation, but these duties did not constitute a reasonable workload. To fill the gap, one was expected to do exactly what I wanted to do, conduct research. As I moved to become Professor and Head of Department, first in Aberystwyth and then at University College London, other duties, principally administrative, crowded in upon me and there was less time for research. But still it got done, because I wanted it to get done, often in conjunction with good, graduate students.

Research, at least in my case, consists of taking questions that interest one and to which you feel you might, given enough time and effort, be able to find an answer; working on them, producing an answer, which often turns out to be quite different from the form originally anticipated, and publishing the results for others to read. There are many aspects to this creative work but the one to be emphasized here is that the questions I chose to answer were selected by me. There was no superior, as there would have been in industry, posing me problems and expecting answers. There was no deadline to be met. This was freedom of thought in its true sense, requiring little more than a comfortable office, a good library, and, most important of all, time in which to think deeply about what interested you. Good answers produce rewards in promotion and more money but that is not the real motivation, which comes instead from the excitement of the chase, to explore where no one has been before, to think deeply, and to come up with something that is genuinely new. And all this free from the interference of others except those you wish to consult. That is true academic freedom that dictators hate so much.

At least during the first 20 years of my researches, I do not recall ever asking myself or being asked by others, whether what I was doing was worthwhile. Society paid me a salary that provided a comfortable living for myself and my family, giving me enough time to think and write, yielding appreciation from the few people who bothered to read my answers. I suppose if someone had asked me to justify my salary, I should have mumbled something about the training in statistics I had given to many students and the value of statistics in society. But nobody did ask and my conscience did not bother me; it was the chase that mattered. Later, however, as I began to sit on committees and come into more contact with life outside the university, I did wonder about the relevance to society of the answers I had given to questions I had chosen and, more widely, about the value of statistical ideas and methods produced by others. When I thought about this, the answers were not terribly encouraging, for admittedly the discovery of the harmful effects of smoking was mostly due to statistical analysis, and statisticians had played an important role in the breeding of new plants and animals, but I had had little to do with these activities and few had attempted to use the answers my research had provided, let alone succeeded. It had been a good life for me, but had it been a worthwhile one from the viewpoint of society?

Research, especially in disciplines that use a lot of mathematics, is a young person's game and after early retirement I did little research but began to read more widely and consider problems that had not seriously entered into my comfortable research world. And I made a discovery. There were people out there, like politicians, journalists, financiers, lawyers, and managers, who were, in my opinion, making mistakes; mistakes that could have been avoided had they known the answers to the questions pondered in my ivory tower. In other words, what I had been doing was not just an exercise in pure thought, but appeared to have repercussions in the world that could affect the activities of many people and ultimately all of us. This is a phenomenon that has been observed repeatedly; namely that if people are given the freedom and opportunity to use their reasoning abilities to explore without any application in mind, what is termed pure research, they often come up with results that are applicable. Ivory towers can yield steel and concrete, produce food and shelter. This book is an attempt to explain in terms that motivated, lay persons can understand, some of the discoveries about uncertainty made in academe, and why they are of importance and value to them, so that they might use the results in their lives. In a sense, it is a justification for a life spent in academe.

The preceding paragraphs are too personal and for clarification it is necessary to say something more about scientific research. Research is carried out by individuals and often the best research is the product of one person thinking deeply on their own. For example, relativity is essentially the result of Einstein's thoughts. Yet, in a sense, the person is irrelevant, for most scientists feel that if he had not discovered relativity, then someone else would; that relativity is somehow "out there" waiting to be revealed, the revelation necessarily being made by human beings but not necessarily by that human being. This may not be true in the arts. For example, if Shakespeare had not written his plays, it would not follow that someone else would have produced equivalent writing. Science is a collective activity, much more so than art, and although some scientists stand out from the rest, the character of science depends to only a very small extent on individuals and what little effect they have disappears over time as their work is absorbed into the work of others. There are two lessons to be learnt from this as far as this book is concerned. First, my contribution to the results described herein is very small and is swamped by the work of others. It is as if I had merely added a brick or two to the whole building. Second, I have not thought it advisable in a book addressed to a general audience to attribute ideas to individuals. Our concern with individual scientists is often misplaced, because it is the collective wisdom that is important. The situation is made worse by the fact that the ideas are often attributed to the wrong individual. The ideas with which this work is usually associated are termed Bayesian, after Thomas Bayes, who had hardly anything to do with them. Generally, there is Stigler's law of eponymy that says that a scientific notion is never attributed to the right person; in particular, the law is not due to Stigler. Some scientists are named in the book because results are universally named after them—Bayes rule, for example, or de Finetti's theorem.

Here is a book about uncertainty, showing how it might be measured and used in your life, especially in decision making and science. It tells the story of great discoveries made in the twentieth century that merit dispersal outside the narrow community where they were developed. New ideas need new forms of exposition, so after a collection, in Chapter 1, of examples of where uncertainty impinges on our lives, Chapter 2 is concerned with certain stylistic questions, including the thorny subject of mathematics; it is only in Chapter 3 that the discoveries really begin.

CHAPTER 1

Uncertainty

1.1 INTRODUCTION

There are some statements that you know to be true, others that you know to be false, but with the majority of statements you do not know whether they are true or false; we say that, for you, these statements are *uncertain*. This book is about understanding uncertainty in this sense, about handling it, and, above all, about helping you to live comfortably with uncertainty so that you can better cope with it in your everyday life.

There are two comments that need to be made immediately. The first arises from the fact that the set of statements that you know to be true differs from my set, for you know things that I do not. Equally, things that are uncertain for you may be known to me; but there is more to it than that, for if we take a statement about which we are both uncertain, you may have more confidence that it is true than I do; we differ in our degrees of uncertainty. The upshot of these considerations is that uncertainty is a personal matter; it is not *the* uncertainty but *your* uncertainty. Admittedly, there are some situations where almost all agree on the uncertainty but these are rare and confined to special scenarios, for example, some aspects of gambling. Statements of uncertainty are personalistic; they belong to the person making

Understanding Uncertainty, Revised Edition. Dennis V. Lindley.
© 2014 John Wiley & Sons, Inc. Published 2014 by John Wiley & Sons, Inc.

them and express a relationship between that person and the real world about which a statement is being made. In particular, they are not objective in the sense that they express a property that is the same for all of us. It follows that throughout this book we will be referring to a person, conveniently called "you", whose uncertainty is being discussed; it may sometimes be appropriate for you, the reader, to interpret it as referring to yourself but generally it applies to some unidentified person, or group of persons expressing a common opinion. You are uncertain about some aspect of the world and that uncertainty does not refer solely to you, or solely to the world, but describes a relationship between you and that world.

The second comment is to note that for any of us, for any "you", the number of statements about which you are uncertain is vastly in excess of the number of statements for which their truth or falsity is known to you; thus all statements about the future are uncertain to some degree. Uncertainty is everywhere, so it is surprising that it is only in the twentieth century that the concept has been systematically studied and, as a result, better understood. Special types of uncertainty, like those arising in gambling, had been investigated earlier but the understanding of the broad notion, applicable to everyday life, is essentially a modern phenomenon. Because uncertainty is everywhere and affects everyone, a proper appreciation of it is vital for all persons, so this book is addressed to everyone who is prepared to listen to a reasoned argument about an ubiquitous concept. This book is for you, whoever you are. We begin with a collection of examples of uncertainty designed to demonstrate how varied, important, and numerous are statements where you genuinely do not know the truth.

1.2 EXAMPLES

EXAMPLE 1. IT WILL RAIN TOMORROW

For all of us who live in climates with changeable weather, this statement is uncertain. It has become almost a classic example of uncertainty because weather is of interest, even importance, to many of us; because meteorologists have seriously studied the question of how to make forecasts like this;

and because it is a statement whose uncertainty will be removed after tomorrow has passed, so that it is possible to check on the quality of the statement, a feature of which meteorologists are very conscious and which will be discussed in §5.12. Notice too, that you can change the degree of your uncertainty about rain by looking out of the window, by consulting a barometer, or by switching on the TV, and we will see in Chapter 6 just how this change may be effected.

A careful discussion here would require clarification of what is meant by "rain"; will a trace suffice, or is at least 0.01 cm in the rain gauge needed before rain can be said to have fallen? Which place is being referred to and where will the gauge be placed? What is meant by "tomorrow"—from midnight to midnight, or 24 hours from 7 A.M., as might be administratively more convenient? In this chapter we deal with illustrative examples and can be casual, but later, when more precision is introduced, these matters will assume some importance, for example, when the skills of meteorologists in predicting the weather are being assessed, or when the quality of mercy in a court of law is described. Again we return to the point in §5.12.

EXAMPLE 2. THE CAPITAL OF LIBERIA IS MONROVIA

The first example, being about the future, is uncertain for everyone living in a variable climate, but with Liberia the personal nature of uncertainty is immediately apparent, as many, but not all of us, are unsure about African politics. Your ignorance could easily be removed by consulting a reference source and, for this reason, such statements, commonly put in the form of a question, are termed almanac questions. The game of Trivial Pursuit is built around statements of this type and exploits the players' uncertainties.

EXAMPLE 3. THE DEFENDANT IS GUILTY

This is uncertainty in a court of law, and "guilt" here refers to what truly happened, not to the subsequent judgment of the court. Although Example 1 referred to the future and Example 2 to the present, this refers to the past. In the two earlier examples, the truth or falsity of the statement will ultimately be revealed; here it will usually remain forever uncertain, though the primary function of the court is, by the provision of evidence, to remove much of that uncertainty with the court's decision. The process of trial in a court of law will be discussed in §§6.6 and 10.14.

EXAMPLE 4. THE ADDITION OF SELENIUM TO YOUR DIET WILL REDUCE YOUR CHANCE OF GETTING CANCER

This is typical of many medical statements of interest today; in another example, selenium may be replaced by vitamin C and cancer by the common cold. Generally a treatment is held to affect a condition. Some medical statements you believe to be true because they are based on a large body of evidence, whereas others you may consider false and just quackery; but most are uncertain for you. They refer to topics that might come within the purview of science, where a scientist might rephrase the example in a less personal way as "selenium prevents cancer". This last statement is a scientific hypothesis, is uncertain, and could be tested in a clinical trial, where the scientist would additionally be uncertain about the number of cancers that the trial will expose. Contrary to much popular belief, science is full of uncertainty and is discussed in Chapter 11. Scientific experiments and the legal trial of Example 3 are both methods for reducing uncertainty.

EXAMPLE 5. THE PRINCES IN THE TOWER WERE MURDERED ON THE ORDERS OF RICHARD III

Richard III was the king of England and mystery surrounds the deaths of two princes in the Tower of London during his reign. Much of what happened in history is uncertain and this statement is typical in that it deals with a specific incident whose truth is not completely known. The arguments to be presented in this book are often thought to be restricted to topics like gambling (Example 7), or perhaps science (Example 4), but not relevant to cultural matters like history, art (Example 6), or the law (Example 3). In fact, they have the potential to apply wherever uncertainty is present, which is everywhere. Admittedly historians are rarely explicit about their doubts but one historian, in accord with the thesis to be developed here, said that his probability, that the above statement about the princes was true, was 98%.

EXAMPLE 6. MANY EIGHTEENTH CENTURY PAINTERS USED LENSES AND MIRRORS

Until recently this was thought unlikely to be true but recent studies have produced evidence that strongly supports the idea. Science and art are not necessarily hostile; aside from optics and paint, they come together in the uncertainty that is present in them both.

EXAMPLE 7. A CARD DRAWN FROM A WELL-SHUFFLED PACK WILL BE AN ACE

This example is typical of those that were discussed in the first systematic studies of uncertainty in the seventeenth century, in connection with gambling, and differs from the previous ones in that the degree of uncertainty has been measured and agreed by almost everyone. Because there are four aces in a pack of 52 cards, the chance of an ace is 4 divided by 52, or 1 in 13. Alternatively expressed, since there is one ace for every 12 cards of other denominations, the odds are 12 to 1 against an ace. ("Odds" and "chance" are here being used informally; their precise meaning will be discussed in §3.8.) It is usual to refer to *the* chance but, once you accept the common value, it becomes *your* chance. Some people associate personal luck with cards, so that for them, their chance may not be 1 in 13.

EXAMPLE 8. THE HORSE, HIGH STREET, WILL WIN THE 2:30 RACE

Horse racing is an activity where the uncertainty is openly recognized and sometimes used to add to the excitement of the race by betting on the outcome. Notice that if High Street is quoted at odds of 12 to 1, so that a stake of 1 dollar will yield 12 if High Street wins, this largely reflects the amount of money placed on the horse, not any individual's uncertainty; certainly not the bookmaker's, who expects to make a profit. Your own odds will help you decide whether or not to bet at 12 to 1. The distinction between betting odds and odds as belief is explored in §3.8. Betting is discussed in §14.5.

EXAMPLE 9. SHARES IN PHARMACEUTICAL COMPANIES WILL RISE OVER THE NEXT MONTH

The buying and selling of stocks and shares are uncertain activities because you do not know whether they will rise or fall in value. In some ways, the stock exchange is like the race course (Example 8), but there is a difference in that the odds are clearly displayed for each horse, whereas the quantitative expression of doubt for the stock can only be inferred from its price now and how it has moved in the past, together with general information about the market. Gambling in the stock market differs from that at the casino (Example 7) because the chances at the latter are generally agreed whereas the existence of buyers and sellers of the same stock at the same time testifies to lack of agreement.

EXAMPLE 10. INFLATION NEXT YEAR WILL BE 3.7%

Statements of this type, with their emphatic "will be", often appear in the media, or even in specialist publications, and are often called either predictions or forecasts (as with the weather, Example 1). They are surely uncertain but the confident nature of the statement tends to disguise this and makes the 3.7% appear firm, whereas everyone, were they to think about it, would realize that 3.8%, or even 4.5%, is a serious possibility. The assertion can be improved by inserting "about" before the figure, but this is still unsatisfactory because it does not indicate how much variation from 3.7% is anticipated. In general, predictions or forecasts should be avoided, because they have an air of spurious precision, and replaced by claims of the form "inflation next year will most likely be between 3.1% and 4.3%", though even here "most likely" is imprecise. Exactly how uncertainty statements about a quantity, here an inflation index, should be made will be discussed in Chapter 9. Many people are reluctant to admit uncertainty, at least explicitly.

EXAMPLE 11. THE PROPORTION OF HIV CASES IN THE POPULATION CURRENTLY EXCEEDS 10%

At first glance this example appears similar to the previous one but notice it is not an assertion about the future but one concerning the present, the uncertainty arising partly because not every member of the population will have been tested. It improves on Example 10 by making a claim about a range of values, above 10%, rather than a single value. People are often surprised by how little we know about the present, yet at the same time, do not want the uncertainty removed because the only method of doing so involves an invasion of privacy, here the testing for HIV. Uncertainty arising from an inability to question the whole population is considered in Chapter 9.

EXAMPLE 12. IF AN ELECTION WERE TO BE HELD TOMORROW, 48% WOULD VOTE DEMOCRAT

There are two main causes for the uncertainty here, both of which are frequently commented upon and thought by many to make polls unsatisfactory. The first is the recognition that in reaching the 48% figure the pollsters only asked very few people, perhaps thousands in a population of millions; the second is caused by people either not telling the truth or changing their views between the question being posed and the action of voting. Methods for handling the first issue have been developed, and the polling firms are among the most sophisticated handlers of uncertainty in the world.

EXAMPLE 13. THERE WILL BE A SERIOUS NUCLEAR ACCIDENT IN BRITAIN NEXT YEAR

The uncertainty here is generally admitted and discussed. Two important features are the extreme seriousness of the statement if true, and the very small chance that it will be true. The balance between these two aspects is not easy to resolve and is of very real concern in a society where people are more comfortable with small risks of moderate chance like road accidents, than with accidents of a nuclear type. Methods are developed to handle this in §5.5.

EXAMPLE 14. JESUS WAS THE SON OF GOD

For at least some Christians, this statement is not uncertain, nor is it for atheists, whereas for agnostics, it is uncertain. It is included here because some people hold that the certainty felt by believers here is different in kind from the certainty they feel about Monrovia being the capital of Liberia (Example 2), at least after the almanac has been consulted, one being based on faith, the other on facts. This is a sensible distinction, for it is unsatisfactory to equate faith with checking an almanac. Nevertheless, some of the ideas to be considered in this book may be relevant to discussions concerning faiths.

Incidentally, it was said in the first sentence of the last paragraph that the statement was "not uncertain". The double negative is deliberate because "certain" is an ambiguous word. It can mean "sure", as would be apt here, but it can also mean "particular". Uncertain does not have this ambiguity, "unsure" being a near synonym.

EXAMPLE 15. THE BRITISH SHOULD REDUCE THE AMOUNT OF SATURATED FAT IN THEIR DIET

This example is similar to that concerning selenium (Example 4) but is expressed in terms of a recommendation and comes with some authority from a government via the Ministry of Health, who also explain the reasoning, claiming it will reduce your chance of death from heart disease. Nevertheless, there is some uncertainty about it if only because people in some parts of France consume more saturated fat than some people elsewhere, yet have a lower rate of death from heart disease. Chapter 10 considers the incorporation of uncertainty into action, where statements like this one about fat can affect one's actions and where other considerations, such as enjoyment of butter, cream, and cheese, need to be balanced against possible health effects.

EXAMPLE 16. THE PLANTING OF GENETICALLY MODIFIED (GM) CROPS WILL DAMAGE THE ENVIRONMENT

Most people consider this statement uncertain, while others are so sure it is true that they are prepared to take action to destroy any GM crops that are planted. Indeed, some will go so far as to destroy those grown to provide information about them and thereby remove, or at least reduce, the uncertainty. Others recognize the value of GM rice in improving the diets of some people in the third world. Issues concerning genetic modification are complex because they can affect both our health and the environment and also have economic consequences. The ideas to be developed in this book are designed to fit uncertainties together and to combine them with our objectives, thus providing some assistance in balancing the many features of an issue to reach an acceptable conclusion. We have first to develop concepts appropriate for a single uncertainty, but our real emphasis has to be on combining uncertainties, and combining them with considerations necessary to implement reasonable actions in the face of uncertainty.

EXAMPLE 17. THE FLIGHT WILL ARRIVE IN LONDON TOMORROW MORNING

This is a typical, uncertain statement about transportation. Whenever we set off on a journey from one place to another, whether on foot, by bicycle, car, bus, train, boat or plane, there is uncertainty about whether we shall reach our destination without mishap and on time, so that it becomes important to compare uncertainties. It is sometimes said that travel by air is the safest form of transport, which is true if the measurement is by number of fatal accidents per thousand miles; unfortunately aviation accidents mostly occur at the start or finish of the journey, so are concentrated into relatively short periods of time. Takeoff is optional; landing is compulsory. What are needed are sensible ways of measuring and comparing uncertainties, and this is what we try to provide in this book. People repeatedly find it hard to compare one risk with another, so that there is need for a way of assessing risks that will help us understand how the risk of car travel compares with that of planes: how the risk from Alzheimer's disease compares with that from serious indulgence in sporting activities. To achieve this it is necessary to measure uncertainty.

EXAMPLE 18. MRS. ANDERSON WAS ANASTASIA, DAUGHTER OF THE LAST TSAR OF RUSSIA

Mrs. Anderson was thought by some to be the daughter who others thought had been killed in the revolution. This historical statement was, until recently, uncertain, yet of so much interest that several books and a film were devoted to the mystery. A few years ago I made a study of the available evidence that led me to think that the statement was probably true, largely because Mrs. Anderson knew things that it was unlikely anyone but the Princess would have been expected to know. Later DNA evidence has virtually removed the uncertainty, demonstrating not merely that she was not the Princess, but establishing exactly who she was. The mystery having been destroyed, people have lost interest in Anastasia, demonstrating that uncertainty can sometimes be enjoyed.

EXAMPLE 19. THE SUN WILL RISE TOMORROW AT THE TIME STATED

Technically this statement is uncertain for you, because it is possible that some disturbance will affect our solar system; yet that possibility is so remote that it is sensible for you to act as if you knew it to be true. We shall have occasion later to return to the topic of statements that you believe to be true without totally firm evidence. A relation of mine was sure of her age but when, in her 50s, she needed a passport for the first time in her life and, as a result, needed to get her birth certificate to establish her citizenship, she was astounded to find she was a year younger than she had thought. Statements of pure logic, like $2 \times 2 = 4$, are true, but little else has the solidity of logic.

EXAMPLE 20. THE SKULL IS 7 MILLION YEARS OLD AND IS THAT OF A HOMINID

Even for palaeontologists, this is uncertain and there are different opinions that arise, not because people can be quarrelsome, but because there are understandable difficulties in fitting the pieces of fossil evidence together. In the early stages of a study, even when conducted using sound, scientific principles, there is, as discussed in Chapter 11, a lot of uncertainty. One aspect has been discussed statistically, namely the assignment of dates, so that a respectable body of evidence now exists for which the uncertainty has been, if not removed, at least lessened.

1.3 SUPPRESSION OF UNCERTAINTY

The long list of examples demonstrates how common is the phenomenon of uncertainty. Everything about the future is uncertain, as is most of the past; even the present contains a lot of uncertainty, due to your ignorance, and uncertainty is everywhere about you. Often the uncertainty does not matter and you will be able to proceed as if tomorrow will be just like today, where the sun will rise, the car will start, the food will not be poisoned, the boss will be her usual self. Without this certainty, without this assurance of continuity, life as we know it would be impossible. Nevertheless, we all encounter situations where you have to take cognizance of uncertainty and where decisions have to be made without full knowledge of the facts, as in accepting a job offer or buying a new house, or even on deciding whether to have a picnic.

Despite uncertainty being all about us, its presence is often denied. In Britain, though not in the United States, the weather forecast will state categorically that "it will rain" (Example 1) and then sometimes look foolish when it does not. Economists will predict the rate of inflation (Example 10) and then get it wrong, though because the time scale is different from the meteorologist's, we sometimes do not notice the error. This is slightly unfair because, as mentioned in the example, economists are mending their ways and quoting intervals, thereby recognizing the uncertainty. Newspapers can report an HIV rate (Example 11) as if it were true, or cite the numbers at a demonstration as fact even though the police and participants differ. Television executives hang desperately onto audience ratings, largely ignoring the errors present. People in the humanities rarely mention uncertainty (Example 5). Even the best historians, who are meticulous with their sources, can blur the borderline between facts and opinions. Lawyers (Example 3) do admit uncertainty and use language like "beyond reasonable doubt" or "the balance of probabilities"; nevertheless, at the end of the trial the jury has to ignore the uncertainty and pronounce the defendant "guilty" or not. Politicians are among the worst examples of people who deny any uncertainty, distorting the true scenario to make their view appear correct. There are places like the casino (Example 7) or the race course (Example 8) where the uncertainty is openly admitted and exploited to add to the excitement.

One reason for the suppression is clear: People do not like to be unsure and instead prefer to have everything sharply defined. They like to be told emphatically that the sun will shine, rather than to hear that there might be the chance shower to spoil the picnic, so they embrace the false confidence of some weather forecasts, though they are annoyed when the forecast is incorrect. But if some uncertainty is present, and we have seen that uncertainty is almost everywhere, it is usually better to face up to it and include it in your thoughts and actions, rather than suppress it. Recognition of the uncertainty in investing in stocks, or taking out a pension contract, is valuable because it helps to guard against things going wrong. Suppression of uncertainty can cause trouble, as the law has found when it claims to have removed the uncertainty by the jury announcing a verdict of guilty. To go to appeal or have a case reviewed can be difficult, partly because no one likes to admit they were wrong, but partly because the uncertainty lay unrecognized. Scientists, who are more open about uncertainty than most, still cling to their beloved theories and have trouble in accepting the maverick worker, partly because they are reluctant to entertain uncertainty. There is a clear and beautiful example of the misplaced dislike of uncertainty in the Ellsberg paradox discussed in §9.11.

Part of the thesis of this book is that, instead of neglecting or, worse still, suppressing uncertainty, it is better to recognize its presence everywhere, bringing it out into the open and discussing the concept. Previously this has not been done, partly because it is no use exposing something if, when you have done so, you do not know how to handle it, like opening a Pandora's box of misery. The past and present neglect and suppression therefore have sense behind them, but recently a change has taken place and the purpose of this book is to tell you about it. What has changed is that we now know how to handle uncertainty; we know what the rules are in Pandora's box. Beginning with the study of uncertainty in games of chance, the net has widened to the appreciation that the simple rules discovered there, and they are truly simple, just controlled addition and multiplication, apply beyond gambling to every uncertain situation, so that you can handle beliefs nearly as assuredly as facts. Early sailors had difficulty going out of the sight of land but when the rules of navigation became better understood, with the use of the stars and accurate clocks, voyages across oceans became practicable. Today we travel the seas, the air, and even space, because of our understanding of the rules; so I contend that now the rules of uncertainty have been understood, we no longer need to neglect or suppress it but can live comfortably even when we do not know.

1.4 THE REMOVAL OF UNCERTAINTY

If uncertainty is such a common feature of our lives, and yet we do not like it, the obvious thing to do is to remove it. In the case of the capital of Liberia (Example 2), this is easily done; one just goes to an almanac and checks that indeed Monrovia is the capital, though it would be as well to bear in mind that the almanac may be out of date or even wrong, or that an error can be made in consulting it, so that some uncertainty remains, but at least the uncertainty will be lessened. The removal of uncertainty is not usually as easy as it is with almanac questions. The court of law is a place where a serious attempt is made to reduce, if not remove, uncertainty. Some places use an adversarial approach, which allows both sides to present facts that they think are relevant, in the hope that the jury will feel convinced one way or the other about the defendant's guilt. Both these examples show that the usual way to remove or reduce uncertainty is by the production of facts; these are statements that are essentially free of uncertainty, like the almanac, or are much more likely to be accepted as true than the original statement. A major task of this book is to show exactly how this reduction takes place. The legal process is considered in §10.14.

The adversarial method is not the only way to obtain and process facts. Scientists collect data and perform experiments, which are assembled to infer general rules that are often deterministic and involve little uncertainty, like Newton's laws of motion. Careful measurements of the motions of the heavenly bodies led eventually to accurate calculation of their orbits so that, for example, an eclipse ceased to be uncertain but could be predicted with great accuracy. Scientific facts differ from legal facts in that they are repeatable, whereas legal evidence is not. If a scientist reports the results of an experiment, then it is an essential feature of the scientific method that other scientists be able to repeat the experiment and obtain the same result, whereas the witness's statement that he was with the defendant at the time of the crime is not capable of repetition. The repeatability aspect of science, with its consequent removal of almost all uncertainty, often leads people to think that all science is objective, as it virtually is after there has been a lot of confirmatory repetition, but active science is full of uncertainty, as healthy disagreement between scientists testifies. Science is discussed in Chapter 11.

One of our examples (Example 14) differs in style from the rest in that the agnostic's uncertainty about Jesus being the son of God is difficult to change since no further facts about Jesus are likely to be obtained. The most plausible way to change is to accept the statement as an article of faith, essentially removing the uncertainty altogether. This would ordinarily be done in connection with other features of the faith, rather than by facts. This is not to say religions do not themselves change in response to facts. The Catholic Church moved from thinking of the Earth as the center of our part of the universe, to a view that centered on the Sun; this in response to astronomical data.

Whether the ideas presented in this book, and especially the three basic rules, apply to faiths is debatable. The wisest advice is perhaps that offered by Oliver Cromwell to the Church of Scotland, "believe it possible you may be mistaken". Acceptance of this advice would lessen tensions between different faiths. Cromwell's rule for probability is discussed in §6.8.

1.5 THE USES OF UNCERTAINTY

So far the emphasis has been on our dislike of uncertainty and methods taken to avoid the phenomenon, yet there are situations in which you actually enjoy the uncertainty and without it life would be duller. Examples are provided by mysteries where you do not know the solution, as with Mrs. Anderson in Example 18; once the mystery has been cleared up, the story loses its interest. A difference between a puzzle and, say, uncertainty about your health lies in the fact that the consequences that could flow from the removal of the uncertainty are not experienced by you in the first case, but will be in the second. Once you know she was not Anastasia, you shrug your shoulders and pass onto the next puzzle; once you are diagnosed as having cancer you have to live with the unpleasantness. So perhaps it is not that we dislike uncertainty; rather we are concerned about possible outcomes. Perhaps it is not the uncertainty about the rain (Example 1) that concerns us but rather the thought of the spoiled picnic.

Yet this cannot be the whole story, as there are uncertainties that many of us enjoy, where we do have to experience the results, some of which may, if we overindulge, be most unpleasant. The obvious ones are gambling with cards (Example 7) or betting on the horses (Example 8). Here we can, and often do, lose our money, yet we gamble because of the excitement found in the activity. Our study will reveal how this enjoyment, quite apart from monetary considerations, can be combined with the rules mentioned earlier to provide a reasoned account of gambling.

Here is a serious example of the benefits of uncertainty. In Chapter 8 we shall discuss clinical trials, that is, experiments in which patients are given a treatment or a drug to investigate whether it improves their health. In order to assess the drug's effectiveness, it is necessary to take other, similar patients and give them a placebo, something that is outwardly like the drug but in fact contains only some innocuous material. Comparing the changes in the patients on the drug with those receiving the placebo, it is possible to measure the value of the drug. In order that the conclusions from a trial be reliable, it has to be conducted with care and one precaution is to ensure that the patients do not know whether they are receiving the drug or the placebo. To anticipate a term to be introduced in §3.2, the patients on the drug are selected at random from a pool of patients, so that every participant in the trial is uncertain about what they are taking. It is also desirable to ensure that the clinician conducting the trial is equally uncertain, as we shall see when discussing Simpson's paradox in §8.2. Many experiments today actively encourage an element of uncertainty, by selecting at random, in order to make the results more reliable than they would be were it not present.

There is another merit of uncertainty that appears whenever a competitive element is present, as in sport or the conduct of war. If you are competing against an opponent, then it is to your advantage to increase their uncertainty, for example, by creating the impression that you are about to do one thing when you intend to do another. There will be little in this book about the bluffing aspect of uncertainty because we are concerned with a single person, the "you" of the language introduced in §1.7, and there are real difficulties in extending the calculus to two "yous" that are in competition. A famous, simple example of this is the prisoner's dilemma, mentioned in §5.11. We develop a calculus for "you"; there does not exist an entirely satisfactory calculus for two or more competitors and, in my view, this omission presents a serious, unsolved problem.

Notice that in the competitive situation it is not so much that you want your opponent to be uncertain, or even wrong, but that you want to have information that they do not have. You know when you are going to attack, they do not. It is your information that matters, information to be kept from them. Information is power, which is why politicians, when in power, hate the open government that they espoused when in opposition. One of our principal tasks will be to see how information can be used to your advantage. The concept of information within the calculus is treated in §6.12.

1.6 THE CALCULUS OF UNCERTAINTY

In this book uncertainty is recognized and accepted as an important part of our lives. No attempt is made to disguise or deny it; rather it is brought out into the open and we learn to handle it as confidently as we do those features about which we are sure. We learn to calculate with uncertainty, much as a card-player calculates the situations in a game of bridge. Indeed, the rules of calculation are essentially those that operate in cards or roulette.

In most circumstances that operate in cards, more than one feature is uncertain and the various uncertainties need to be combined. Similarly, a juror hearing witnesses will be uncertain about their veracity and need to meld it with the doubts concerning the defendant's guilt. A scientist performing an experiment may be uncertain about the pressure used, the purity of the material, as well as about the theory under investigation. In reacting to the offer of a job, you will be uncertain about the move involved, the nature of the work, and many other features. A doctor will need to combine appreciation of the uncertain symptoms in order to reach an overall diagnosis. In every one of these cases, many uncertainties have to be amalgamated to produce the overall judgment, so that a central task is for us to see how to put several uncertainties together.

There are things that combine very easily: numbers. Addition and multiplication are so easy that even a computer can perform them, a computer being only as wise as its programmer. One day we may have artificial intelligence but today most computers can only perform the logic they have been taught. If then, we could measure uncertainty, in the sense of attaching numbers to the statements, just as we did above with the ace drawn from the pack of cards, then the combination would present fewer difficulties and involve only the rules of arithmetic. This will be done; we will measure uncertainty, and then develop the three wonderful rules of combination. It is in the appreciation of the rules, and the ability to use them, that the strength of this book resides. We shall calculate with uncertainties and the machinery to do this is called the calculus of uncertainty.

Scientists already use statistical methods, developed from these rules, to help them interpret their data. It will be sometime before jurors have their computer with them to assess the uncertain guilt, but the beginning of the idea can be seen in the treatment of forensic science in §6.6. One day the historian will calculate the odds against Richard III being the culprit (Example 5) rather than plucking a number out of the air as the historian quoted might have done.

It is an unfortunate fact of life that many people, especially those working in the arts or the media, have a strong dislike of numbers and are unhappy using them. Although there is likely to be genuine variation in the ease with which numbers are handled, my personal belief is that almost all can be taught to manipulate with figures and, just as important, appreciate the power that such a facility can bring. Here we shall calculate but I have tried to expound the mechanics in a simple manner. All that I ask is a willingness on the reader's part to cooperate by showing some motivation to learn, genuinely to want to understand uncertainty.

1.7 BELIEFS

We have seen that uncertainty involves a statement, whose truth is contemplated by a person. It is now convenient to introduce the standard language that is used in the calculus of uncertainty. Instead of "statement", we refer to an "event", thus the event of rain tomorrow or the event of selenium affecting cancer. Sometimes "event" will seem a strange nomenclature, as when referring to the event that Monrovia is the capital of Liberia, but it is usually apt and experience has shown that it is useful as a standard term. Thus an event is uncertain for you if you do not know whether it is true or not.

We also need to have a term for the person facing the uncertainty for, as we have seen, one person's uncertainty can be different from another's. As already mentioned, the term "you" will be used and we will talk about your uncertainty for the event. In many cases you, the reader, can think of it as a reference to yourself, while in others it may be better to think of someone else.

A term is needed to describe what it is that you feel about the event. The phrase usually employed is "degree of belief"; and we will talk about your degree of belief in the truth of the event, so that you have the highest belief when you think it is true, and least when false. Belief is a useful word because it does emphasize that the uncertainty we are talking about is a relationship between you, on the one hand, and an event, on the other. Belief does not reside entirely with you because it refers to the world external to you. Belief is not a property of that world because your degree of belief may reasonably be different from mine. Rather belief expresses a relationship between you and the world, in particular between you and an event in that world. The word that will be used to measure the strength of your belief is probability, so that we talk about your probability that an event is true, or more succinctly, your probability for the event. One of the greatest experts on probability, having written a two-volume work on the topic, calling it simply *Theory of Probability*, wanted an aphorism to include in his preface that would encapsulate the basic concept expressed therein. He chose:

Probability does not exist.

It was intended to shock, for having written 675 pages on a topic, it did not seem sensible to say the topic did not exist. But having brought it to your attention by the shock, its meaning becomes apparent; probability does not exist as a property of the world in the way that distance does, for distance between two points, properly measured, is the same for all of us, it is objective, whereas probability depends on the person looking at the world, on you, as well as on the event, that aspect of the world under consideration. Throughout this book we will refer to *your* probability, though the use of *the* probability is so common in the literature that I may have slipped into the false usage unintentionally.

Our task in this book is to measure beliefs through probability, to see how they combine and how they change with new information. This book is therefore about your beliefs in events. It is not about what those beliefs should be, instead it is solely about how those beliefs should be organized; how they need to relate, one to another. An analogy will prove useful, provided it is recognized that it is only an analogy and cannot prove anything but is merely suggestive. Suppose that this was a book about geometry, then it would contain results about the shapes of figures, for example, that the angles of a plane triangle add to 180 degrees, but it would not tell you what the angles have to be. In fact they can be anything, provided they are positive and add to 180 degrees. It is the same with the beliefs described here, where there will be results, analogous to the sum of the angles of a triangle being 180 degrees, that provide rules that beliefs must obey. We shall say little about what the individual beliefs might be, just as little is said about the individual angles. If you have high belief that the Earth is flat, then there is nothing in our rules to say you are wrong, merely that you are unusual, just as a triangle with one angle only a fraction of a degree is unusual. We claim that the rules provided are universal and should not be broken, but that they can incorporate a wide range of disparate opinions.

Before writing these words, I had heard an argument on the radio between a representative of a multinational corporation and another from an environmental organization. The arguments presented in this book have little to say about who is correct but they have a lot to say about whether either of the participants had organized their beliefs sensibly. It is my hope that correct organization, combined with additional information, will help in bringing the speakers together.

1.8 DECISION ANALYSIS

We all have beliefs and in this book we try to show how they should be organized, but not what they should be. There is, however, a basic question that we need to answer:

What is the point of having beliefs and why should we organize our opinions?

The answer is that we have beliefs in order to use them to improve the way in which we run our lives. If you believe that it will rain tomorrow, you will act on this and not go on with the picnic, but go for an indoor entertainment instead. Action is not essential for beliefs and most of us will not be influenced in our actions by our beliefs concerning the Princes in the Tower (Example 5), but if action is contemplated, as with the picnic, then our beliefs should be capable of being used to decide what the action should be.

This attitude toward beliefs is pragmatic in the sense that it assesses them by how they perform as a guide to action, and it leads from the sole consideration of your attitude toward an uncertain world, to how you are to behave in that world. Some hold that belief is inseparable from action, while we prefer to develop the calculus of belief first, and then extend it to embrace action. The relationship here is asymmetric: actions require beliefs, but beliefs do not necessitate action.

The topic that deals with the use of beliefs in action is called "decision analysis", and it analyzes how you might decide between different courses of action, without saying what the decisions should be, only how they should be organized. The passage from belief to action will introduce a new concept that needs to be blended with the beliefs in order to produce a recommended action. Example 13 supplies an illustration, where the seriousness of the nuclear accident needs to be blended with the small belief that it will happen, in order to decide whether to build more nuclear power plants. The subject is covered in Chapter 10.

In summary, this book is about your approach to uncertainty, how your beliefs should be organized, and how they need to be used in deciding what to do. Before we embark on the program, it is necessary to comment on the method used to tackle these problems. These commentaries form the content of the next chapter and only in Chapter 3 will the development proper begin.

CHAPTER 2

Stylistic Questions

2.1 REASON

The approach adopted, at least at the beginning of this book, is based firmly on reason, the wonderful facility that human beings possess, enabling them to comprehend and manipulate the world about them; and only later will emotional and spiritual aspects of uncertainty be considered. "Reason centers attention on the faculty for order, sense, and rationality in thought" says Webster's dictionary, going on to note that "reason is logic; its principle is consistency: it requires that conclusions shall contain nothing not already given in their premises". A contrasting concept is emotion "the argument which is not an argument, but an appeal to the emotions".

The program that will be adopted is to state some properties of uncertainty that seem simple and obvious, the premises mentioned in the second quotation above, and from them to deduce by reasoning other, more complicated properties that can be usefully applied. As an example of a premise, suppose you think it is more likely to rain tomorrow than that your train today will be late; also that the latter

---

Understanding Uncertainty, Revised Edition. Dennis V. Lindley.
© 2014 John Wiley & Sons, Inc. Published 2014 by John Wiley & Sons, Inc.

event is more likely than that your car will break down on traveling to the railway station; then it is necessary that you think rain is more likely than the breakdown. The references to rain, trains, and accidents are not important; the essential concept is contained in an abstraction. Recalling our use of "you", "event", and "belief" as described in § 1.7, the premise is that if you have stronger belief in event A than in event B; and, at the same time, stronger belief in event B than in event C, then necessarily you have stronger belief in A than in C, the exact meanings of A, B, and C being irrelevant. Starting from abstract premises like this, pure reasoning in the form of logic will be used to deduce other properties of uncertainty that can then be applied to concrete situations to give useful results. Thus, abstract A becomes "rain", B refers to the train, and C refers to the breakdown. This premise is discussed in some detail in §12.10.

There are two points to be made about the premises. Firstly, they are intended to be elementary, straightforward, and obvious, so that no justification is needed and, after reasonable reflection, you will be able to accept them. Secondly, they should be judged in conjunction with the results that flow from them by pure reasoning. It is the package of premises and results that counts, more than the individual items, for if one of the premises is false, then all the consequences are suspect. If you, the reader, find one of the premises unacceptable, as you might that given above, then I would ask you to bear with it and follow through the argument to see where reason takes you; and only then to reach a final judgment. I know of no conclusion that follows by pure reason from the premises adopted here, which appears unsound. Although we shall meet conclusions that at first surprise, further reflection suggests that they are correct and that our common sense is faulty. Indeed, one of the merits of our approach is that it does produce results that conflict with common sense and yet, on careful consideration, are seen to be sound. In other words, it is possible to improve on common sense. The whole package will be termed a calculus, a method of calculating with beliefs.

There is an additional reason for thinking that the conclusions are sound, which rests on the fact that different sets of premises lead to the same conclusions. For example, the premise cited above can be avoided and replaced by another that some find more acceptable, without altering the whole structure. Though only one line of argument will be used in this book, mention will be made of other approaches, the important result being that all lead to the same calculus. It is like several people starting out from different places but finding that all roads lead to Rome. The metaphor is a happy one since one of the leaders in developing a proper understanding of uncertainty, Bruno de Finetti, was a professor in Rome and stood in an election there. Other writers have used premises that do not lead to Rome, while others have dispensed with premises and suggested a calculus that differs from ours. Some of these will be considered from §5.7 onward, but for the moment I ask you to go along, at least temporarily, with the premises and the logic, to see where they lead and how you feel about the construction as a whole. Remember that Newton's premises, his laws of motion, might appear to be abstract, but when they enable the time of an eclipse at a site to be predicted years in advance, they become real.

People are often very good at raising objections to even simple, direct statements. This is no doubt, on occasions, a useful ability, but objections alone are worthless; they must be accompanied by constructive ideas, for otherwise we are left with the miasma that uncertainty presents to us. For many years, I, and many others, had used a premise that appeared eminently sensible and led to apparently excellent results, only to have three colleagues come along with a demonstration that the premise led to an unacceptable conclusion but, at the same time, they showed how a change in the premise avoided the unsound result. This was good, constructive criticism. Our psychology makes us reluctant to admit errors, especially when the errors destroy some of our cherished results, but it has to be done and the amended results are strengthened by my colleagues' perspicacity. So if you think one of the premises used in this book is unsound, be constructive and not merely destructive.

The role of reasoning in appreciating uncertainty has been emphasized because reasoning does not play an important role in some books, so that ours will appear different in some regards from others. To appreciate some of the lines of argument taken here, let us look at the lack of reason in other places.

2.2 UNREASON

Literature

Reasoning, quite sensibly, plays but a small role in literature. Some literature has the straightforward aim of telling a tale, of entertaining, and save for detective novels, few make a pretence of reasoning. Other literature tries, often successfully, to develop insights into the way people and society behave and, to use a term that will occur later, are essentially descriptive. Because people, either individually or collectively, do not use much reasoning, so neither does the description. For example, there is little reasoning in Othello's behavior as he lets his emotions reign with disastrous results. No criticism of Shakespeare is implied here for he does provide us with insights into the workings of the human mind.

Advertising

Whatever reasoning goes on in advertising agencies (and much of it must be good to judge from the effectiveness of the results), the final product is lacking in reason. An advertisement for beer will develop a macho image or a catchy phrase but will fail to mention the way the product is made or the effects that over-consumption might have. The advertisements for lotteries concentrate on the jackpot and fail to mention either the tax element or the profits, let alone the odds. The barrage of advertising that surrounds us does not encourage the faculty of reason; indeed, much of it is deliberately designed to suppress reason, as in the encouragement we receive to eat junk food. Many advertisements persuade us to buy the product, not by reasoning about its qualities but by associating it with an image that we regard favorably. Thus a car that might be attractive to a man has a beautiful woman in the advertisement but makes no mention of its cost. This method of inveigling you into a purchase is unfortunate but a more serious consequence of the continual repetition of this form of persuasion may cause you to abandon reason generally. For instance, you may be led to vote for one party in an election, in preference to another, because its image seemed more attractive, rather than because its policies were better. Spin overcomes substance and bad thinking drives out the good. It is sensible to claim that some advertising makes a contribution to the ills of society, by driving out logical approaches and thereby increasing the possibilities for serious errors.

Politics

In a democratic society with opposing parties, there is an element of conflict because the parties use different premises and the reasoning that flows from them, though these features are often not spelt out honestly. In their simplest form, seen in Europe, these are the premises of capitalism, with its emphasis on the individual: and in opposition, those of socialism with social considerations to the fore. The effect of the existence of at least two sets of arguments means that much of the political process consists in one party trying to convince the other that it is wrong, conviction gets involved with emotion so that the discussion becomes emotional and reason is displaced. This is in addition to the element of conflict mentioned in §1.5. The lack of reasoning is more recently emphasized by the use of spin.

Law

Good law is good reasoning but, in court, where the adversary system is used, emotion sometimes replaces reason. A lawyer, needing to show that the conclusions of this book, as applied to forensic science, were unsound and being unable to do so, resorted to defaming the scientist by referring to the more disreputable aspects of gambling, thereby using emotions to overcome the lack of reason.

Television

Most television programs are for entertainment and cannot be expected to deal with reason. But there are "serious" programs, such as those devoted to science, where reason, which is at the basis of scientific thinking, might be expected to be present, though sometimes it is not. The dominant view is that science must be presented as entertainment, the screen must be full of pretty images, and the scene must shift with great frequency lest the viewer becomes bored; graphs of considerable ingenuity, and in bright colors, are presented without any hint as to what the axes are. This travesty of science arises because the programs are being viewed as entertainment and are primarily developed by entertainers who are not familiar with the scientific mode of thought. Of course science needs to be presented in an interesting way, but the entertainment level should always be subservient to the reasoning.

In Western societies today, and certainly those of Britain and the United States with which I am familiar, there is a tendency to disparage reason and place an emphasis on emotions, as we have seen in literature and advertising. One reason for this is the lack of balance between what C.P. Snow called the two cultures, of the arts and science, one predominately emotional, the other mainly logical. Both cultures are valuable and there is no suggestion that one is right, the other wrong, but rather that the balance has shifted too far toward emotional appeals. We will return to this point in §2.8, but in the meantime, I would ask readers to be prepared for a surfeit of reason when they have been used to one of emotion.

2.3 FACTS

Although this book is about your not knowing the truth about events, there are some events that you do know to be true, or would accept as true were you to have the information. You know that the capital of Liberia is Monrovia, or will know when you have consulted the almanac. You know your age, though recall my relative in §1.2; you know that the Sun is 93 million miles from the Earth, on average; you may know that Denmark voted to join the European Union. Such events will be termed facts. While philosophers will sensibly debate exactly what are facts, some suggesting only logical truths, like two plus two equals four, most of us will recognize a fact when we encounter one. It is not surprising that in talking about uncertainty we should lean heavily on facts, just as the court of law does when interrogating witnesses. Facts form a sort of bedrock on which we can build the shifting sands of uncertainty.

Yet many people do not like facts. This is especially true when the facts go counter to the opinions they hold. There are many examples of governments that have tried to suppress facts that speak against their policies. There is an old adage: "do not bother me with facts, my mind is made up". We have all experienced discomfort when we have to admit as true a fact that conflicts with our opinions, yet would embrace a truth that supports it. A newspaper editor had the right idea when he remarked that "facts are sacred, comment is free". It is this view that will be adopted here and we will study how facts ought to be used to influence our beliefs, which are free.

The development here will be based firmly on reason and much of the reasoning will be with facts, because it is the observation of facts that is a major feature in changing your beliefs about events that are uncertain for you and therefore not facts. Thus, you are uncertain about rain tomorrow, so before retiring you look out of the window and see storm clouds gathering, a fact that changes your belief about tomorrow's rain.

2.4 EMOTION

Although this book will be firmly based on reason and facts, an approach that consists entirely in logical development from premises and the incorporation of facts will be boring and, worse still, irrelevant to a real world that has a richness that owes much to other things beyond reason. Consider these two features, boredom and irrelevance.

It need not be true that reasoning is boring, for if it is accompanied by illustrations that interest the reader, then the ideas can leap into life and have a reality that reason alone lacks. I cannot say how well this has been done here, but concepts like Simpson's paradox in §8.2 seem to most people to be full of interest and lead to a valuable understanding of how some apparently sensible conclusions from facts can be erroneous.

The charge of irrelevance is more serious and to treat beliefs and decision making without reference to emotion would mean that the ideas would be irrelevant to a world that is rightly full of emotion, and would regard us as nothing more than calculators. Fortunately the charge of irrelevance can easily be rebutted, for the path of pure reason leads to a surprise.

There will come a point in the development of the calculus of uncertainty (§10.7) where pure reason will insist that something new has to be introduced and where beliefs alone call out for extra ideas. An additional concept has to be included, a concept to which we are led by pure reason. When we try to interpret this new idea, we see that it deals with emotion. At the funeral service for Princess Diana in Westminster Abbey, there was some music by Verdi, followed by some by Elton John, as a result of which I felt uplifted by the magnificent singing in the glorious building, only to have it shattered by the sound produced by a pop star. It was tempting to say that I believe Verdi is better than John, but this is not true in the same sense that I believe it will rain tomorrow, for the rainfall can be measured, whereas there is no clear, impersonal meaning to Verdi's offering being better than John's. No, my preference for Verdi over John is just that, a preference, and my feelings expressed themselves with emotion, not reason. So it is with our reasoned development about uncertainty. The calculations we need to do cry out for preferences, in addition to beliefs, that depend on emotion, not on reason alone. It would be good to establish a reasoned relationship between Verdi and John (it sounds better if the Italian's name is translated and we use John's real name, and contrast Joe Green with Reggie Dwight), but it cannot be done and I must be content with my preferences and let Dwight ruin the occasion, just as Green may have spoilt it for others.

What the reasoned approach reveals is that emotional considerations must be considered and that, just as we measure belief, so we need to measure our preferences. Emotion is included, not because we feel it desirable, but because reason demands it. The motive for measuring emotional preferences is exactly the same as that advanced for the measurement of belief in §1.6 and enlarged upon later in §3.1. Your beliefs will be measured by probabilities; your preferences by the rather unemotional word, utility, so that my utility for Verdi exceeds that for John, both concepts being personal. We shall not abandon that element of life that provides so much interest but incorporate it into our reason; indeed, incorporate it in a way that makes the two fit together like pieces in a good jigsaw—sometimes so well that they cannot easily be separated. One example has already been met in §1.5 when it was pointed out that when gambling, people take into account the excitement of the gamble in addition to the monetary experiences, here expressed by saying that their utility depends on cash and thrill. Utility is the emotion pleading to be let into the house of pure reason and thereby enriching it.

2.5 NORMATIVE AND DESCRIPTIVE APPROACHES

The claim is therefore made that the approach to uncertainty developed here incorporates many aspects of human endeavor; and this despite the strict adherence to reason. It enables you to both incorporate your beliefs and include your emotional and spiritual preferences. It does not tell you what to believe nor what to enjoy but merely says how you should organize your beliefs and preferences in a reasoned way, leading to a reasoned action. It is for all: for atheist and believer, for manager and hedonist, for introvert and extrovert. It is for everyone. Yet there is something that it is not.

In society people believe and act in ways that have been recorded in literature and studied by psychologists and sociologists. What emerges from these studies is a *description* of the way people act and believe. Literature is mainly a description as when Shakespeare describes what Othello did and thought in reaction to Iago. Psychologists describe peoples' actions through observations and experiments, proceeding to explain the results in general terms. Advertising agencies have exploited the way people behave to present their products. All these approaches start from the observation of people and how they behave in reaction to circumstances, some behavior seeming, on reflection, to be sensible; others, in contrast, to be perverse or even stupid.

The approach here is somewhat the reverse, in that we begin by considering what is sensible in very straightforward circumstances, the premises already referred to, and then use reason to extend simple sense to more complicated scenarios. We use reason to provide what is termed a *normative* or *prescriptive* approach, where the methods of organizing beliefs provide a norm or a prescription against which the descriptive material can be contrasted. If, in this contrast, of prescriptive and descriptive, the normative view is found to be wanting, then it must be abandoned. I know of no case where the normative view can unequivocally be held to be poor in comparison with what happens in the description of reality. Often one needs care in applying the norm, but it can either be made to fit, as with behavior over gambling, or the actuality can be seen to be wrong, as with Ellsberg's paradox in §9.11.

Here is an example of a clash between descriptive and normative modes. We shall discuss the scientific method in Chapter 11 within the framework of probabilities and utilities, going on to discover how scientists ought to analyze their experimental results. But if we look at the way the scientists actually behave, we shall find that although they generally do fit into the normative framework, at least approximately, there are many occasions when they do not. Real scientists do not always behave like the normative scientist. Some of the criticisms that have been leveled against science have aspects of the descriptive viewpoint and are irrelevant for the normative attitude. Scientists are human beings, not mere calculators, and all we claim is that they can be assisted by the methods described here; not that they must use only these methods. Genius does not operate according to rules.

A claim for the normative method is that, if implemented, it should result in better decisions. For example, scientists sometimes use methods for assessing the uncertainty of their hypotheses that can be shown to be unsound, like the tail-area significance tests described in §§11.10 and 14.4. Scientists ought, according to the normative viewpoint, to assess their hypotheses according to a result named after its discoverer, Bayes, and were they to do so, their analyses would, we claim, be more efficient. The normative analysis describes some aspects of how one ought to behave, not how one does behave.

The normative theory is sometimes criticized because it does not describe how the world actually works. It has been said that some of the results are without value because people do not, or even could not, obey them. To that, my reply is how could people obey the normative conclusions when they are not aware of them and, even if they were, have received no training in their use. People cannot calculate without training in arithmetic or in the use of calculators. Why should they be able to use the normative ideas here presented without instruction? Many psychological studies of human decision making are as irrelevant to logical decision making as would be a similar study of people doing multiplication who have received no training in arithmetic.

One field in which the distinction between normative and descriptive approaches has been recognized is economics. Important parts of this discipline are based upon the prescription that people base their behavior on rational expectation, a notion that will be extended and formulated precisely in the concept of maximization of expected utility (MEU) in §10.4. However, there is much evidence that people are not rational, in the economist's sense; nor do they take into account expectation, in the precise interpretation of that word. As a result economic theory often does not correspond with what happens in the market. Some would argue that we need descriptive economics. I would argue that all should be taught about probability, utility, and MEU and act accordingly.

The fact that the normative and descriptive results are so often different is most encouraging, for suppose they were typically the same, then all the arguments in this book, all the probabilities and utilities, would merely serve to show you were right all along and my only reward would be to give a boost to your confidence. That the normative and descriptive results are different, and when they are the normative is better, suggests that the tools for handling uncertainty here developed would, if used, be of benefit to society.

2.6 SIMPLICITY

The analysis will begin by making some assumptions, or premises, that may seem to you to be too simple, and therefore unacceptable, or at best only approximations. This aim for simplicity is deliberate, for simple things have considerable advantages over the complicated. There are people who rejoice in the complicated saying, quite correctly, that the real world is complicated and that it is unreasonable to treat it as if it was simple. They enjoy the involved because it is so hard for anyone to demonstrate that what they are saying can be wrong, whereas in a simple argument, fallacies are more easily exposed. Yet it is easy to show by example that they can be wrong through being complicated.

Consider our solar system with its planets, sun, moons, asteroids, and comets. It is truly complicated for one planet has life on it, another is hot, another has rings, and they appear to move across the sky in complicated ways. Yet forget life and heat and other complications and think of planets as point masses, surely a gross simplification. Newton was able to show that if this were done it was possible to account for the movements by means of a few simple laws, which were so successful that today we can foretell an eclipse with great accuracy and even evaluate the tides. Our claim is that the rules that will be developed here for uncertainty are, in this respect, just like Newton's. Our rules are few in number (three), simple and capable of great development, and they deal with uncertainty in the way that Newton dealt with motion.

There are two great merits to simplicity. The first is that if an idea is simple it is much easier to develop it, and produce new results, than is possible with complications. To return to Newton again, from his simple ideas it was possible for him, and other scientists, to predict many phenomena in the physical world, which, when checked against reality, were seen to be correct. Contrast this with the complicated ideas of others that were incapable of being extended to other situations. The second advantage of simplicity is that of ease of communication, for simple concepts presented by one person are more easily understood by another than are complicated ones. It is no good having a simple idea from which, because of its simplicity, many ideas flow, if they are found to disagree with reality. It is an interesting observation, one sensibly discussed by philosophers, why nature does often appear to us to be so simple. Quantum electrodynamics, with its few premises, explains all of physics except the nucleus and gravitation. The genetic code holds promise of explaining a lot of biology. Simplicity, always checked against facts, is a wonderfully successful idea, so if our description of uncertainty at first feels too simple, even naïve, please bear with us and see what happens. I hope that you will not be disappointed.

2.7 MATHEMATICS

Anyone taking this volume off the shelves of a bookstore and flipping through the pages will see some mathematical symbols, which may frighten them and lead to their replacing it on the shelves unread and unpurchased. Perhaps there should be a health warning attached—"Danger, this book contains mathematics". This would be a pity for it would better read—"The germ of mathematics contained herein is harmless". There is some mathematics here, so let me try to explain why and of what it consists, for it really is harmless, indeed it is positively therapeutic.

For reasons that mathematicians find hard to understand, many people have an aversion to math, castigating it as boring or irrelevant, escaping from instruction in it as soon as possible. The blindness of the humanities to mathematics is unfortunate but must be recognized, hence the words of explanation that follow. The first aspect of mathematics should cause no real problem; it is just another language and is no more formidable than any foreign language. Because our discourse is limited to uncertainty, only a small part of the language will be needed, such as might be contained in a simple phrase book for tourists and the translation, from the English language to the mathematical, thereby eased. When, later on, we write $p(A|B)$, it is merely a translation into mathematical language of the English phrase: "your probability that the defendant committed the crime, given the evidence that has been presented in court". (For the moment, let us not concern ourselves with how the translation is effected.) One advantage of the mathematical form is apparent, it is much shorter; indeed, it is a shorthand.

A second aspect of mathematics is its ability to deal with abstractions. Many people have difficulty in handling general concepts, preferring to think in terms of special cases. Thus, when I remark to someone that "smoking is a cause of lung cancer", they are quite likely to reply that their uncle smoked like a chimney and lived to 85, failing to notice that their single case weighs but little against the tens of thousands in the trials that led to the generalization. Mathematics handles generalizations with ease. We have already done a little mathematics, perhaps without you even noticing it, for when in §2.1 a possible premise was described, it was put in the form "if you have stronger belief in event A than in event B and, at the same time, stronger belief in event B than in event C, then necessarily you have stronger belief in A than in C". There is the abstraction, for A, B, and C can be any events satisfying the first two conditions, the premise asserting that they must necessarily satisfy the third. We had an example with A dealing with rain, B with the train, and C with your car, but the principle expressed in the premise is general and therein lies the abstraction. The only real novelty is the use of capital letters, which provide the required generality. The same idea holds with $p(A|B)$ above, for A and B are two events, in the "translation" A is the defendant's guilt, B the evidence; but we want to talk about probabilities generally and this can be done using letters. The use of $p(|)$ will be discussed later. One major reason why we use mathematics is to achieve this abstraction, to be able to talk about any uncertainty without restriction to a topic like weather or guilt. Mathematics is a language of abstract ideas, which is perhaps why some people find it difficult, but it can be enlivened by examples, which is partly why we began with so many in Chapter 1. Of course the abstraction is enhanced if it can be applied, and it is amazing how many abstract concepts developed by mathematicians without any reference to reality, have proved to be relevant to the real world. I have just read an article about whales, which uses the concept of a Borel field, an abstract idea developed by a French mathematician. My knowledge of whaling is not sufficient to judge its usefulness, but at least the Whaling Commission thought it so by publishing the article in its journal.

Language and abstraction are the two key aspects of mathematics that are used here, and my hope is that they will be of little trouble to you, but there is a third aspect that could be the cause of great difficulties. Having got the language and the symbols for the generality, the mathematician uses an enormous battery of devices to manipulate the language, thereby creating new results. This is the technical side of the subject and one that the general reader cannot be expected to handle, so what is to be done? The procedure adopted here is to give the technical procedure whenever I feel that it is simple, which it fortunately is in most of the problems that will be met, but merely to indicate the new results in more difficult cases. Why not, do I hear you say, omit all the technical problems? The reason is that, in my opinion, it helps enormously to know why something is true, rather than being told it is true, for why should you believe me? Never believe anything on the authority of a single person but seek confirmation—and reason is the best confirmation. For example, we shall meet Bayes rule, one of the most important results in our appreciation of the uncertain world, fit to rank with those of Einstein or Newton in the physical world. I want to convince you that what the rule says must be correct; it has got to be that way. The best way to convince you is to prove it to you, and that is what will be done in §6.3.

If we are going together into technicalities, we need a little preparation, so the final section of this chapter, §2.9, contains what little preparatory mathematics I feel you need to know before reading what follows. At least you need to have the translation from English, and essentially §2.9 provides the phrase book you require before entering the foreign country of the mathematician. So come and explore this strange country. Mathematics is a universal language because it deals with reason, which is common to all of us, unlike religion or literature. I can speak to a Pakistani statistician almost as easily as to my British colleagues simply because we share $p(A|B)$.

2.8 WRITING

The primary purpose of this book is to convey information about uncertainty to you, the reader, and the book should mainly be judged by how well you understand the concepts on completion of your reading. Sound content and elegant clarity are my objectives. As a result, this book differs in style from much modern writing, where the conveying of information does not have high priority, and style matters as much as content. The differences in objectives between science writing and much of modern literature call for a few comments on style. Writing is a linear procedure in that it effectively occupies a single line from first word to last, only physical necessity breaking it up into separate lines and pages. The clearest expression of this linearity used to be found in the early days of computers where the information was fed in on a tape, an unbroken sequence of symbols. This linearity is a nuisance when reason is employed because reasoning is not linear but has connections both backwards and forwards, connections that can be described on tape only by special devices, such

38   STYLISTIC QUESTIONS

as "go to". Consequently, this book has been divided into sections, so that it is possible to refer back and forth to the text that is related to the matter under immediate discussion. The unfortunate result of this is that the reader cannot turn over the pages in sequence but must necessarily search other pages to experience the complete argument. Each section is numbered in the form $a.b$, where a is the number of the chapter and b that of the section in that chapter. Thus, you are now reading section (denoted by§) 2.8, being the eighth section in Chapter 2.

A writer is often urged to avoid repetition of words or phrases in order, correctly, to improve the style. Unfortunately, when reason is employed it is not just confusing to do this, it is blatantly wrong. A journalist recently remarked that a quarter of the people in one place supported an idea, whereas 40% did in another. Presumably she was trying to avoid too much use of percentages but the juxtaposition of the fractional system with the decimal one results in confusion for the reader. Another writer mentioned a person, weighing 18 stone, who had gone to summer camp and lost ten kilos. The use of two different scales and two typographies is ridiculous. An example that will bother us involves the uses of the words "probable", "likely", "chance" and words derived from them. The three are nearly synonymous but in our study they will be given precise meanings that make, for example, probability different from likelihood. So if probability is meant, then it has to be repeated and the variation to likelihood would change the meaning. Doubtless you have already experienced overuse of "uncertainty" but there are no synonyms in the English language, except, so Fowler tells us, "whin", "furze", and "gorse", so precision implies repetition, sorry. After the mathematics in the next section, our preparations will be complete and we will be ready to go on our journey into the uncertain world.

2.9   MATHEMATICS TUTORIAL

Everyone knows a little mathematics, even if it is only arithmetic, with its basic operations of addition $+$, subtraction $-$, multiplication $\times$, division $\div$, and equality $=$, with their associated symbols. Thus

$$5 + 3 = 8,$$

saying that five added to three, on the left-hand side of the equality, equals eight, on the right, and from which it follows that

$$5 = 8 - 3,$$

or, reading backwards, the subtraction of three from eight equals five. The displayed equalities are translations of the English phrases that follow them. Also

$$5 \times 3 = 15,$$

saying that five multiplied by three equals fifteen, with the consequence that

$$5 = 15 \div 3,$$

or the division of fifteen by three equals five. Notice that the second arithmetic equality above follows from the first by subtracting three from each side, for if there are two equal things and the same operation is applied to both sides, the results are also equal: on the left the $+3$ is omitted, on the right -3 is included. Similarly the fourth follows from the third by dividing each side by three: on the left $\times 3$ is omitted, on the right $\div 3$ is included. Operations like these, in which we do the same thing to both sides of an equality, yielding another equality, will find frequent use throughout this book.

Arithmetic becomes more mathematical when we use symbols to replace the numbers. You already appreciate that "three" and "3" are two representations of the same thing; in mathematics other symbols may be used and, for example, letters of an alphabet may be employed but with the difference that a, for example, may be used for any number. Thus, you know

$$5 + 3 = 3 + 5,$$

because the order in which numbers are taken to be added does not matter. We could equivalently write

$$a + b = b + a,$$

where *a* replaces 5 and *b* replaces 3. However, the last equality is much more powerful than that involving the mere addition of 3 and 5 because it holds for any pair of numbers *a* and *b*; it says that if you add *b* to *a*, you get the same result as adding *a* to *b*. Here we have an example of the abstraction mentioned in §2.7 that enables one to make general statements, here about any pair of numbers, in a form convenient for manipulation; in particular, when we introduce probability, it will be possible to make general statements about probability. Incidentally, if you feel that $a+b = b+a$ is trivial and obvious, notice that it is not true that $a - b = b - a$, nor $a \div b = b \div a$. As an example of manipulation with letters in place of numbers, consider the statement

$$a + b = c,$$

which says that the number *a*, when added to *b*, yields *c*. This is not true for any numbers, but if it is true, then

$$a = c - b,$$

on subtracting *b* from both sides of the equality. (Compare the case $5 + 3 = 8$ above.) We will follow the mathematical style and refer to "equation" rather than "equality", that word being used in more informal contexts. Notice that the symbols *a*, *b*, *c*, are printed in italics so that there is a clear distinction between a cow and *a* cows. The Greek alphabet will be used in addition to the Roman, but the Greek will be explained when needed.

We shall follow the standard, mathematical practice and not normally use the arithmetical symbols × for multiplication, nor ÷ for division. For multiplication, no symbol is used, the two numbers being run together; thus *ab* replaces $a \times b$. This could not be done with the numeric description where 23 could mean 2×3 or twenty-three, but it is useful and economical when the representation is by letter. If a number is multiplied by itself, *aa*, we abbreviate to a^2, the index 2 indicating two *a*'s in the multiplication. a^2 is called the square of *a*. If $a^2 = b$, then *a* is called the square root of *b*. Thus $3 \times 3 = 9$, so that 9 is the square of 3, and 3 is the square root of 9. We often write $\sqrt{b}$ for the square root of b though the notation $b^{1/2}$ is preferred because the superscripts add: $b^{1/2} \times b^{1/2} = b^{1/2+1/2} = b$. For division, $a \div b$ is replaced by a/b, using the solidus / instead of $\div$. The solidus is called "forward slash" in computer terminology. a is termed the numerator and b the denominator. Sometimes it is typographically more convenient to rotate the solidus to the horizontal and, in so doing, carry the denominator with it, so that a/b becomes $\dfrac{a}{b}$. If $a = 1$, then $1/b$ is called the reciprocal of b. One half is the reciprocal of 2. There is one other mathematical convention we will need, that involving brackets, usually round ones ( ), which are needed to distinguish, for example,

$$a/b + c \quad \text{from} \quad a/(b+c).$$

The first expression means take the number a, divide it by b and add the result to c; the second says add b to c and divide a by the result. Try it with some numbers: where

$$10/2 + 3 = 8,$$

yet

$$10/(2+3) = 2.$$

Generally operations within brackets, here addition, take precedence over those outside, here division. Most pocket calculators use brackets, though the better ones use the superior reverse-Polish notation, which avoids them.

There is one result involving brackets, that we shall frequently use, which says

$$a(b+c) = ab + ac.$$

In words, if two numbers, b and c, are added and the result multiplied by a, the final result is the same as multiplying a by b, then multiplying a by c, and adding the products together. Try it with some numbers

$$6(2+3) = 6 \times 5 = 30,$$

or alternatively,

$$6 \times 2 + 6 \times 3 = 12 + 18 = 30.$$

Be careful though, this works for multiplication but not division, where $a/(b+c)$ is not $a/b + a/c$. Multiplication being harder than addition, $a(b+c)$ is easier to evaluate than $ab + ac$.

There are two other symbols we need, $>$ and $<$, signs of inequality. If a and b are two numbers, we write $a > b$ to mean that a is larger than b, as $5 > 3$. The symbol is easy to understand and remember because it is larger on the left and smaller on the right, where it reduces to a single point, just as a on the left is larger than b on the right. Similarly $a < b$ means that a is smaller than b. Clearly if $a > b$, then $b < a$. If a and b are any two numbers, then one, and only one, of the following three statements must be true: $a < b$, $a = b$, or $a > b$. If both $a > b$ and $b > c$, then necessarily $a > c$. The reader may like to explore the similarity between this result and the premise described in the second paragraph of §2.1; lower-case letters replace the upper-case there and $>$ replaces "is believed more strongly than". If a is positive, we may write $a > 0$. If $a > b$, then $a - b > 0$.

We often want to use several quantities and, rather than using different letters like a, b, c, it is often convenient to number them. Thus, we write a_1, a_2, a_3, the numbering being presented through subscripts. Subscripts are much used. Often we want to employ several quantities without saying how many there are, in which case we might say there are n of them, without specifying the value of n, 3 in the example. In that case, the quantities are listed as $a_1, a_2, \ldots a_n$, where the dots indicate the omitted values between the second, a_2, and the last a_n. It is purely a convention to write the first two and the last, filling the gap with dots.

Some other conventions will be used. Often it will be useful to display a statement of equality, usually called an equation, by writing it in isolation, centered on a line, as has been done above with

$$a(b + c) = ab + ac. \tag{2.1}$$

Often it is necessary to refer to the equation later in the text, so it is numbered in round brackets at the end of the line. We can then, as here, refer to Equation (2.1). The reference is an example of the nonlinearity mentioned in the last section. Equation (a.b) means equation b of chapter a; thus, here (2.1) is the first equation in this, the second chapter. There are two arithmetical conventions that will be used. A number is given to two, say, *significant figures* when other figures are ignored. Thus, 0.2316 becomes 0.23, or 2.316 becomes 2.3. People often fail to understand the difference between 0.3 and 0.30; the latter is given to two significant figures and the second 0 is just as significant as any other digit. A decimal is given to two, say, *decimal places* if only the first two are provided. Thus, 0.2316 becomes 0.23 and 0.02316 becomes 0.02 to two decimal places.

That is all the mathematics you need to start you off; so please try it and see the advantages that it brings. If, in glancing through the book, your eye catches the formulae in §14.4, do not be put off. They are only there for people who sensibly want to verify my calculations.

CHAPTER 3

Probability

3.1 MEASUREMENT

In this chapter the systematic study of uncertainty begins. Recall that there is a person "you", contemplating an "event", and it is desired to express your uncertainty about that event, which uncertainty is called your "belief" that the event is true. The tool to be used is reason (§2.1) or rationality, based on a few fundamental premises and emphasizing simplicity (§2.6). The first task is to measure the intensity of your belief in the truth of the event; to attach to each event a number that describes your attitude to the statement. Many people object to the assignment of numbers, seeing it as an oversimplification of what is rightly a complicated situation. So let us be quite clear why we choose to measure and what the measurement will accomplish. One field in which numbers are used, despite being highly criticized by professionals, is wine-tasting, where a bottle of wine is given a score out of 100, called the Parker score after its inventor, the result being that a wine with a high score such as 96 commands a higher price than a mere 90. Some experts properly object that a single number cannot possibly capture all the nuances that are to be found in that most delectable of liquids. Nevertheless, numbers do have a role to play in wine-tasting,

---

Understanding Uncertainty, Revised Edition. Dennis V. Lindley.
© 2014 John Wiley & Sons, Inc. Published 2014 by John Wiley & Sons, Inc.

where a collection of different wines is tasted by a group of experts, the object being to compare the wines, which naturally vary; variation, as we shall see in §9.1, gives rise to uncertainty. In addition to the wines being different, so are the tasters, and in a properly conducted tasting, it is desirable to sort out the two types of variation and any interaction between wines and tasters, such as tasters of one nationality preferring wines of their own country. If tasting results in comments like "a touch of blackcurrant to a background of coffee with undertones of figs", sensible comparisons are almost impossible. A useful procedure is for each taster to score each wine, the usual method employing a score out of 20 devised at the University of California at Davis. It is then possible by standard statistical methods to make valuable judgments both about the wines and their tasters.

The point here is that, whether it is the Parker or Davis score that is employed, the basic function is to compare wines and tasters. Whether a wine with an average score of 19 is truly better than one with an average of 16, will depend on the variation found in the tasting. (Notice that there are uncertainties here, but wine tasters do not always mention them.) What is not true is that the scores for different wines are combined in any way; a Chablis at 17 is not diluted with a claret at 15 to make a mixture at 16. The numbers are there only for comparison; 17 is bigger than 15. The situation is different with uncertainty where, in any but the simplest scenario, you have to consider several uncertainties and necessarily need to combine them to produce an overall assessment. A doctor has several beliefs about aspects of the patient, which need to be put together to provide a belief about the treatment. It is this combination that makes measurement of uncertainty different from that of wine, where only comparison is required. Now numbers combine very easily and in two distinct ways, by addition and by multiplication, so it is surely sensible to exploit these two simple procedures by associating numbers with your beliefs. How else is the doctor to combine beliefs about the various symptoms presented by the patient?

We aim to measure separate uncertainties in order to combine them into an overall uncertainty, so that all your beliefs come together in a sensible set of beliefs. In this chapter, only one event will be discussed and the combination aspect will scarcely appear, so bear with me while we investigate the process of measurement itself for a single event, beginning with some remarks about measurement in general. A reader who is unconvinced may like to look at §6.4, which concerns the uncertainty of someone who has just been tested for cancer. Without numbers, it is hard to see how to persuade the person of the soundness of the conclusion reached there.

Take the familiar concept of a distance between two points, where a commonly used measure is the foot. What does it mean to say that the distance is one foot? All it means is that somewhere there is a metal bar with two thin marks on it. The distance between these two marks is called a foot, and to say that the width of a table is one foot means only that, were the table and the bar placed together, the former would sit exactly between the two marks. In other words, there is a standard, a metal bar, and all measurements of distance refer to a comparison with this standard. Nowadays the bar is being replaced by the wavelength of krypton light and any distance is compared with the number of waves of krypton light it could contain. The key idea is that all measurements ultimately consist of comparison with a standard with the result that there are no absolutes in the world of measurement. Temperature was based on the twin standards of freezing and boiling water. Time is based on the oscillation of a crystal, and so on. Our first task is therefore to develop a standard for uncertainty.

Before doing this, one other feature of measurement needs to be noticed. There is no suggestion that, in order to measure the width of the table, we have to get hold of some krypton light; or that to measure temperature, we need some water. The direct comparison with the standard is not required. In the case of distance, we use a convenient device, like a tape measure, that has itself been compared with the standard or some copy of it. The measurement of distances on the Earth's surface, needed for the production of maps, was, before the use of satellites, based on the measurement of angles, not distances, in the process known as triangulation, and the standard remains a conceptual tool, not a practical one. So do not be surprised if you cannot use our standard for uncertainty, any more than you need krypton light to determine your height. It will be necessary to produce the equivalent of tape measures and triangulation, so that belief can be measured in reality and not just conceptually.

In what follows, extensive references will be made to gambles and there are many people who understandably have strong, moral objections to gambling. The function of this paragraph is to assure such sensible folk that their views need not hinder the development here presented. A gamble, in our terminology, refers to a situation in which there is an event, uncertain for you, and where it is necessary for you to consider both what will happen were it true, and also were it false. Webster's dictionary expresses our meaning succinctly in the definition of a gamble as "an act . . . having an element of... uncertainty". Think of playing the game of Trivial Pursuit and being asked for the capital of Liberia, with the trivial outcomes of an advance on the board or of the move passing to your opponents. Your response is, in Webster's sense and ours, a gamble. The examples of §1.2 show how common is uncertainty and therefore how common are gambles in our sense. We begin by contemplating the act, mentioned by Webster, and it is only later, when decision analysis is developed in Chapter 10, that action, following on from this contemplation, is considered. In §14.5 we have a little to say about gambling, in the sense of monetary affairs in connection with activities such as horse racing, and will see that the moral objections mentioned above can easily be accommodated using an appropriate utility function.

3.2 RANDOMNESS

The simplest form of uncertainty arises with gambles involving physical objects such as playing cards or roulette wheels, as we saw in Example 7 of §1.2. The standard to be used is therefore based on a simple type of gamble. Take an urn containing 100 balls that, for the moment, are as similar as modern mass-production methods can make them. There is no significance in 100; any reasonably large number would do and a mathematician would take n balls, where n stands for any number, but we try to avoid unnecessary math. An urn is an opaque container with a narrow neck, so that you cannot see into the urn but can reach into it for a ball, which can then be withdrawn but not seen until it is entirely out of the urn.

Suppose that the balls are numbered consecutively from 1 to 100, with the numbers painted on the balls, and imagine that you are to withdraw one ball from the urn. In some cases you might feel that every ball, and therefore every number, had the same chance of being drawn as every other; that is, all numbers from 1 to 100 had the same uncertainty. To put it another way, suppose that you were offered a prize of 10 dollars were number 37 to be withdrawn; otherwise you were to receive nothing. Suppose that there was a similar offer, but for the number 53. Then if you are indifferent between these two gambles, in the sense that you cannot choose between them, you think that 37 is as uncertain as 53. Here your feeling of uncertainty is being translated into action, namely an expressed indifference between two gambles, but notice that the outcomes are the same in both gambles, namely 10 dollars or nothing, only the circumstances of winning or losing differ. We shall discuss later in Chapter 10 the different types of gambles, where the outcomes differ radically and where additional problems arise.

There are circumstances where you would not exhibit such indifference. You might feel that the person offering the gamble on 37 was honest, and that on 53 a crook, or you might think that 37 is your lucky number and was more likely to appear than 53. Or you might think that the balls with two digits painted on them weighed more than those with just one, so would sink to the bottom of the urn, thereby making the single-digit balls more likely to be taken. There are many occasions on which you might have preferences for some balls over others, but you can imagine circumstances where you would truly be indifferent between all 100 numbers. It might be quite hard to achieve this indifference, but then it is difficult to make the standard meter bar for distance, and even more difficult to keep it constant in length. The difficulties are less with krypton light, which is partly why it has replaced the bar.

If you think that each number from 1 to 100 has the same chance of being drawn; or if a prize contingent on any one number is as valuable as the same prize contingent on any other, then we say that you think the ball is *taken at random*, or simply, *random*. More formally, if your belief in the event of ball 37 being drawn is equal to your belief in the event of ball 53, and similarly for any pair of distinct numbers from 1 to

100, then you are said to believe that the ball is drawn at random. This formal definition avoids the word "chance", which will be given a specific meaning in §7.8, and embraces only the three concepts, "you", "event", and "belief".

The concept of randomness has many practical uses. In the British National Lottery, there are 49 balls and great care is taken to make a machine that will deliver a ball in such a way that each has the same chance of appearing; that is, to arise at random. You may not believe that the lottery is random and that 23 is lucky for you; all we ask is that you can imagine a lottery that is random for you. Randomness is not confined to lotteries, thus, with the balls replaced by people, not in an urn but in a population, it is useful to select people at random when assessing some feature of the population such as intention to vote. We mentioned in §1.5, and will see again in §8.5, how difficulties in comparing two methods can be avoided by designing some features of an experiment at random, such as when patients are randomly assigned to treatments. Computer scientists have gone to a great deal of trouble to make machines that generate numbers at random. Many processes in nature appear to act randomly, in that almost all scientists describe their beliefs about the processes through randomness, in the sense used here. The decay of radioactive elements and the transfer of genes are two examples. There is a strong element of randomness in scientific appreciation of both the physical and the biological worlds and our withdrawal of a ball from the urn at random, although an ideal, is achievable and useful.

3.3 A STANDARD FOR PROBABILITY

We have an urn containing 100 balls, from which one ball is to be drawn at random. Imagine that the numbers, introduced merely for the purpose of explaining the random concept, are removed from the balls but instead that 30 of them are colored red and the remaining 70 are left without color as white, the removal or the coloring not affecting the randomness of the draw. The value 30 is arbitrary; a mathematician would have r red, and $n - r$ white, balls. Consider the event that the withdrawn ball is red. Until you inspect the color of the ball, or even before the ball is removed, this event is uncertain for you. You do not know whether the withdrawn ball will be red or white but, knowing the constitution of the urn, have a belief that it will be red, rather than white.

We now make the first of the premises, the simple, obvious assumptions upon which the reasoned approach is based, and measure your belief that the random withdrawal of a ball from an urn with 100 balls, of which 30 are red, will result in a red ball, as the fraction 30/100 (recall the mathematical notation explained in §2.9) of red balls, and call it your probability of a red ball being drawn. Alternatively expressed, your belief that a red ball will be withdrawn is measured by your probability, 30/100. Sometimes the fraction is replaced by a percentage, here 30%, and another possibility is to use the decimal system and write 0.3, though, as explained in §2.8, it pays to stay with one system throughout a discussion. There is nothing special in the numbers, 30 and 100; whatever is the fraction of red balls, that is your probability.

Reflection shows that probability is a reasonable measure of your belief in the event of a red ball being taken. Were there more than 30 red balls, the event would be more likely to occur, and your probability would increase; a smaller number would lessen the probability. If all the balls were red, the event would be certain and your probability would take its highest possible value, one; all white, and the impossible event has the lowest value, zero. Notice that all these values are only reasonable if you think that the ball is drawn at random. If the red balls were sticky from the application of the paint, and the unpainted, white ones, not, then the event of being red might be more likely to occur and the value of 0.3 would be too low.

In view of its fundamental importance, the definition is repeated with more precision. If you think that a ball is to be withdrawn at random from an urn containing only red and white balls, then your *probability* that the withdrawn ball will be red is defined to be the fraction of all the balls in the urn that are red.

The simple idea extends to other circumstances. If a die is thrown, the probability of a five is 1/6, corresponding to an urn with six balls of which only one is red. In European roulette, the probability of red is 18/37, there being 37 slots of which only 18 are red. In a pack of playing cards, the probability of a spade is 13/52, or 1/4, and of an ace, 4/52 = 1/13. These considerations are for a die that you judge to be balanced and fairly thrown, a roulette wheel that is not rigged and a pack that has been fairly shuffled, these restrictions corresponding to what has been termed random.

The first stage in the measurement of uncertainty has now been accomplished; we have a standard. The urn is our equivalent of the metal bar for distance, perhaps to be replaced by some improvement in the light of experience, as light is used for distance. Other standards have been suggested but will not be considered here. The next stage is to compare any uncertain event with the standard.

3.4 PROBABILITY

Consider any event that is uncertain for you. It is convenient to fix ideas and take the event of rain tomorrow (Example 1 of §1.2), but the discussion that follows applies to any uncertain event. Alongside that event, consider a second event that is also uncertain for you, namely, the withdrawal at random of a red ball from an urn containing 100 balls, of which some are red, the rest white. For the moment, the number of red balls is not stated. Were there no red balls, you would have higher belief in the event of rain than in the impossible extraction of a red ball. At the other extreme, were all the balls red, you would have lower belief in rain than in the inevitable extraction of a red ball. Now imagine the number of red balls increasing steadily from 0 to 100. As this happens, you have an increasing belief that a red ball will be withdrawn. Since your belief in red was less than your belief in rain at the beginning, yet was higher at the end with all balls red, there must be an intermediate number of red balls in the urn such that your beliefs in rain and in the withdrawal of a red ball are the same. This value must be unique, because if there were two values, then they would have the same beliefs, being equal to that for rain tomorrow, which is nonsense as you have greater belief in red with the higher fraction. So there are two uncertain events in which you have the same belief: rain tomorrow and the withdrawal of a red ball. But you have measured the uncertainty of one, the redness of the ball; therefore, this must be the uncertainty of the other, rain. We now make a very important definition:

Your *probability* of the uncertain event of rain tomorrow is the fraction of red balls in an urn from which the withdrawal of a red ball at random is an event of the same uncertainty for you as that of the event of rain.

This definition applies to any uncertain event, not just to that about the weather. To measure your belief in the truth of a specific uncertain event, you are invited to compare that event with the standard, adjusting the number of red balls in the urn until you have the same beliefs in the event and in the standard. Your probability for the event is then the resulting fraction of red balls.

Some minor comments now follow before passing to issues of more substance. The choice of 100 balls was arbitrary. As it stands, every probability is a fraction out of 100. This is usually adequate, but any value between 0 and 1 can be obtained by increasing the total number of balls. When, as with a nuclear accident (Example 13 of §1.2) the probability is very low, perhaps less than 1/100, yet not zero, the number of balls needs to be increased. As we have said, a mathematician would have r red, $n-r$ white, balls and the probability would be r/n.

The following point may mean nothing to some readers, but some others will be aware of the frequency theory of probability, and for them it is necessary to issue a warning: there is no repetition in the definition. The ball is to be taken once, and once only, and the long run frequency of red balls in repeated drawings is irrelevant. After its withdrawal, the urn and its contents can go up in smoke for all that it matters. Repetition does play an important role in the study of probability (see §7.3) but not here in the basic definition.

Some writers deny the existence, or worth, of probability. We have to disagree with them, feeling convinced by the measuring techniques just proposed. Others accept the concept of probability but distinguish between cases where the probability is known, and those where it is unknown, the probability in the latter case being called *ambiguous*. For example, the probability of a coin falling heads is unambiguous at $1/2$, whereas the probability of your candidate winning the election is ambiguous. (Rather than referring to "the" probability, we would prefer "your" probability.) In the descriptive mode, the distinction is important because, as Ellsberg's paradox (§9.11) shows, people make different choices depending solely on whether the probability is ambiguous or not. In the normative mode adopted here, the distinction is one of measurement, an ambiguous probability being harder to measure than one with no ambiguity. The paradox disappears in the normative view, though your action may depend on how good your measurement processes are. As will be seen in Chapter 13, they are not as good as one would wish.

3.5 COHERENCE

In the last section, we took a standard, or rather a collection of standards depending on the numbers of red balls, and compared any uncertain event with a standard, arranging the numbers such that you had the same beliefs in the event and in the standard. In this way, you have a probability for any uncertain event.

You immediately, and correctly, respond, "I can't do it". You might be able to say that the number of red balls must be at least 17 out of a 100 and not more than 25, but to get closer than that is impossible for most uncertain events, even simple ones such as rain tomorrow. A whole system has been developed on the basis of lower (17/100) and upper (25/100) probabilities, both of which go against the idea of simplicity and confuse the concept of measurement with the practice of measurement. Recall the metal bar for length; you cannot take the table to the institution where the bar is held and effect the comparison. It is the same with uncertainty, as it is with distance; the standard is a conceptual comparison, not an operational one. We put it to you that you cannot escape from the conclusion that, as in the last section, some number of red balls must exist to make the two events match for you. Yes, the number is hard to determine, but it must be there. Another way of expressing the distinction between the concept and the practice is to admit that reasoning persuades you that there must exist, for a given uncertain event, a unique number of red balls that you ought to be able to find, but that, in practice you find it hard to determine it. Our definition of probability provides a norm to which you aim; only measurement problems hinder you from exactly behaving like the norm. Nevertheless, it is an objective toward which you ought to aim.

When it comes to distance, you would use a tape measure for the table, though even there, marked in fractions of inches, you might have trouble getting an accuracy beyond the fraction. With other distances, more sophisticated devices are used. Some, such as those used to determine distances on the Earth's surface between places far apart, are very elaborate and have only been developed in the last century, despite the concept of distance being made rigorous by the Greeks. So please do not be impatient at your inability, for we do not as yet have a really good measuring device suitable for all circumstances. Nevertheless, you are entitled to wonder how such an apparently impossible task can be accomplished. How can you measure your belief in practice? Much of the rest of this book will be devoted to this problem. For the moment, let me try to give you a taste of one solution by means of an example.

Suppose you meet a stranger. Take the event that they were born on March 4 in some year. You are uncertain about this event but the comparison with the urn is easy and most of you would announce a probability of 1/365, ignoring leap years and any minor variations in the birth rate during the year. And this would hold for any date except February 29. The urn would contain 365 balls, each with a different date, and a ball drawn at random. Now pass to another event that is uncertain for you. Suppose there are 23 unrelated strangers and consider the uncertain event that, among the 23, there are at least two of them who share the same day for the celebration of their births. It does not matter which day, only that they share a day. Now you have real difficulty in effecting the comparison with the urn. However, there exist methods, analogous to the use of a tape measure with length, that demonstrate that your probability of a match of birthdays is very close to 1/2. These methods rely on the use of the rules of the calculus of probability to be developed in later chapters. Once you have settled on 1/365 for one person, and on the fact that the 23 are unrelated, the value of 1/2 for the match is inevitable. You have no choice. That is, from 23 judgments of probability, one for each person, made by comparison with the standard, you can deduce the value of 1/2 in a case where the standard was not easily available. The deduction will be given in §5.5.

56 PROBABILITY

The principle illustrated here is called *coherence*. A formal definition of coherence appears in §5.4. The value of 1/2 coheres with the values of 1/365. Coherence is the most important tool that we have today for the measurement of uncertainty, in that it enables you to pass from simple, measurable events to more complicated ones. Coherence plays a role in probability similar to the role that Euclidean geometry plays in the measurement of distance. In triangulation, the angles and a single distance, measured by the surveyor, are manipulated according to geometrical rules to give the distance, just as the values of 1/365 are manipulated according to the rules of probability to give 1/2. Some writers use the term "consistent", rather than "coherent", but it will not be adopted here. The birthday example was a diversion; let us return to the definition of probability in §3.4.

3.6 BELIEF

The definition of probability holds, in principle, for any event, the numerical value depending not only on the event but also on you. Your uncertainty for rain tomorrow need not be the same as that of the meteorologist, or of any other person. Probability describes a relationship between you and the world, or that part of the world involved in the event (see §1.7). It is sometimes said to be *subjective*, depending on the subject, you, making the belief statement. Unfortunately, subjectivity has connotations of sloppy thinking, as contrasted with objectivity. We shall therefore use the other common term, *personal*, depending on the person, you, expressing the probability. Throughout this book, probability expresses a relationship between a person, you, and the real world. It is not solely a feature of your mind; it is not a value possessed by an event but expresses a relationship between you and the event and is a basic tool in your understanding of the world. There are many uncertainties upon which most people agree, such as the 1/365 for the birthday in the last section, though there is no complete agreement here. I once met a lady at a dinner party who, during the course of the evening, in which birth dates had not been mentioned, turned to me and said, "you are an Aries". She had a probability greater than 1/365 for dates with that sign, a value presumably based on her observation of my conversation. She is entitled to her view and considered alone, it is not ridiculous although in combination with other beliefs she might hold, she may be incoherent. Note: I am not an Aries.

Similarly, there are events over which there is a lot of disagreement. Thus, the nuclear protester and the nuclear engineer may not agree over the probability of a nuclear accident. One of the matters to be studied in §§6.9 and 11.6 is how agreement between them might be reached, essentially both by obtaining more information and by exposing incoherence.

Probability therefore depends both on the event and on you. There is equally something that it should not depend on—the quality of the event for you. Consider two uncertain events: a nuclear accident and winning a lottery. The occurrence of the first is unpleasant, that of the second highly desirable. These two considerations are not supposed to influence you in your expressions of belief in the two events through probability. This is important, so let us spell it out.

We suppose that you possess a basic notion of belief in the truth of an uncertain event that does not depend on the quality of the event. Expressed differently, you are able to separate in your mind how plausible the event is from how desirable it is. We shall see in Chapter 10 that plausibility and desirability come together when we make a decision, and strictly it is not necessary to separate the two. Nevertheless, experience seems to show that people prefer to isolate the two concepts, appreciating the advantages gained from the separation, so this view is taken here.

To reinforce this point, consider another method that has been suggested for comparing your uncertainty of an event with the standard. In comparing the nuclear accident with the extraction of a red ball from an urn in order to assess your probability for the former, suppose that you were invited to think about two gambles. In the first, you win $100 if the accident occurs; in the second, you win the same amount if the ball is red. The suggestion is that you choose the number of red balls in the urn so that you feel the two gambles are equivalent. The comparison is totally different from our proposal because the winning of $100 would be trivial if there were an accident and you might not be alive to receive it, whereas the red ball would not affect you and the prize could be enjoyed. In other words, this comparability confuses the plausibility of the accident with its desirability, or here, horror. Gambling for reward is not our basis for the system and where it was mentioned in §3.2, the rewards were exactly the same in all the gambles considered, so desirability did not enter. Some aspects of gambling are raised in §14.5.

3.7 COMPLEMENTARY EVENT

Consider the event of rain tomorrow (Example 1 of §1.2). Associated with this event is another event that it will not rain tomorrow; when the former is true, the latter is false and vice versa. Generally, for any event, the event that is true when the first is false, and false when it is true, is called the event that is *complementary* to the first. Just as we have discussed your belief in the event, expressed through a probability, so we could discuss your probability for the complementary event. How are these two probabilities related? This is easily answered by comparison with the withdrawal of a red ball from an urn. The event complementary to the removal of a red ball is that of a white one. The probability of red is the fraction of red balls in the urn and similarly, the probability of white is the fraction of white balls. But these two fractions always add to one, for there are no other colors of ball in the urn; if 30 are red out of 100, then 70 are white. Hence, the standard event and its complement have probabilities that add to one. It follows by the comparison of any event with the urn that this will hold generally. If your belief in the truth of an event matches the withdrawal of a red ball, your belief in the falsity matches with a white ball. Stated formally, it means the following:

Your probability of the complementary event is one minus your probability of the original event. If your probability of rain tomorrow is 0.3, then your probability of no rain tomorrow is 0.7.

This is our first example of a rule of probability; a rule that enables you to calculate with beliefs and is the first stage in developing a calculus of beliefs. Since calculation is involved, it is convenient to introduce a simple piece of mathematics, effectively rewriting the above statement in another language.

3.7 COMPLEMENTARY EVENT

Instead of using the word "event", it is often useful to use a capital letter of the Roman alphabet. E is the natural one to use, being the initial letter of event, and thereby acting as a mnemonic. Later it will be necessary to talk about several events and use different letters to distinguish them, thus E, F, G, and so on. When we want to state a general rule about events, it is not necessary to spell out the meaning of E, which can stand for any event; whereas in an application of the rule, we can still use E, but then it will refer to the special event in the application. Your probability for the event E is written $p(E)$. Here the lower-case letter p replaces probability and the brackets encompass the event, so that in a sense they replace "of" in the English language equivalent. Some writers use P or Pr or *prob* but we will use the simple form p. Notice that p always means probability, whereas E, F, G, and so on refer to different events and $p(E)$ is simply a translation of the phrase "your probability for the event E". It might be thought that reference should also be made to "you" but since we will only be talking about a single person, this will not be necessary, see §3.6.

Let us have a bit of practice. If R is the event of rain tomorrow, then the statement that your probability of rain is 0.3 becomes $p(R) = 0.3$. If C is the event of a coincidence of birthdays with 23 people (see §3.5), then $p(C) = 1/2$ to a good approximation. Notice that R is an event, r a number of balls, and mathematicians make much more use of a distinction between upper- and lower-case than does standard English. If E is any event, then the event that is complementary to E is written E^c, the raised c standing for "complement" and again, the initial letter acting as a mnemonic. Complement being such a common concept, many notations besides the raised c are in use.

With this mathematical language, the rule of probability stated above can be written as follows:

$$p(E^c) = 1 - p(E),$$

this being a mathematical translation of the English sentence "The probability of the complementary event is equal to one minus the probability of the event". Mathematics has the advantage of brevity and, with some practice, has the benefit of increased clarity. Notice that in stating the rule, we have not said what the event E is since the statement is true for any event.

Let us perform our first piece of mathematical calculation and add $p(E)$ to both sides of this equation (see §2.9) with the result

$$p(E) + p(E^c) = 1,$$

or in words, your probability for an event and your probability for the complementary event add to one. Several important rules of probability will be encountered later, but they have one point in common—they proscribe constraints on your beliefs. While you are free to assign any probability to the truth of the event, once this has been done, you are forced to assign one minus that probability to the truth of the complementary event. If your probability for rain tomorrow is 0.3, then your probability for no rain must be 0.7. This enforcement is typical of any rule in that there is great liberty with some of your beliefs, but once they are fixed, there is no freedom with others that are related to them and we have an example of the coherence mentioned in §3.5. You are familiar with this phenomenon for distance. If the distance from Exeter to Bristol is 76 miles, and that from Bristol to Birmingham is 81 miles, then that from Exeter to Birmingham, via Bristol is inevitably 157 miles, the sum of the two earlier distances. Mathematically, if the distance from A to B is x, and that from B to C is y, then the distance from A to C, via B, is $x+y$, a statement that is true for any A, B, and C and any x and y compatible with geography.

3.8 ODDS

Although probability is the usual measure for the description of your belief, some people prefer to use an alternative term, just as some prefer to use miles instead of kilometers for distance, and we will find that an alternative term has some convenience for us in §6.5. To introduce the alternative measure, let us return to any uncertain event E and your comparison of it with the withdrawal at random of a red ball from an urn containing r red and w white balls, making a total of $n = r + w$ in all. Previously, we had 100 balls in total, $n = 100$, purely for ease of exposition. As before, suppose r is adjusted so that you have the same belief in E as in the random withdrawal of a red ball, then your probability for E, $p(E)$, is the ratio of the number of red balls to the total number of balls, $r/(r+w)$. The alternative to probability as a measure is the ratio of the number of white balls to that of red w/r and is called the *odds against* a red ball and therefore, equally the odds against E. Alternatively, reversing the roles of the red and white balls, the ratio of the number of red balls to that of white r/w is termed the *odds on* a red ball or the odds on E. We now encounter a little difficulty of nomenclature and pause to discuss it.

The concept of odds arises in the following way. Suppose that, in circumstances where E is an event favorable to you, such as a horse winning, you have arranged the numbers of balls such that your belief in E equals that in a red ball being withdrawn; then there are w possibilities corresponding to E not happening because the ball was white, and r corresponding to the pleasant prospect of E. Hence, it makes sense to say w against E and r for E, or simply "w to r against", expressed as a ratio w/r. As an example, suppose your probability is a quarter, 1/4, that High Street will win the 2.30 race at Epsom (Example 8 of §1.2), then a quarter of the balls in the matching urn will be red, or equivalently, for every red ball there will be three white; the odds against High Street are 3 to 1, the odds on are 1 to 3; as ratios, 3 against, 1/3 on.

Odds are commonly used, at least in Britain, in connection with betting (§14.5). Odds in betting are always understood as odds against; in the few cases where odds on are used, they say "odds on". Thus "against" is omitted but "on" is included. As a way through this linguistic tangle, we shall always use odds in the sense of odds on. If we do need to use odds against, the latter word will be added. This is opposite to the convention used in betting and is weakly justified by the fact that our probabilities will commonly be larger than those encountered in sporting events, also because a vital result in §6.5 is slightly more easily expressed using odds on. It will also be assumed that you are comfortable with using fractions.

There is no standard notation for odds and we will use $o(E)$, o for odds on replacing p for probability. There is a precise relationship between probability and odds, which is now obtained as follows. $p(E)$

is the fraction $r/(r+w)$ and equally, $p(E^c)$ is the fraction $w/(w+r)$. The ratio of the former fraction to the latter is r/w, which is the odds on, $o(E)$. The reader is invited to try it with the numbers appropriate to High Street. Consequently, we have the general result that the odds on an event is the ratio of the probability of the event to that of its complement. That sentence translates to

$$o(E) = \frac{p(E)}{p(E^c)}.$$

Because a ratio printed this way takes up a lot of vertical space, it is usual to rewrite this as

$$o(E) = p(E)/p(E^c), \tag{3.1}$$

keeping everything on one line, as in English (see §2.9). Recall that $p(E^c) = 1 - p(E)$, so that (3.1) may be written

$$o(E) = p(E)/[1 - p(E)]. \tag{3.2}$$

The square brackets are needed here to show that the whole content, $1 - p(E)$, divides $p(E)$; the round brackets having been used in connection with probability.

Equation (3.2) enables you to pass from probability on the right to odds on the left. The reverse passage, from odds to probability, is given by

$$p(E) = o(E)/[1 + o(E)]. \tag{3.3}$$

To see this, note that $p(E) = r/(r+w)$, so that dividing every term on the right of this equality by w, $p(E) = (r/w)/(1 + r/w)$ and the result follows on noting that $o(E) = r/w$. Thus, if the odds on are 1/3, the probability is 1/3 divided by $(1 + 1/3)$ or 1/4. The change from 3 in odds to 4 in probability, caused by the addition of 1 to the odds in (3.3), can be confusing. Historians have a similar problem where dates in the 16 hundreds are in the seventeenth century; and musicians have four intervals to make up a fifth, so we are in good company. Notice that if your probability is small, then the odds are small, as is clear from (3.2).

Similarly, a large probability means large odds. Probability can range from 0, when you believe the event to be false, to 1, when you believe it to be true. Odds can take any positive value, however large, and probabilities near 1 correspond to very large odds; thus, a probability of 99/100 gives odds of 99.

Odds against are especially useful when your probability is very small. For example, the organizers of the National Lottery in Britain state that their probability that a given ticket (yours?) will win the top prize is 0.000 000 071 511 238, a value that is hard to appreciate. The equivalent odds against are 13,983,815 to 1. There is only one chance in about 14 million that your ticket will win. Think of 14 million balls in the urn and only one is red. Another example is provided by the rare event of a nuclear accident.

In everyday life, odds mostly occur in connection with betting, and it is necessary to distinguish our usage with their employment by bookmakers (§14.5). If a bookmaker quotes odds of 3 to 1 against High Street winning, it describes a commercial transaction that is being offered and has little to do with his belief that the horse will win. All it means is that for every 1 dollar you stake, the bookmaker will pay you 3 dollars and return your stake if High Street wins; otherwise you lose the stake. The distinction between odds as a commercial transaction from odds as belief is important and should not be forgotten. You would ordinarily bet at odds of 3 to 1 against only if your odds against were smaller, or in probability terms, if your probability of the horse winning exceeded 1/4.

3.9 KNOWLEDGE BASE

Considerable emphasis has been placed on simplicity, for we believe that the best approach is to try the simplest ideas and only to abandon them in favor of more complicated ones when they fail. It is now necessary to admit that the concept of your probability for an event, $p(E)$ as just introduced, is too simple and a complication is forced upon us. The full reason for this will appear later but it is perhaps best to introduce the complication here, away from the material that forces it onto our attention. Our excuse for duping the reader with $p(E)$ is a purely pedagogical one of not displaying too many strange ideas at the same time.

Suppose that you are contemplating the uncertain event of rain tomorrow and carrying through the comparison with the balls in the urn, arriving at the figure of 0.3. It then occurs to you that there is a weather forecast on television in a few moments, so you watch this and, as a result, revise your probability to 0.8 in the light of what you see. Just how this revision should take place is discussed in §6.3. So now there are two versions of your belief in rain tomorrow, 0.3 and 0.8. Why do they differ, for they are probabilities for the same event? Clearly because of the additional information provided by the forecast, which changes the amount of knowledge you have about tomorrow's weather. Generally, your belief in any event depends on your knowledge at the time you state your probability and it is therefore oversimple of us to use the phrase "your probability for the event". Instead, we should be more elaborate and say "your probability for the event in the light of your current knowledge". What you know at the time you state your probability will be referred to as your *knowledge base*.

The idea being expressed here can alternatively be described as saying that any probability depends on two things, the uncertain event under consideration and what you know, your knowledge base. It also depends on the person whose beliefs are being expressed, you, but as we have said, we are only thinking about one person, so there is no need to refer to you explicitly. We say that probability depends on two things, the event and the knowledge. Some writers on probability fail to recognize this point, with a resulting confusion in their thoughts. One expert produced a wrong result, which caused confusion for years, the expert being so respected that others thought he could not be wrong. In the light of this new consideration, the definition of probability in §3.4 can be rephrased. Your probability of an uncertain event is equal to the fraction of red balls in an urn of red and white balls when your belief in the event *with your present knowledge* is equal to your belief that a single ball, withdrawn at random from the urn, will be red. The change consists in the addition of the words in italics.

This necessary complexity means that the mathematical language has to be changed. The knowledge base will be denoted by $\mathcal{K}$, the initial letter of knowledge, but written in script to distinguish it from an event. In place of $p(E)$ for your probability of the event E, we write $p(E|K)$. The vertical line, separating the event from your knowledge, can be translated as "given" or "conditional upon". The whole expression then translates as "your probability for the event E, given that your knowledge base is K". In the example, where the event is "rain tomorrow" and your original knowledge was what you possessed before the forecast, $p(E|K)$ was 0.3. With the addition of the forecast, denoted by F, the probability changes to $p(E|F$ and $K)$ and 0.8. Your knowledge base has been increased from K to F and K.

Despite the clear dependence on how much you know, it is common to omit K from the notation because it usually stays constant throughout many calculations. This is like omitting "you" because there is only one individual. Thus, we shall continue to write $p(E)$ when the base is clear. In the example, after the forecast has been received, we shall write $p(E|F)$. Although the knowledge base is often not referred to, it must be remembered that it, like you, is always present. The point will arise in connection with independence in §8.8.

Some people have put forward the argument that the only reason two persons differ in their beliefs about an event is that they have different knowledge bases, and that if the bases were shared, the two people would have the same beliefs, and therefore the same probability. This would remove the personal element from probability and it would logically follow that all with knowledge base K for an uncertain event E, would have the same uncertainty, and therefore the same probability $p(E|K)$, called a *logical* probability. We do not share this view, partly because it is very difficult to say what is meant by two knowledge bases being the same. In particular, it has proved impossible to say what is meant by being ignorant of an event or having an empty knowledge base, and although special cases can be covered, the general concept of ignorance has not yielded to analysis. People often say they know nothing about an event but all attempts to make this idea precise have, in my view, failed. In fact, if people understand what is under discussion, such as rain tomorrow, then, by the mere fact of understanding, they know something, albeit very little, about the topic. In this book, we shall take the view that probability is your numerical expression of your belief in the truth of an event in your current state of knowledge; it is personal, not logical.

3.10 EXAMPLES

Let us return to some of the examples of Chapter 1 and see what the ideas of this chapter have to say about them. With almanac questions, such as the capital of Liberia, the numerical description of your uncertainty as probability would not normally be a worthwhile exercise, though notice how, in the context of "Trivial Pursuit" your probability would change as you consulted with other members of your team and, as a result, your knowledge base would be altered. A variant of the question would present you with a number, often four, possible places that might be the capital, one of which is correct. This multiple-choice form does admit a serious and worthwhile use of probability by asking you to attach probabilities to each of the four possibilities rather than choosing one as being correct, which is effectively giving a value 1 to one possibility and 0 to the rest. An advantage of this proposal in education is that the child being asked could face up to the uncertainty of their world and not be made to feel that everything is either right or wrong. There is a difficulty in making such probability responses to multiple-choice examination questions, but these have been elegantly overcome and the method has been made a real, practical proposal.

The legal example of guilt (Example 3) is considered in more detail in §10.14 but for now, just note how the uncertain event remains constant throughout the trial but the knowledge base is continually changing as the defense and prosecution present the evidence.

Medical problems (selenium, Example 4 and fat, Example 15) are often discussed using probability and therefore raise a novel aspect because you, when contemplating your probability, may have available one or more probabilities of others, often medical experts in the field. You may trust the expert and take their probability as your own but there is surely no obligation on you to do so since the expert opinion has to be combined with other information you might have, such as the view of a second expert or of features peculiar to you. There exists some literature, which is too technical for inclusion here, of how one person can use the opinions of others when these opinions are expressed in terms of probability.

Historians only exceptionally embrace the concept of probability, as did one over the princes in the tower (Example 5) but they are enthusiasts for what we have called coherence, even if their form is less numerate than ours. When dealing with the politics of a period, historians aim to provide an account in which all the features fit together; to provide a description in which the social aspects interact with the technological advances and, together with other features, explain the behavior of the leading figures. Their coherence is necessarily looser than ours because there are no rigid rules in history, as will be developed for uncertainty, but the concepts are similar. Whether probability will eventually be seen to be of value in historical research remains to be seen though, since so much in the past is uncertain, the potential is there.

The three examples, of card play (Example 7), horse racing (Example 8), and investment in equities (Example 9) are conveniently taken together because they are all intimately linked with gambling, though that term seems coarse in connection with the stock market; nevertheless, the placing of a stake in anticipation of a reward is fundamentally what is involved. Games of chance have been intimately connected with probability, and it is there that the calculus began and where it still plays an important role, so that many cardplayers are knowledgeable about the topic. There is a similar body of experts in odds, namely, bookmakers, but here the descriptive results seem to be at variance with the normative aspects of §2.5 (see §14.5). Bookmakers are very skilled and it would be fascinating to explore their ideas more closely, though this is hindered by their understandable desire to be ahead of the person placing the bet, indulging in some secrecy. A descriptive analysis of stockbrokers would be even more interesting since they use neither odds nor probabilities. There is a gradation from games of chance, where the probabilities and rewards are agreed and explicit; to horse racing, where the rewards are agreed but probabilities are not, since your expectation of which horse is going to win typically differs from mine; to the stock market, where nothing is exposed except the yields on bonds.

Some of the examples, but especially that of opinion polls before an election (Example 12), are interesting because the open statement of uncertainty itself can affect an uncertainty. An obvious instance of this arises when a poll says that the incumbent is 90% certain of winning the election, with the result that her supporters will tend not to vote, deeming it unnecessary, and your probability of their victory will drop. The other examples either raise no additional issues, or, if they do, the issues are better discussed when we have more familiarity with the calculus of probability to be developed in the following chapters. Instead, let us recapitulate and see how far we have got.

3.11 RETROSPECT

It has been argued that the measurement of uncertainty is desirable because we need to combine uncertainties and nothing is better and simpler at combination than numbers. Measurement must always involve comparison with a standard and here we have chosen balls in urns for its simplicity. Other standards have been used, perhaps the best being some radioactive phenomena, which seem to be naturally random and, like krypton light, have the reliance of physics behind them. It has been emphasized that the role of a standard is not that of a practicable measuring tool but rather a device for producing usable properties of uncertainty. From these ideas, the notion of probability has been developed and one rule of its calculus derived, namely that the probabilities of an event and its complement add to one.

So what has been achieved? Quite frankly, not much, and you are little better at assessing or understanding uncertainty than you were when you began to read. So has it been a waste of time? Of course, my answer is an emphatic "No". The real merit of probability will begin to appear when we pass from a single event to two events, because then the two great rules of combination will arise and the whole calculus can be constructed, leading to a proper appreciation of coherence. Future chapters will show how new information, changing your knowledge base, also changes your uncertainty, and, in particular, explains the development of science with its beautiful blend of theory and experimentation. When we pass from two events to three, no new rules arise, but surprising features arise that have important consequences.

CHAPTER 4

Two Events

4.1 TWO EVENTS

So far we have only investigated a single event and its complement. We now pass to two events and the relationships between them. To fix ideas, consider these two events that you are contemplating now but that refer to what happens one year in the future:

Inflation next year will exceed 4%, *and*

Unemployment next year will exceed 9% of the workforce.

These are two events of economic importance about which you are uncertain, and which are to be discussed in connection with a fixed knowledge base. It is tedious to have to spell out the whole sentences describing the events each time they are mentioned, so let us simply call the first event *high* and the second *many*. The complementary event, high not happening, will be termed *low*, and the complement of many will be *few*. So you are uncertain about high (inflation above 4%) and about many (unemployed above 9%). By taking the event high on its own, you can proceed as in the previous chapter and assess the

---

Understanding Uncertainty, Revised Edition. Dennis V. Lindley.
© 2014 John Wiley & Sons, Inc. Published 2014 by John Wiley & Sons, Inc.

probability of high using the comparison with red and white balls in an urn. Suppose you think that if there are 40 red balls (and 60 white) in the urn, then the uncertainty of a red ball being drawn at random is the same as that of high inflation, then the probability of high inflation for you is 0.40 (and of low, 0.60).

Now let the same thing be done for the event of many unemployed, but to avoid confusion, in the new urn, also with 100 balls, replace red by spotted and white by plain. Then you will arrange for the number of spotted balls to be selected so that the uncertainty of drawing a spotted ball at random matches the uncertainty of many. Suppose that you select 50 as the number of spotted balls needed for the match, leaving 50 plain, then your probability of many unemployed is 0.50; you think that many and few employed are equally probable.

This has attended to each event separately, but the key question for the economist is likely to concern any possible relationship between the two. How can this be expressed in our scheme? The trick is to combine the two urns in the sense that we still have a single urn with 100 balls, but in this urn, the balls are basically either red or white and, at the same time, either have black spots added or are left plain. Thus, there are four types of ball; spotted red, plain red, spotted white, and plain white. The number of red balls has already been settled at 40, being the sum of the numbers in the first two of these four categories. Similarly, the sum of the numbers in the first and third, the spotted balls, is 50. The situation can conveniently be represented in a table.

		INFLATION		
		High (red)	Low (white)	Total
UNEMPLOYMENT	Many (spotted)			50
	Few (plain)			50
	Total	40	60	100

The rows in the table correspond to unemployment, the columns to inflation. There are two basic rows and two basic columns to which columns and rows of totals have been added. So far, by taking the two events separately, you have settled on the totals, both along the rows and down the columns, but what has not been settled are the entries in the table, we have only the margins of the table not the interior entries. There are several ways of thinking about these but, for the moment, let us concentrate on the number of balls that are both red and spotted, corresponding to both high inflation and many unemployed and to an entry in the top, left-hand corner of the table. This is a new event, in which two other events are both true, and can be contemplated in the same way as before; namely by thinking about how many balls of that type, red and spotted, are needed to make the uncertainty of drawing a red, spotted ball equate to your uncertainty of the event of both high inflation and many unemployed. The value cannot exceed 40 as only 40 red balls are available to receive spots. Suppose that you settle on 10. The table now looks as shown below:

		INFLATION		
		High (red)	Low (white)	Total
UNEMPLOYMENT	Many (spotted)	10		50
	Few (plain)			50
	Total	40	60	100

The table can be completed by simple arithmetic without any extra considerations of uncertainty; thus, in the first column, there are 40 red balls in all, of which 10 are spotted, so there must be 30 that are plain red. Continuing in this way with the rows, the table may be completed and then it looks as shown below:

		INFLATION		
		High (red)	Low (white)	Total
UNEMPLOYMENT	Many (spotted)	10	40	50
	Few (plain)	30	20	50
	Total	40	60	100

On dividing every entry by 100 to obtain the fractions of balls of the different types, we have your probabilities.

		INFLATION		
		High (red)	Low (white)	Total
UNEMPLOYMENT	Many (spotted)	0.1	0.4	0.5
	Few (plain)	0.3	0.2	0.5
	Total	0.4	0.6	1.0

Recall that you reached this table by the consideration of only three events, "high", "many", and "both high and many", but now you have probabilities for many other uncertain events. For example, your probability for the highly desirable event of low inflation and few unemployed is 0.2. (As an aside, recall that desirability does not enter into the numbers so far obtained in the table, which relate purely to questions of uncertainty, see §3.6.) From the three original numbers in the table, you can calculate others such as the one just mentioned. There are many uncertain events in the table and all have their uncertainties determined from the three already given. This is an example of the principle of coherence mentioned in §3.5, for although you were free to choose the original three values, once they are settled, all the others follow by the rules of probability and you have no further choice. They must all cohere. This is an important point and we will repeatedly return to it, but let us now look at some further statements that can be derived from the table and to which you have, perhaps unwittingly, committed yourself.

It is useful to have a term to describe such a table; it is called a *contingency* table because it describes how one event is contingent upon another. Strictly, it is a contingency table of probabilities, here of size 2×2 since there are 2 rows and 2 columns.

4.2 CONDITIONAL PROBABILITY

Let us be a little more mathematical and introduce Roman letters for the two original events and their complements. In each case, the initial letters are used, thus H for high inflation and M for many unemployed.

4.2 CONDITIONAL PROBABILITY

Then for complements, H^c is L, low, and M^c is F, few. The contingency table of probabilities is now as follows:

	H (red)	L (white)	
M (spotted)	0.1	0.4	0.5
F (plain)	0.3	0.2	0.5
	0.4	0.6	1.0

Here the corresponding patterns on the balls have been retained to aid the understanding of what follows. Using the notation introduced in §3.7, we can write, for example, $p(M) = 0.5$ and $p(H) = 0.4$. Recall that a fixed knowledge base is supposed and is omitted from the notation for simplicity, though it should not be forgotten. The event of both high inflation H and many unemployed M, considered earlier, is the event that may be denoted H & M, though it is usual to omit the ampersand and write this simply as HM. Thus the occurrence of both events is the new event written by putting the two symbols together. This is only one way of combining two events; another will be encountered later in §5.1. The consideration above gave $p(HM) = 0.1$ and from the three assessments we saw that $p(LF) = 0.2$, among others that could have been derived.

Let us look at the uncertainty in this table in another way. So far, the two features of inflation and unemployment have been treated symmetrically, whereas another possibility is to think about how one, say inflation, might influence unemployment, and see how the numbers in work are dependent on a change in the value of money. This viewpoint would lead you to think about what would happen to unemployment were inflation low. Notice the use of the subjunctive mood here; it is not known that inflation is low, you are merely thinking about what might happen were it to be low in a year's time. So let us look at your uncertainty of M were L true, and show that your uncertainty here can be expressed as a probability, obtained from the numbers already in the table. If L obtains, the equivalent for the standard is the withdrawal of a white ball; so supposing that L is true corresponds to supposing that the ball is white. If the ball is white, the only uncertainty lies in whether it is spotted, corresponding to many unemployed, M. But of the 60 white balls, 40 are spotted, so the proportion of spotted balls among the white is 40/60 = 2/3, and it is proportions that can be equated to probabilities. Consequently, it is not necessary to think about your probability of M were L to obtain; it has already been found as 2/3 from the table of your original judgments. Expressed as a decimal, to conform with the others, to two decimal places, this is 0.67, and in this style, 0.2 in the table should be 0.20 and the others similarly. A reference back to the end of §2.9, where decimal representation was discussed, may help here.

This probability is written in mathematical notation as $p(M|L)$ and is read "your probability of M were L true". The vertical line has been used in §3.9 to mean "given", but here it means "were". The distinction will be discussed in §4.6. Some writers make a distinction between probabilities such as $p(M)$, which only explicitly refer to one event, and those such as $p(M|L)$, which mention two. The former are called, as here, probabilities, but the latter are termed "conditional probabilities", differing in that although both refer to the uncertainty of M, the latter is conditional on L. The latter term will not be used, except in special circumstances, because we view all probabilities as conditional, on at least the knowledge base K (see §3.9) and the adjective is superfluous. Thus, in our view, $p(M|K)$ and $p(M|L \, \& \, K)$ differ only in the conditions and not on the type of probability.

The value of $p(M|L)$ has been calculated, in terms of the standard of balls in an urn, as 40/60 but it could equally be found in terms of probabilities, which are proportions of balls, rather than numbers. Thus, from the last table, $p(M|L)$ is 0.40/0.60, which is identifiable as the ratio of $p(ML)$ to $p(L)$, thus enabling the standard to be forgotten and all the calculations to be expressed in terms of probabilities. This idea will form the basis of the multiplication rule of probability to be developed in §5.3; for the moment we only need to notice the calculation of the "conditional probability" as the ratio of two probabilities.

What these ideas show is that from your three, original uncertainty judgments, $p(H)$, $p(M)$, and $p(HM)$, many other uncertainties can be deduced, such as $p(M|L)$, by coherence within your belief system. The reader might like to try others that follow from the table above; thus $p(M|H) = 0.10/0.40 = 0.25$. The reverse effect of unemployment on inflation can be found by considering probabilities such as

$p(L|M) = 0.40/0.50 = 0.80$ and $p(H|F) = 0.30/0.50 = 0.60$. All are expressions of how one event is contingent upon another. It is worth emphasizing, at the cost of repetition, that all these probabilities, and many others, have been obtained from the three original ones, $p(H)$, $p(M)$, and $p(HM)$, by coherence. You need not have started from these three. Another popular method is to think about $p(L)$, from which $p(H)$ is immediate as the complement, and then, using the standard urn, consider $p(M|L)$ and $p(M|H)$; taking inflation first and then thinking how unemployment depends on it. The reader will easily be able to obtain all the entries in the table from these three, just as with the original three. The order may be reversed, starting with M (and F) and then considering L. You are free to make what judgments you like about some events, but once having made them, you are no longer free to judge others; their judgments can be calculated from the original values. If you do not like the probabilities obtained by calculation, then your only resource is to return to the original values and change these until overall coherence is attained and you are satisfied that all the numbers reflect your beliefs. This idea will be used as a basis for some probability assessments in Chapter 13. Coherence is our most important tool in the evaluation of our uncertainty, and this book is not about what your uncertainties must be but about how they must cohere.

4.3 INDEPENDENCE

The study of the uncertainty relations, expressed through probability, between unemployment and inflation are continued in this section and the contingency table of probabilities is repeated here for convenience, with the slight change that a second decimal place, always a 0, has been included since the calculations that follow will be done to this precision.

	H	L	
M	0.10	0.40	0.50
F	0.30	0.20	0.50
	0.40	0.60	1.00

$p(M|L) = \dfrac{0.40}{0.60}$

From the final column of this table, we see that your probability of there being many unemployed next year, $p(M)$, is 0.50. But suppose there was low inflation next year; then the previous column shows that your probability of many unemployed has increased to $0.40/0.60 = 0.67$ using the ratio concept of the last section. In symbols, $p(M) = 0.50$ but $p(M|L) = 0.67$. This is a quantitative expression of your belief that low inflation will tend to increase unemployment. (Some economists may not agree with this, so we repeat, the values are not mandatory; insert your own, but be coherent.) Also recall that these are your judgments now about events a year ahead.

Suppose that you did not have this belief but felt that inflation and unemployment next year were unrelated. Were this so, and you still felt that $p(M) = 0.50$, you would have $p(M|L) = 0.50$ as well. What would the entries in the table look like then? Again recognizing $p(M|L)$ as the ratio of $p(ML)$ to $p(L)$, the entry against ML must be 0.30. All the other entries in the table then follow by simple arithmetic as before and the table will look as shown below:

	H	L	
M	0.20	0.30	0.50
F	0.20	0.30	0.50
	0.40	0.60	1.00

$p(M|H) = \dfrac{0.20}{0.40} = 0.50$ $\qquad p(M|L) = \dfrac{0.30}{0.60} = 0.50$

This table was derived on your view that L did not affect M with $p(M) = p(M|L)$. But from the numbers in the table, it can be seen that it is also true that $p(M) = p(M|H)$, so that high inflation would similarly not affect your uncertainty regarding the number of unemployed. Not only this, consider the effect the other way round, of unemployment on inflation. $p(L) = 0.60$ but equally, $p(L|M) = 0.30/0.50 = 0.60$ and similarly, $p(L|F) = 0.60$. Again we have an example of coherence. Once you have decided that L did not affect M, you have decided that no aspect of inflation, high or low, effects unemployment, many or few; nor does unemployment affect inflation. In this case, we say that the two events are independent. Let us make this more formal and, to clarify a further point, recall the knowledge base on which all your judgments of uncertainty have been made. The concept is stated for any two events, E and F, and not just for the specific events of high inflation and many unemployed.

If two events E and F are such that, on knowledge base K, you assert that
$$p(E|K) = p(E|FK),$$
then we say that you judge E and F to be *independent given* K.

The motivation for the use of the word independent is that your uncertainty of E is independent of the inclusion of F. From the considerations just advanced with inflation and unemployment, F^c similarly does not affect E, nor does E or E^c affect F, so that the relationship of independence is symmetric between the two events and you can talk about the independence of them, or that one of the events is not contingent upon the other. To exhibit this symmetry between the two events, the defining equation just displayed can be written as
$$p(EF|K) = p(E|K) \times p(F|K).$$
In words, your probability of both events happening is the product of their separate probabilities. The reader will be able to verify this using the ratio considerations, though the point will be examined more carefully when the multiplication rule is introduced in §5.3.

Independence is an extremely important concept in the study of uncertainty. A glimpse into why this is so can be seen from the contingency table, for if two events, M and L in the example, are independent, you will only have to think about $p(M)$ and $p(L)$, or equivalently, the numbers of spotted and of white balls in the urn. All the other probabilities, or numbers of balls that are both spotted and white, will follow from the admission of independence, and you can obtain the body of the table from the margins. Without independence, many calculations of uncertainties would become prohibitively difficult.

4.4 ASSOCIATION

In your original assessment of the uncertainties involving the two events of low inflation L and many unemployed M, you did not regard them as independent on your knowledge base. Instead, you thought that the probability of many unemployed would be increased were the inflation to be low. In symbols, $p(M|L)$ exceeds $p(M)$, omitting reference to K. The numbers were 0.67 and 0.50, respectively. The same inequality is true if every event is replaced by its complement. Thus, $p(M^c|L^c)$ exceeds $p(M^c)$, or $p(F|H)$ exceeds $p(F)$, the numbers being 0.75 and 0.50, respectively. The same inequality even persists if the events are interchanged. Thus, just as $p(M|L)$ exceeds $p(M)$, $p(L|M)$ exceeds $p(L)$, the numbers being 0.80 and 0.60, respectively. Indeed, the example was constructed to reflect the view of one economist who thought that more people in work (event F) would put money in peoples' pockets and thereby increase inflation (event H); $p(H|F) = 0.60$ exceeds $p(H) = 0.40$. We say that two events M and L are positively associated if the occurrence of one increases the probability of the other, or generally,

If, on knowledge base K, two events E and F are such that

$$p(E|FK) > p(E|K),$$

then the two events are said to be *positively associated* on K.

(Here use has been made of the symbol >, meaning "greater than", explained in §2.9.) It then follows that $p(F|EK) > p(F|K)$ on reversing the roles of the two events, and the same inequalities hold if both events are replaced by their complements.

A senior policeman was quoted as saying that the proportion of members of an ethnic minority among those convicted of mugging was higher than the proportion in the general population. In our language and omitting reference to a fixed knowledge base, $p(E|C) > p(E)$, where E is the event of belonging to the ethnic minority and C conviction for mugging. This association implies that $p(C|E) > p(C)$, the members of the ethnic minority are more likely to be convicted of mugging than is a random member of the population. To some, the second statement sounds more racist than the first, yet they are equivalent. We return to this example in §8.7.

Notice that the inequalities are reversed if one of the events is replaced by its complement but the other is not. Thus, if E and F are positively associated, then $p(E|F^c) < p(E)$. In our example, $p(M|L) > p(M)$, so that M and L are positively associated, whereas, recalling that L^c is H, $p(M|H) < p(M)$, the numbers being 0.25 and 0.50, respectively. The situation will become clearer when the multiplication rule of probability has been introduced in §5.3. If, in the definition of positive association just displayed, $>$ is replaced by $<$, or in words, "exceeds" is replaced by "is less than", the events are said to be *negatively associated on K*. The upshot is that the positive association between E and F implies positive association between their complements, but a negative association between E and F^c or between E^c and F. In your judgment, low inflation is positively associated with many unemployed, but is negatively associated with few unemployed. If E and F are independent on the knowledge base, $p(E|F) = p(E)$ and there is zero association. Numerical measures of the amount of association have been introduced, and are much used, but need not concern us here.

4.5 EXAMPLES

Some of the examples in §1.2 are extended here to illustrate the ideas of independence and association. If, to the event of rain tomorrow (Example 1), we add the event of rain on the day after tomorrow, then you would not ordinarily regard them as independent because in many places, weather on successive days tends to be similar and you would ascribe positive association between them. Many events that occur in time sequence exhibit this positive association, as high inflation one year tends to be followed by high inflation the following. Negative association within successive members of a sequence is rare.

Two almanac events (Example 2), would usually be independent. Thus, the capital of Liberia being Monrovia is totally unconnected with the population of France being above 60 million, in that being told the truth of one would have no effect on your uncertainty of the other. Of course, if they were both events concerning France, there might be some association.

In legal trials (Example 3), independence is often appropriate. For example, evidence about the defendant being at the scene, and evidence about the possession of a weapon, may be independent in your view. However, there can be subtle connections that result in associations that are hard to handle coherently. For example, if the same witness is involved in the scene and weapon evidence, there may be reason to think of an association. This is discussed further in §10.14.

The medical examples of selenium (Example 4) and saturated fat (Example 15) remind us that association is a problem that often arises, especially in the treatment of patients, because the effect of one medicine may be influenced by other medical aspects. Thus it is possible that the beneficial effect of selenium may be reduced if saturated fat is removed from the diet. Drug companies have to be aware of the possible interactions of one drug with another and the possibilities become even more complex when more than two events are involved, as will be seen in Chapter 8. The lack of independence has bedeviled much medical research so that special techniques have been developed to overcome it.

The event that a card will be an ace (Example 7) is not independent of the event that a second card, drawn at random from the same pack without replacement of the first ace, will also be an ace. For if the first event occurs, there are only 3 aces left in the pack, now of 51 cards, and your probability of an ace is reduced from 4/52 to 3/51 and there is negative association. The event that High Street will win the 2.30 race (Example 8) is certainly not independent of another horse winning the same race, but it may be reasonable to judge it independent of Congress winning the 3.15.

The electricity industry in Britain, rather than the nuclear (Example 13) provides a good example of an unsound judgment of independence. The electricity grid, carrying supplies around the country, is designed so that a failure in one part of it can be compensated by rerouting the electricity, with the intention that no place suffers an interruption to the supply. Even two failures can be allowed for, with a third route available. In the original calculations of supply, failures were supposed to be independent and the calculations of the small probability of a place having no supply were based on this assumption. However, one cause of failure is the trees fouling the electricity cables, especially in a high wind in the spring when growth of the trees is vigorous. Thus a storm in April can cause several failures, revealing that independence of interruption events is not a sensible assumption. Once this is recognized, the common cause can be introduced, and under a different knowledge base incorporating the storm, independence recovered, reminding us how important the base is in the definition, for E and F may be dependent under K, but independent under GK, for some G (see §8.8).

History (Example 5), as was seen in §3.10, is affected by a nonnumerate form of coherence and is similarly conscious of association in a more literary guise than that presented here. Politics (Example 12) is replete with associations such as those between social class and voting. Independence plays an important role in science (Examples 15 and 20) because the ability to repeat scientific experiments is a key concept, as explained in §11.11, and the repetitions are typically independent because they are performed by different scientists in different environments. Most experiments, especially those in the biological sciences, have an element of uncertainty in them that has to be handled with a statistical analysis that almost always involves basic assumptions of independence. Indeed, some writers have been so overwhelmed by independence that it dominates their understanding of probability to such an extent that a popular statistics textbook hardly mentions conditional probability. Scientific procedures for handling association within an experiment are discussed in §11.4.

4.6 SUPPOSITION AND FACT

In §3.9, the vertical line in the mathematical expression for probability was used in the sense of "given". Thus, $p(E|K)$ was your probability that the event E was true, given your knowledge base K, whereas in §4.2, the same line has been used in the sense of supposition, using the subjunctive. Thus, $p(E|F)$ was your probability that the event E was true, were another event F true. There is a vast difference between knowing something to be so, as with K, and supposing it to be so, as with F, so is this a case of sloppy mathematical language? We think not and explain here the apparent liberty of using the same symbol for apparently different concepts.

The notation $p(E|F)$ omits reference to the knowledge base; so let us temporarily introduce some more complicated, but accurate notation and write $p(E|F:K)$ for your probability of E on the supposition that F is true, and knowing K; the colon separating the supposition, on the left, from knowledge, on the right. Now contemplate the two probabilities

$$p(E|FG:K) \quad \text{and} \quad p(E|F:GK),$$

where G is a third event. The only difference between these two probabilities is that in the one on the left, the truth of G is mere supposition, being to the left of the colon, whereas in the probability on the right, G is known to be true, being to the right of the colon, while all the other elements in the two expressions are the same. Consequently, the change from the left-hand probability to the right-hand one is entirely accounted for by your learning that G is not just supposition, but a fact. We now make the assumption that the change from supposition to fact does not alter your uncertainty of E, and therefore your probability; in other words, the two probabilities displayed above are equal.

At first sight, this assumption looks wrong. If you have learned that inflation was low, rather than merely supposing it to be low, you might well appreciate the uncertainty about unemployment differently. By contrast, we saw in §4.5 that the probability of a second ace when one had already been drawn was 3/51, and the value would surely be the same whether you were merely making a supposition and thinking about the situation before any card had been taken, or whether you had actually seen the first ace. So the assumption is sometimes reasonable. Why the difference between inflation and the aces?

The answer is that when you learn about inflation, you almost certainly learn about something else as well. Recall that it is inflation next year that is under discussion, and so the event can become a fact only after the passage of a year. During that year you will have learned many other things. In other words, your knowledge base will have changed, apart from the extra knowledge of inflation. But in the two probabilities displayed above, about which the assumption was suggested, they had the same base, K. So the assumption amounts to saying that if G passes from supposition to fact, and nothing else happens, then the two uncertainties are the same, as with the aces. In this form it is often found acceptable. If it is, then there is no need to make the distinction between supposition and fact, and the vertical line can be used for either or both purposes. Hence, the notation $p(E|F)$ is adequate and there is no need for the complication.

4.7 SEEING AND DOING

Having developed the concept of your probability of an uncertain event E, given that you either know or suppose another event F to be true, a concept that has been denoted $p(E|F)$, we will, in the next chapter, develop the rules that such probabilities must obey, essentially the rules governing coherence between uncertainties. These rules are really simple and, with a little experience, are not hard to use. Because of the comparison with the standard, they are equivalent to the rules that obtain for the proportions of balls of different types in an urn. If the calculus of probability, at least at the level at which it will be used here, is as simple as the balls in an urn, then in contrast, there is an aspect of probability that really is difficult, so difficult that even experts in the field can make errors more frequently than they care to admit. The difficulty lies in the relationship between the probabilities, on the one hand, and the real uncertain world you are attempting to describe on the other. There is a difficulty in translating your opinions about inflation and employment into probabilities, and then, after calculating, translating the results of the calculations back into reality. It does not stretch language too far to say that the science is easy but the art is difficult. Examples of this arise in many places in this book, but here we explore the notion of association, as developed here, and an apparently similar notion of causation.

Let us return to the example of unemployment and inflation and consider carefully what you mean by $p(H|F) = 0.6$. You are contemplating now what the economy will look like a year ahead, in particular, on the two events of high inflation and few unemployed then, thinking of how they are associated and saying that if few are unemployed, then the probability of high inflation is 0.6, greater than it would be with many unemployed. Suppose that you, the person making this statement, are a politician concerned about unemployment in the country and are proposing to increase public expenditure, creating new jobs, and hence reducing the number out of work. You might think that high inflation would possibly result and even think that many people at work cause inflation because they have more money to spend. This could be incorrect since the original statements of association refer to a passive situation in which you are merely contemplating, whereas causation here reflects an intervention, by raising public expenditure. The contrast has been happily expressed by distinguishing between "seeing" and "doing". Association says that if you see F at the end of the year, then you will expect to see H.

Causation says that if you do something to make F true, you will get H. There is a lot of difference between seeing and doing because the latter involves intervention in the system, whereas the former merely demands observation.

Here is another example: a chemical engineer was disturbed by the variability of the product being made, and launched an investigation to find the cause of the variability. An obvious result of the study was that there was a critical temperature for the chemical reaction involved, either too hot or too cold and the product was unsatisfactory. In our language, an association was established between quality and temperature. The process being used had no real control over the temperature of the vessel in which the product was made, so it was decided to redesign the vessel so that it could be kept near to the critical temperature revealed by the study. This was done, at a considerable expense, only to discover that the quality of the product was markedly less than in the study. The reason was that the new design had affected other features of the production process, which were not previously thought important. The investigation had been concerned with seeing what happened. The new process was the result of doing something. Again we see that there can be a real difference between seeing and doing. There is a well-established association between heavy consumption of saturated fat and heart disease (Example 15) found by observing people, but it does not follow that reducing the amount of fat, doing something, will reduce deaths from heart disease, although more recent evidence suggests that it may be true. The association between smoking and lung cancer was established before it was shown that smoking was a cause of lung cancer.

We shall have little to say about causation but the distinction between seeing and doing will often arise, especially in connection with Simpson's paradox in §8.2. The general point to be made here is that in making your probability statements, you need to be alert to the precise interpretation of the events involved.

CHAPTER 5

The Rules of Probability

5.1 COMBINATIONS OF EVENTS

It has been shown how your uncertainty of an event E, when you know or suppose event F to be true, can be described by a number $p(E|F)$, termed your probability of E, given F. This is for a knowledge base that will be supposed fixed throughout the following discussion and therefore mostly omitted from the notation. In almost all practical cases, several uncertainties, or probabilities, are involved and it is necessary to combine them to reach an overall measure. Probabilities combine according to rules, and the aim of this chapter is to explain the rules so that you can perform the necessary calculations. There are three basic rules from which all others can be derived; one of them is slight and the other two are developed from the two ways in which events may be combined. This chapter begins with a study of these two ways.

We have already seen in §4.2 one way in which two events can be combined. If E and F are any two events, then the event that is true if, and only if, both events are true, was written $E\ \&\ F$, or more succinctly, EF. It is called the *conjunction* of the two events. If E is the event of rain tomorrow, Saturday, and F is the event of rain on Sunday, then EF

Understanding Uncertainty, Revised Edition. Dennis V. Lindley.
© 2014 John Wiley & Sons, Inc. Published 2014 by John Wiley & Sons, Inc.

is the event that it rains on both days. If E is the event that a person is white, and F is the event that the same person is male, then EF is the event that the person is a white male. If E is the event that the ball, taken at random from the standard urn, is red, and F is the event that it is spotted, then EF is the event that the ball is both red and spotted.

There is another way in which two events can be combined. Consider the event that is true provided at least one of E and F is true and is therefore only false if both events are false. It is called the *disjunction* of E and F and will be written E or F. (It is sometimes written $E+F$ but this can be misleading since the plus sign with events does not obey the same rules as it does with addition in arithmetic.) When E is the event of rain on Saturday, and F that of rain on Sunday, E or F is the event of rain at some time during the weekend. If E is the event of a ball being withdrawn from the urn and found to be red, and F the event of a similar withdrawal having black spots added, then E or F is the event of the ball being decorated, in the sense of having color or spots applied. If E is the event of high inflation and F that of many unemployed, then E or F is the event that the government will experience unpopularity either from the inflation or from the unemployed.

It may be helpful to distinguish between these two methods of combining events by presenting the rules of combination in the form of a truth table that describes the truth of a combination in terms of the truths of the original events. At the same time, recall that we have already seen another way of creating a new event, by means of the complement E^c in §3.7. Here is the truth table for all the three methods, conjunction, disjunction, and complement in which, for any row of the table, the status of the first two events determines the status of the other three. For example, in the first row, if both events are true, then so are the conjunction and the disjunction but the complement is false.

E	F	EF	E or F	E^c
True	True	True	True	False
True	False	False	True	False
False	True	False	True	True
False	False	False	False	True

There are other ways in which events can be combined, including combinations involving more than two events, but they can all be expressed in terms of the three given in the truth table.

The rule that governs the relationship between your probabilities of an event and its complement was considered in §3.7, where it was seen that your probability of the complement was one minus your probability of the event; or equivalently that your probabilities of an event and of its complement add to one. In symbols,

$$p(E^c) = 1 - p(E).$$

We now consider the rules that apply to the methods of combining two events by disjunction and conjunction. They are necessarily more complicated than that for the complement, involving, as they do, two events. The rules will be developed by the device of comparison with the standard of balls withdrawn at random from an urn, since what holds for the standard must also hold for the general concept of probability. Essentially, the rules of probability are just those of balls of different types in an urn.

5.2 ADDITION RULE

We first look at the way probabilities behave when two events are combined by disjunction, and begin by taking the standard urn that has been used before with balls that are either red or white, and simultaneously, either spotted or plain. Let R be the event that a ball, drawn at random from the urn, is red; and S the event that it is spotted. The combination R or S is then the event that the ball is decorated, either by color or spots, or both.

Now recall that probability is just the fraction of relevant balls in the urn, so, out of 100, we have merely to count the numbers to obtain the probabilities. A first reaction might be to say that the number of balls that are either red or spotted is the number that are red plus the number spotted. But a moment's reflection will show that this is false, for in so doing, the balls that are both red and spotted will have been counted twice, once as red, once as spotted. The following is the true state of affairs:

✳   The number of balls that are decorated is the number that are red, plus the number that are spotted, less the number that are both red and spotted.

In that sentence, the first number is related to the disjunction of the events, R or S, the next two to the individual events, and the last to the conjunction of the events, RS. Recalling that probability is the fraction of balls of the relevant type, we can divide every number in the statement above by 100 and interpret the results as probabilities, so that the statement is equivalent to:

The probability that a withdrawn ball is decorated, either by color or by spots or by both, is the probability that it is red, plus the probability that it is spotted, less the probability that it is both red and spotted.

Let this be written in mathematical language. We have

$$p(R \text{ or } S) = p(R) + p(S) - p(RS).$$

Since any two events E, F admit comparison with the urn, this result holds for any two events, so that

$$p(E \text{ or } F) = p(E) + p(F) - p(EF). \tag{5.1}$$

Equation (5.1) is the general rule for calculating the probability of either of the two events occurring, the disjunction. It is not a happy result since it involves not just the individual probabilities of the two events, but also the probability of the event that arises from the other method of combination, conjunction, EF. There is an important, special case where the result simplifies.

Two events E and F are *exclusive* if it is impossible for them to occur simultaneously. Alternatively expressed, the conjunction is impossible. The obvious case is where F is the complement of E, for an event cannot be both true and false. Here are some examples of exclusive events.

1. Inflation next year will exceed 6%. Inflation next year will be below 3%.
2. The defendant was at the scene of the crime in the club. The defendant was at home.

3. High Street will win the 2.30. Gladiator will win the 2.30.
4. Your neighbor will vote Republican. Your neighbor will vote Democrat.

If E and F are exclusive, then you will assess the impossible event, EF to have probability zero, $p(EF) = 0$. Equation (5.1) then simplifies to produce the result that

$$p(E \text{ or } F) = p(E) + p(F). \tag{5.2}$$

This result is much simpler than the general form (5.1) but it applies only to exclusive events and is therefore limited in scope. For obvious reasons, (5.2) is called the addition rule of probability, though we shall use the same term for the general Equation (5.1). In view of its fundamental importance, the addition rule of probability is now stated in full generality and, in particular, we recall that all uncertainties are relevant to a knowledge base K.

If E and F are any two events, uncertain for you on knowledge base K, then

$$p(E \text{ or } F \mid K) = p(E \mid K) + p(F \mid K) - p(EF \mid K). \tag{5.3}$$

This is the *addition rule of probability*. If E and F are exclusive on K, then

$$p(E \text{ or } F \mid K) = p(E \mid K) + p(F \mid K).$$

Although this result may seem, at first, a little complicated, though simpler when it relates to exclusive events, recall that it is only an expression about fractions of balls in the standard urn. When considering an example, it is often useful to think of the calculations in terms of fractions of balls.

5.3 MULTIPLICATION RULE

The addition rule deals with the combination E or F, the disjunction of two events, though, in general, it involves the conjunction EF as well. We now turn to this last form of combination and develop another rule, using

90 THE RULES OF PROBABILITY

the same device, as in the last section, of comparison with the standard urn. If, as before, R refers to the event of the ball being red, and S to the event of being spotted, RS is the event of being both red and spotted. The red balls are either spotted or plain, so the number that are both red and spotted is equal to the number that are red multiplied by the fraction of the red ones that are spotted. Dividing by the total number of balls, 100, it can be seen that the fraction of balls that are both red and spotted is the fraction that are red times the fraction of the red that are spotted. (It may clarify the statement to insert actual numbers of balls; for example, those in the third table of §4.1.) Replacing each of the fractions by probabilities, we have the following:

Your probability that a ball, withdrawn at random from the urn, is both red and spotted is your probability that it is red, multiplied by your probability that it is spotted, given that it is red.

Here the concept of conditional probability, explored in §4.2, has been used, equating the fraction of red that are spotted with your probability of being spotted, given that a ball is red. Finally, we turn the literary form into mathematical language and replace the special events, R and S, by general events, E and F, to obtain the result

$$p(EF) = p(E) \times p(F \mid E).$$

It is usual to omit the multiplication sign, as explained in §2.9, and write

$$p(EF) = p(E)p(F \mid E). \qquad (5.4)$$

As with the addition rule, let us next restate it in full generality including reference to the knowledge base.

For any two events E and F that are uncertain for you on knowledge base K,

$$p(EF \mid K) = p(E \mid K)p(F \mid EK). \qquad (5.5)$$

This is the *multiplication rule of probability*. *Product rule* is an alternative term.

Like the addition rule, it is merely an expression of a result concerning fractions of balls, and it can be useful to think in terms of these when calculating. Notice one important feature of the multiplication rule; it involves two knowledge bases, unlike the addition rule that had only one, K; for here, in addition to K, there is K augmented by E. This feature will play a vital role in describing how your uncertainties change when new information, E, is acquired.

In the case of the addition rule, it was seen that it simplified if the two events were exclusive, and that then the result could be expressed in terms of your individual probabilities of the two events, without the inclusion of the other combination EF. The multiplication rule can similarly be simplified and involve only the individual probabilities. This happens, not when the events are exclusive, but when they are independent, see §4.3. Recall that E and F are independent, given K, if $p(F \mid EK) = p(F \mid K)$ and using this in (5.5), we have the following result:

If E and F are independent, given K, then

$$p(EF \mid K) = p(E \mid K)p(F \mid K). \qquad (5.6)$$

The disjunction E or F is sometimes called the *sum* of the two events, and the conjunction EF the *product*. With this terminology, the last form of the multiplication rule reads that your probability of the product of two independent events is the product of their separate probabilities, a result that is attractive because it is easy to remember. Unfortunately, it is true only if the events are independent; otherwise it is wrong, and often seriously wrong. Similarly, (5.2) reads that your probability of the sum of two events is the sum of their separate probabilities. Again, this is true only under restrictions, but this time the restriction is not independence but the requirement that the events be exclusive. Simple as these special forms are, their simplicity can easily lead to errors and are therefore best avoided unless the restrictions that made them valid are always remembered throughout the calculations. The desire for simplicity has often been emphasized, but here is an example where it is possible to go too far and think of the addition and multiplication rules in their simpler forms, forgetting the restrictions that must hold before they are correct. Notice that the restriction, necessary for the simple form of the addition rule, that the events be exclusive, or the disjunction impossible, is a logical restriction, having nothing to do with uncertainty, whereas independence, the restriction with the multiplication rule, is essentially probabilistic. It is perhaps pedantic to point out that the simple form of the addition rule is correct if you judge the disjunction to have probability zero, rather than knowing it is logically impossible, but we will see in §6.8 that it is dangerous to attach probability zero to anything other than a logical impossibility.

5.4 THE BASIC RULES

There are now two rules that your probabilities have to obey: addition and multiplication. To these we add a third, one that is so simple that we have passed it by as obvious, for it merely says that any probability lies between the limits of 0 and 1 and that an event that you know to be true has probability at the upper limit of 1. This is strangely called the *convexity* rule. The three rules are now stated together:

Convexity Rule. For any event E with knowledge base K, your probability of E, given K, $p(E \mid K)$, is a number between 0 and 1. If, on K, you know E to be true, then your probability is 1.

Addition Rule. For any two events, E and F, with knowledge base K,

$$p(E \text{ or } F \mid K) = p(E \mid K) + p(F \mid K) - p(EF \mid K).$$

Multiplication Rule. For any two events, E and F, with knowledge base K,

$$p(EF \mid K) = p(E \mid K)p(F \mid EK).$$

It is a fact that can hardly be emphasized too strongly that these three rules encapsulate everything about probability, and therefore everything about your uncertainty measurement, in the sense that although the rules have been obtained through comparisons with a standard, all other properties of probability can be deduced from these three, and the standard forgotten. Although we have used the standard of balls drawn at random from an urn, it will be seen in §5.7 that other standards lead to the same three rules. All that you need to know about your uncertainty measurements, all the results that experts have obtained (and some are very sophisticated) are contained within these three rules. It is now possible to give a formal definition of the concept of coherence mentioned in §3.5. A person's beliefs are *coherent* if, when those beliefs are expressed in terms of probabilities, they obey the three rules just stated. In §5.7, it is shown that an incoherent person is potentially capable of losing money for sure.

As an undergraduate, I once attended a course of 24 lectures on Newtonian mechanics. The first lecture was devoted to a careful mathematical formulation of Newton's laws and, at the end, the lecturer explained that the remaining 23 lectures would merely consist of calculations based on these laws and went on to say that, since you can calculate, in a sense, the lectures are redundant. A similar feature obtains here, so that once you have understood the three laws of probability just stated, you can calculate for yourselves and not read further. Of course, the undergraduates continued with the lectures to gain experience in calculation and, more importantly, to see how to apply Newton's method. So I hope that you will continue with this book, but I do sincerely suggest that you ensure that the rules are thoroughly understood before proceeding.

All the properties of probability follow from the three rules, but equally, the term probability is used only in the sense of something that obeys the three rules. It sometimes happens that probability is employed to mean any number that measures belief, lying between 0 and 1; that is, merely obeying the first, convexity, rule. Here we will follow Humpty Dumpty in making probability mean exactly what we say, that is, obeying all three rules, not just a subset.

The fact that the three rules have been derived from assumptions, and not just invented, is not always appreciated. One cannot sit down in one's ivory tower of the Prologue and invent rules. This is because there are uncertain events of some simplicity (we have chosen to use balls in an urn) where convexity, addition and multiplication do hold. One school of thought replaces the last two by rules that use maxima and minima. These rules may be suitable in some contexts, but they do not obtain with balls in urns or other simple situations that we will meet later. People may sensibly reject our comparison of general uncertainty with balls in urns, or betting at the casino, but some justification for alternative rules, such as maxima and minima, is required.

Another way of expressing the ideas of the previous paragraphs is to say that the calculations of several uncertainties, forming a calculus of probabilities, are based entirely on addition and multiplication. What is also remarkable about this is that probabilities combine in two different ways: addition and multiplication. Most common concepts only combine in one way. Lengths add, but they do not multiply; when they do, they produce something different, area. You may add sums of money but it makes no sense to multiply 3 dollars by 7 dollars. Even human beings can combine in only one way to produce another human being, by the addition of sperm to egg, though there are now many ways of effecting the addition. Probability is so rich because of its two methods of combination, corresponding to the two ways that events may combine.

Despite this richness, recall that the three rules are only results concerning fractions of balls in an urn, expressed in a different language. If you have trouble thinking about some of the results in this book, do not be ashamed to think in terms of fractions of balls in an urn, if you find that convenient. Nevertheless, experience shows that it is better to forego the habit and calculate directly in terms of probabilities. Notice that the rules have been expressed in terms of probabilities and not in terms of the odds (§3.8) with which some people are more familiar. This is because the rules are easier to comprehend, and to use, in probability form. The interested reader might like to translate the addition rule into odds; the result is a mess. Bookmakers are familiar with the multiplication rule but sometimes do not understand the addition rule.

After the above had been written, a little voice in my ear said that I was not quite correct and that the addition rule in the form stated above is not adequate for all that had been claimed. "Don't forget conglomerability" it said, as if one could forget a word as long as that. The objection is sound; most probabilists do introduce a conglomerability rule, which cannot be justified by reference to the standard, and use it extensively to produce deep results. My contention is that conglomerability is just a mathematical device for handling infinities and not a basic property relevant to the finite situations that we shall encounter. We shall not need to use sophisticated tools such as integration or differentiation but can make do with simple devices that are adequate for a layperson's understanding. To assuage the curious, and to encourage curiosity, we will mention conglomerability in §12.9 when more experience of using the rules has been gained.

There is another important (some might say outrageous) claim for the three rules. Most of us like to think of ourselves as logical, though we often have difficulty in living up to the ideal. Now logic deals with the truth and falsity of events and the rules are principally captured in the truth table of §5.1; thus if E and F are both true, then it logically follows that EF, the conjunction, is also true. Probability deals with all events, principally uncertain ones, but true and false events are included with $p(E) = 1$ if E is true and 0 if false. I claim that probability is the unique extension of logic and that your ideal should not be to be logical, but to be coherent in the sense of the three rules. The grand assertion is that you must see the world through probability and that probability is the only guide you need. "Understanding Uncertainty" means knowing the three rules of probability. The language of life is that of probability. Probability is as essential as ethics, religion, physics, genetics, and politics. Probability should operate everywhere and is a feature of all these topics because uncertainty is present in all of them.

5.5 EXAMPLES

One result in probability was met in §3.7 but does not appear in the basic rules just listed; so to support our claim that all results in probability follow from the basic ones, let us derive the earlier result from these. There it was seen that your probability of an event and your probability of its complement necessarily added to one; yet this does not appear in the three rules and the complement is not even mentioned. To establish the correctness of this result, take any event E and its complement E^c. These two events are exclusive (§5.2) since they cannot both be true. The addition rule may therefore be applied in its simpler form of (5.2) to provide

$$p(E \text{ or } E^c) = p(E) + p(E^c).$$

96 THE RULES OF PROBABILITY

The event on the left-hand side of this equation, E or E^c, must be true since E is either true or false, and by the convexity rule, a true event has probability 1. Hence, the above result reads

$$1 = p(E) + p(E^c),$$

and the earlier result is obtained, thereby substantiating the claim that the earlier result follows from the three rules. Further rules that follow from the basic ones will be developed later.

In §3.5, a problem about birthdays was mentioned to illustrate the idea that, from some easily assessed probabilities, others that were harder to think about, could be found. Let us see how the rules achieve this and begin with three people. For each of them, you state that your probability that they were born on March 4 is 1/365, and similarly for any other date. You further believe that knowledge of the birthday of any of them would not affect your probability for any other; in the language of §4.3, on your knowledge base, the birth dates for different people are independent. The rules are now used to calculate the probability that at least two of the three share the same, unstated birthday.

To do this, we calculate the probability of the complementary event that none of them share a birthday, when the required result will be one minus this value, by the general result just obtained. Take the three people in order. The first person will have some birthday and for the second to have a different day, it must be among the 364 other days, so your probability is 364/365 by the addition over the 364 exclusive possibilities. Now take the third; their birthday is restricted to the remaining 363 days, so your probability that the day will be different from the first two is 363/365. By the multiplication rule, and using your assumed independence, your probability that all three will differ is 364/365 times 363/365. This is 0.9918 to four decimal places. It follows that your probability that at least two of them share a birthday is one minus this, at 0.0082. This is small, less than 1%.

In the original example, there were 23 people, not 3, but the general method of calculation is the same. Having 364 days available for the second, 363 for the third, there will be 362 for the next, 361 for the next, decreasing by one each time. For 23, your probability for all

23 having different birthdays will be

$$\frac{364}{365} \times \frac{363}{365} \text{ and so on until } \times \frac{343}{365},$$

the last fraction and product corresponding to the twenty-third person. A calculator enables this to be found to be about $\frac{1}{2}$. Hence, your probability of the complement, that at least two people share their birthdays, is also about $\frac{1}{2}$, as stated. Thus, from probabilities that are easily thought about, such as 1/365, others that are far from easy can be calculated. This use of the probability rules is typical.

As another example of the rules, consider the probability of a nuclear accident in Example 13 of §1.2. The following simplified version will illustrate the ideas of how this might be assessed. Suppose that an accident can occur only if two things simultaneously happen: first, a fuel rod jams, an event R; second, the cooling thermostat fails, an event C. Then you are interested in the event RC. The probability of this can be evaluated by the multiplication rule as $p(R)p(C|R)$. Suppose your probability of jamming is 0.01 and, if the rod jams, the probability that the thermostat will fail to respond to the consequent overheating is 0.04. In symbols, $p(R) = 0.01$ and $p(C|R) = 0.04$. Then the probability of an accident is the product $0.01 \times 0.04 = 0.0004$. Notice that although the two separate probabilities are modest, the product is small. In reality, more than two things will have to occur simultaneously for there to be an accident and as each involves multiplication by a probability, necessarily less than one, the accident probability decreases with each multiplication and its very small value can be assigned, provided the individual probabilities, of modest size, can be found. Again, in practice, there will be many ways for an accident to arise; each can have its probability found by this use of the multiplication rule and the addition rule used to combine them. Thus, if in addition to the failure method mentioned above with probability 0.0004, there was another method, exclusive of the first, with probability 0.0007, the total probability would be 0.0011. Two words of caution need to be included: first, the uncertainties refer to a fixed period of time, say over a year, and second, they refer to a fixed knowledge base.

5.6 EXTENSION OF THE CONVERSATION

The three rules are the basic ones but many others can be derived from them, a particularly useful one having the delightful name of the extension of the conversation. Although the proof of the rule involves some technical mathematics, which we said would be avoided in general, it is given here because the technicalities are very simple, and because it will perhaps show the reader, by example, how these simple ideas can be used to extract many other results from three basic rules. The reader prepared to embark on the journey of discovery should recall that probabilities are only numbers and that the manipulations that follow are essentially a combination of arithmetic and the basic rules; the language, such as $p(E\,|\,F)$, covering many possible arithmetical interpretations, one of which is given in the cancer example that follows. The reader who is not interested can proceed directly to the result, Equation (5.7) helped by the literary equivalent in the following paragraph.

Take any event E whose probability you wish to find, and let F be another event. From these, two other events can be constructed, EF and EF^c. The latter is the event that is true if, and only if, both E is true and F is false. The events EF and EF^c are exclusive since F cannot be both true and false. The addition rule in its simpler form (5.2) gives

$$p(EF \text{ or } EF^c) = p(EF) + p(EF^c).$$

The event on the left is just E since the truth of F is irrelevant. So

$$p(E) = p(EF) + p(EF^c).$$

Next, apply the multiplication rule to each of the two terms on the right. Thus, $p(EF) = p(E\,|\,F)p(F)$, and similarly with F^c. The result is

$$p(E) = p(E\,|\,F)p(F) + p(E\,|\,F^c)p(F^c). \tag{5.7}$$

This is the rule of the *extension of the conversation* and it holds for any two events on a common knowledge base. The reason for the terminology is that you are considering, or conversing, about E on the left of (5.7), and extending the conversation on the right to include event F.

Like all the results, this can be thought of in terms of ratios of balls in the standard urn. Let E correspond to red and F to spotted. Then the rule says that the proportion of red balls is equal to the proportion of red among the spotted, times the proportion of spotted, plus the proportion of red among the plain, times the proportion of plain. The last sentence is the literary equivalent of (5.7).

An immediate question is what use is this rule; why extend the conversation to include another event, making life more complicated? The reason is that the conditional probabilities of E that appear on the right-hand side of (5.7) are often easier to think about than those of E on its own. Here is an example that we meet again in §6.4. Suppose there is a clinical test for cancer that can yield either a positive + or negative − result but is not perfectly reliable, thus introducing an element of uncertainty. Suppose that all the patients showing a positive result have to go for a more extensive analysis. It then becomes important to know how probable it is that a person will show + on the test. With a clinical test, it is usual to know how good it is in the sense that the probabilities of false positives and false negatives are agreed and known. That is, if C is the event of having cancer, $p(-\mid C)$ for the false negatives, and $p(+\mid C^c)$ for the false positives, are known. Here C^c is, as usual, the complement, not having cancer. If the conversation is extended from + (playing the role of E) to include C (playing that of F) then, from (5.7),

$$p(+) = p(+\mid C)p(C) + p(+\mid C^c)p(C^c). \qquad (5.8)$$

Finally, with $p(+\mid C) = 1 - p(-\mid C)$ and $p(C^c) = 1 - p(C)$, your required probability of a positive outcome, $p(+)$, has been expressed in terms of the known falsity rates and your probability that, before the test, a patient has cancer, $p(C)$.

Here is a numerical example. The rates are low for a good test, so suppose

$$p(-\mid C) = 0.01 \text{ and } p(+\mid C^c) = 0.05.$$

The first value means that a patient who truly has cancer has only one chance in 100 of slipping past the test undetected and therefore is almost certain, probability 0.99, of being detected, $p(+\,|\,C) = 0.99$. The second value means that a cancer-free patient still has probability 0.05 of a positive test result. If only 2% of the population has cancer, and you consequently take $p(C) = 0.02$, your probability of a positive result is, by (5.8),

$$0.99 \times 0.02 + 0.05 \times 0.98 = 0.0688.$$

So you assess the probability of a positive result at almost 7%, much greater than the 2% of the patients who truly have cancer.

The analysis just advanced recognizes that a positive test result can arise from one of two causes; a patient with cancer can be correctly diagnosed, $p(+\,|\,C)$, or a healthy patient can respond incorrectly, $p(+\,|\,C^c)$. If all patients had cancer, only the first case would apply and $p(+) = p(+\,|\,C)$; if none, then only the second operates and the errors are experienced, $p(+) = p(+\,|\,C^c)$. The general result (5.8) is a combination of these two cases and the formula for the extension of the conversation reflects this, adding a proportion $p(C)$ of the first value, and the complementary proportion $p(C^c)$ of the second. We say that the conditional probabilities have been *mixed* and (5.7) reflects this mixture of $p(E\,|\,F)$ and $p(E\,|\,F^c)$ in proportions $p(F)$ and $1 - p(F) = p(F^c)$.

It has repeatedly been emphasized that probabilities behave like proportions of balls in urns, so let us redo the cancer calculations in those terms without any of the mathematical apparatus. Mathematicians rather frown on this form of argument but it is useful for those who find the symbolism too abstract and, most importantly, it is correct; its serious disadvantage is that it only handles special, numerical cases and does not, unlike the general Equation (5.7), reveal the structure of the edifice that is your logical way of thinking about uncertainty.

Instead of our usual urn of 100 balls, let this one have 10,000 balls, otherwise the numbers will be uncomfortably small. With 2% of the patients having cancer, there will be 200 balls labeled "cancer" and the remaining 9,800, "cancer free". The falsity rate for the former was 1%, so 2 will register negative, leaving 198 positive. The falsity rate for the latter was 5%, 1 in 20, so 490 will register positive. Hence, the total number of positives is $198 + 490 = 688$ out of 10,000, exactly as before. Notice that the high rate of positives, nearly 7%, compared with the cancer rate of 2% is mostly due to the 490 healthy patients that the test got wrong. An alternative way of laying out the arithmetic is by the use of a contingency table, as was done with unemployment and inflation in §4.1.

5.7 DUTCH BOOKS

Let us see where we have got to in the argument from where we began with examples of uncertainty in Chapter 1. As a result of some assumptions about the measurement of uncertainty involving comparison with a standard of balls in an urn, we have demonstrated that the measurement, called probability, has to obey three rules, convexity, addition and multiplication, which form the basis of a calculus that can be used to make your appreciation of uncertainty coherent. This derivation of the rules from a standard has been used because it is perhaps the simplest and the least free from objections. However, some readers may not be convinced and their doubt is not at all unreasonable. You may feel unhappy using a single number to describe something as subtle as not knowing, or you may be concerned at a comparison that in many cases, for example with inflation, you would have difficulty in making. In this and the following sections, we discuss how other, quite different approaches lead to the same rules. That is, whatever way uncertainty is approached, probability is the *only* sound way to think about it. The alternative approaches will only be dealt with in outline, hopefully enough for you to appreciate their main ideas. If probability is the only sound way to think about uncertainty, it is valuable to have many derivations of the rules, thereby strengthening your confidence in the rules.

One derivation has been briefly mentioned in §3.6 that involved gambling. As usual, let E be an uncertain event and suppose a gamble on E is offered by you at odds of 5 to 1 against, meaning that, for a stake placed by another, you will pay out five times that stake if E is subsequently found to be true, and return the stake. If E is not true, then you will retain the stake. The importance of numbers, as has already been explained, lies in their abilities to combine easily, so let us suppose you contemplate a second gamble, but this time on the complement of E, denoted, as usual, E^c. With two gambles, something interesting happens. Here is an example.

Suppose that you offer a gamble on E at odds against of 1 to 1, commonly called "evens", and, at the same time, offer one on E^c at odds against of 2 to 1. Next suppose that I come along and place a stake of 3 on the first gamble and a stake of 2 on the second. What will happen? If E is true, then you will lose 3 on the first and gain the stake of 2 on the second—a total loss of 1. Suppose E is false, E^c true, then you will keep the stake on the first, a gain of 3, but will lose 4 on the second, because you will have to pay out twice (2 to 1) the stake (of 2). So your total loss will be 1. Hence, whatever happens, whether E is true or false, you will lose 1. You might just as well give me 1 and forget about the gambles.

If, as here, I can choose the stakes such that I will win for sure, and you will lose for sure, we say that I have made a *Dutch book* against you. It has just been shown that if you give odds of evens against an event and odds of 2 to 1 against its complement, then you will lose money for sure with an intelligent placing of stakes. (The stakes of 3 and 2 were selected deliberately.) Clearly, you want to avoid the possibility of a Dutch book and the question is how can this be done? The answer is to turn the odds into probabilities, as with equation (3.3) in §3.8, and arrange the probabilities to add to 1. In the example, your equivalent probability for E was 1/2, that for E^c was 1/3. These do not add to 1. There is more discussion of this in §14.5.

The method of the example may easily be extended to prove that the avoidance of Dutch books implies one of the rules of probability, namely,

$$p(E) + p(E^c) = 1,$$

as obtained in §3.7. Using more complicated combinations of events and their associated gambles, it is possible to derive all three rules of probability. Hence, we have here a quite different approach to uncertainty, employing gambles, which leads to exactly the same results, thereby adding to our confidence that the results are correct. There are three reasons why the gambling approach has not been employed here. First, many people, understandably, object to gambling. Second, there are difficulties with uncertainty being confused with desirability, see §3.6. The third reason is that the proofs are more complicated than those presented here. Notice that in the urn approach, all the rules of probability have been proved, not merely presented as plausible. A proof seems to me to be the best way of convincing you that the rules are correct, indeed, inevitable.

One final remark about Dutch books before we leave them. If you go to a race meeting and investigate the odds offered by a bookmaker against the horses in a single race, you will find that it is always possible for you to arrange your stakes such that you will *lose* money for sure. That is, the bookmaker has arranged his odds such that he has the potentiality for a Dutch book against unwary gamblers. Of course, he cannot guarantee this, since he does not control the stakes, but this is how he makes his money. Turn the bookmaker's odds against into probabilities and you will find they always add to more than 1. Again See §14.5.

5.8 SCORING RULES

Some people reasonably object to the derivation of the rules used here because they feel that the standard is not usable, or operational, though we attempted to overcome this objection in §3.5. Here is a method of deriving probability that is operational, though unfortunately it has been little used.

Suppose that I ask you to give me a number that describes your uncertainty for an event E, where you are free to use whatever process you like, even one that merely provides a number that keeps this annoying inquisitor quiet. But you are told that, if the event is subsequently found to be true, you will be given a score that is the square of the difference between your number and 1. If it is false, you will be scored by the square of the difference between your number and 0. For an explanation of "square" see §2.9. Thus, if you say 0.7 and the event is true, you score $(1 - 0.7)^2 = 0.3^2 = 0.09$; if false, you score $0.7^2 = 0.49$. In practice, the scores are multiplied by 100, giving a modest 9 if the event is true but a more substantial 49 if false. In symbols, if you provide a number x you will score $(1-x)^2$ or x^2 according to whether the event is true or is false. What number x will you give? The scores are to be thought of as penalty scores so that you aim to make them as small as possible. Furthermore, you may be asked to provide uncertainty numbers for other events, in which case the scores for the various events will be added to produce a total score.

It is easy to see, and so easy that it is left as an exercise for the reader, that your score will be smaller, and hence better, if you obey the convexity rule of probability. That is, the number provided must lie between 0 and 1, and that if you know the event to be true, it must be 1. Using these scores for several events, it is possible to prove the addition and multiplication rules of probability. That is, the numbers that you give must satisfy those rules or else you will necessarily receive a larger penalty score than you need have done. The proofs here are not so easy and are omitted, which is a pity since the proof of the multiplication rule is one of the most beautiful pieces of modern, simple mathematics. The use of a scoring rule based on squares therefore leads to the same rules of probability. It is called a *quadratic* rule.

Nevertheless, an objection will occur to many of you; why use those scores, other scores might have given different results; for example, those based on maxima and minima mentioned in §5.4. What happens if scores that are not quadratic are used? There are two possibilities. The first type of score means that effectively you will think of any event as either "true" or "false" with no shades of meaning in between. Alternatively expressed, you will, to any event, assign one of two numbers, one number can be interpreted as "true", the other as "false". The second type of scoring rule will lead to a range of numbers, which may not be probabilities, but will be capable of being transformed into probabilities. For example, one scoring rule leads to your stating odds. These, as we have seen in §3.8, can be easily transformed into probabilities. The first type of rule is unsatisfactory because it does not distinguish between different strengths of beliefs in an uncertain event. Nevertheless, a pupil under instruction is often forced by their teacher to say "true" or "false" when uncertain, so denying their uncertainty. (The comments in §3.10 are relevant here, as is the problem in §12.5.) There is no general agreement on which, among the second type of scoring rule, is best. The choice between them may depend on other factors. For example, odds are preferred by some people, probabilities by others. An important conclusion is that there are no scoring rules that lead to the maxima and minima rules mentioned in §5.4. Scoring rules necessarily lead to probability, or something equivalent to it, such as odds.

5.9 LOGIC AGAIN

A modern computer operates according to the rules of logic. Each statement is regarded as either true or false and the calculations operate within the rules to produce other statements that are similarly either true or false. The computer does not deal with uncertainty directly but can only handle uncertainty by operating with probabilities, or other numbers, which themselves obey the rules of arithmetic. Suppose we were to think of a machine that dealt with uncertainty directly rather than with just truth and falsity; what rules would it have to obey? The person "you" that has been used previously is replaced here by a machine and we are enquiring what rules this machine would have to use in a generalization of logic (compare §5.4). This is a complicated matter but let me try to convey an outline of the ideas involved.

As before, take two events, E and F. There are several uncertainties here such as those of

$$E, \quad F, \quad E|F, \quad F|E, \quad EF,$$

discussed in Chapter 4. Thus, there is the uncertainty of one event, were the other known to be true, exemplified by $E|F$. Some basic requirements establish that there must be relations between these five uncertainties, in the form that some must be functions of the others. What functions could these be? Here the mathematics becomes a little involved, but the result is that the relationships are just those of the probability rules and the functions are just those of addition and multiplication. In other words, we are back to the familiar territory, even though the starting point and the route have both been different. In many ways this is the best way of deriving the rules because it assumes so little and because it exhibits the powerful feature of probability, namely that it is a generalization of logic. It has the pedagogical objection that the proofs are hard, much harder than those offered here with balls in urns, which is not too serious provided the reader has trust in the mathematician. It also has the more cogent objection, namely that it encourages the idea that a computer can measure uncertainty. This is not so; all the material does is to provide the rules and it does not say what the probabilities should be, only how you (or the computer) should manipulate those it has. This point has been made in §1.7: you are free to believe what you wish within the bounds of the rules prescribed by probability, but these you must never offend. Unfortunately, many writers have not appreciated this point and tried to develop a machine concept of ignorance, from which other uncertainties could be derived. This is unsound. A machine cannot know what an uncertainty is any more than it can tell whether an event is true or false, except in comparison with other events that it has been told about.

5.10 DECISION ANALYSIS

There is one other way of justifying the rules of probability. It ignores the concept of uncertainty directly and instead inquires how you should *act* in the face of uncertainty. Forget about the uncertain events; only consider whether you should act this way or that when faced with a situation in which you do not know all the facts. The emphasis is on the action, rather than on thinking about the events, as we have done. Alternatively expressed, you need to decide what to do, so the topic is called "decision analysis". We shall not have anything to say about this here, because in Chapter 10 it is demonstrated how the results we already have, enable you to act in the face of uncertainty. With this before us, it will be easier to understand the advantages and disadvantages of decision analysis. We shall also see how decision analysis leads us back to probability, in that coherent actions necessitate uncertainty being so described and the three rules used.

The upshot of the material in the last four sections is that we have five, rather distinct methods of establishing the rules of probability. These are

1. Comparison with a standard.
2. Avoidance of Dutch books in gambling.
3. The use of scoring rules.
4. An extension of logic.
5. Action in the face of uncertainty.

These can be thought of as five pillars supporting the same edifice, the edifice of probability, and whichever way you look at uncertainty, the end result is the same. It is possible to use other approaches that lead to different rules, for example, involving upper and lower probabilities, see §3.5. These can be dismissed on grounds of simplicity, or for confusing the idea of measurement with the practice. By contrast, there are no approaches that lead to rules of comparable simplicity to those of probability. We can therefore go forward in the real confidence that our rules are the proper ones to use. Recall that although they may appear, especially in their use, to be complicated, in reality they are only expressions of simple properties of proportions of balls in urns.

We next go on to develop, from the three rules, another rule of great importance; so important that it deserves a chapter to itself. But before doing so, consider two words of caution lest my enthusiasm for probability becomes too gross.

5.11 THE PRISONERS' DILEMMA

Our cautionary tale concerns two prisoners, Ann and John, where each has separate decisions to make, whether or not to confess to a crime. Their dilemma is described in the contingency table below, though it differs from the tables encountered in Chapter 4

108   THE RULES OF PROBABILITY

in that the entries are not probabilities, but the consequences of their decisions.

		JOHN'S DECISION	
		Confess	Not Confess
ANN'S	Confess	0,0	2, −2
DECISION	Not Confess	−2,2	1,1

Ann controls the rows, the upper corresponding to her confessing, the lower applying to when she does not. John similarly controls the columns, on the left confessing, on the right not confessing. The entries in the table are each a pair of numbers, the first being the reward to Ann, the second that to John. For example, if neither confesses, the right hand, lower entry, they both get a reward of 1; whereas if Ann confesses with John still not confessing, she would increase her reward to 2, at the cost of John losing 2; the entry 2, − 2 in the top, right. The rewards have been selected for simplicity, rather than as an accurate reflection of prison conditions. The problem is how they should act when they cannot communicate, so that each is uncertain about what the other will do. The uncertainties here, caused by one person not knowing what the other will do, suggest probability, so that Ann might consider her probability that John will confess, while he evaluates his probability of her confessing. But consider the following argument:

Ann thinks what she should do were John to confess, when the left-hand column of the table is relevant. Recalling that the first entry in each pair refers to her reward, she should also clearly confess for, although she will get no reward (value 0) as a result, it is better than the loss of 2 (value −2) that will arise if she does not confess. Similarly if John were not to confess, the right-hand column, confession (value 2) is preferable for her, yielding more than not confessing (value 1). So Ann argues that whatever John does, it is better for her to confess, and her uncertainty about his choice is irrelevant. Similarly when John considers what Ann might do, it is always better for him to confess (0 instead of −2 if Ann confesses, 2 instead of 1 if she does not), and his uncertainty about her is also irrelevant, with the upshot that they both decide to confess, ignoring their uncertainties and both ending up with no reward, 0,0.

This conclusion is strange, since, had neither confessed, they would each have increased their reward by 1. The difficulty with both not confessing is that, had John not confessed, Ann could have improved her position by confessing, increasing her reward to 2, and similarly with their roles reversed. No such improvement is possible when both confess. The dilemma therefore poses a real problem for which several resolutions have been proposed, none of which is entirely satisfactory. To report on the real progress that has been made would take us too far from the main thesis. The point that concerns us here is that although both participants face uncertainty, the expression of that uncertainty in the form of probability does not help in the resolution of the dilemma. The difficulty appears to be this: the whole treatment of uncertainty presented here concerns a single person, you, facing uncertainty; whereas the dilemma involves two persons who are not able to cooperate. It is possible to consider two people, with their individual uncertainties, by means of probability, provided there is an element of cooperation between them. For example, if John announces that his probability of rain tomorrow is 0.7, then it is possible to evaluate the effect his statement of probability might have on Ann when she contemplates the possibility of rain tomorrow. It is the lack of cooperation, and the consequent separation of their roles, that seems to be the cause of the trouble. The cautionary tale of Ann and John reminds us that the treatment of uncertainty offered here is not universally applicable.

What seems to be true is that if only one person, or one group of persons acting in cooperation is involved, then the probability calculus is satisfactory. At the other extreme, if there is a complete lack of cooperation, as in the prisoners' dilemma, or with two armies in a battle, probability may fail. There are intermediate cases, such as a company marketing a product for sale to consumers, where there is no cooperation between producer and consumer but equally no hostility. Here the probability calculus appears to be helpful and typically produces a sensible resolution for both the producer and the consumer. Games, where there is competition between two players, can cause real problems to both practitioners and mathematicians that have not been satisfactorily resolved.

5.12 THE CALCULUS AND REALITY

The calculus of probability, based on the three rules, is, at least in the situations discussed in this book, rather easy to use, being essentially, as its name suggests, merely a method of calculation. What can be more difficult is to relate the calculus to the real, uncertain world. The problem will repeatedly arise as we progress through the book. Here we anticipate some of the difficulties, using the specific example of weather forecasting. We suppose that a meteorologist has to forecast tomorrow's weather in Arizona. Future weather (Example 1 of §1.2) is uncertain, so probability should be incorporated into the forecast. Actually, British weather forecasts rarely include probabilities, preferring emphatic statements such as "it will rain tomorrow". One argument put forward in defense of this policy is that people will not understand probabilities, to which my response is that people will not until they are used. No, the difficulties with probability lie deeper.

Many people have a low opinion of weather forecasts, a view that often stems from the emphatic nature of the forecasts. They recall the occasions when the statement "it will rain" has been followed by a dry day, forgetting the days when it did rain as forecasted. According to our thesis, the emphatic statement should be replaced by one like "the probability of rain tomorrow is 0.8." A minor advantage of this style is that the meteorologist is less obviously wrong since, even when the weather is dry, he is covered by the 0.2 possibility of dry weather. Recall that the probability reflects the meteorologist's belief that the event, rain tomorrow, will be true. (I once heard a forecaster in Florida, where probability is used, say that it meant that 80% of you would get wet.) But what does it mean that "rain tomorrow" is true? Does it mean it will rain somewhere in Arizona, or everywhere in the state? Are the inhabitants of Tucson entitled to criticize the forecast if their city is dry when most are wet? These considerations suggest that the probability form has unsatisfactory features. A way out of the difficulty is indicated by looking at the practice of bookmakers. A popular activity in England is to bet on having a white Christmas. There has to be a precise definition of "white Christmas" in order that there is no argument about when the payout takes place. A typical definition is that at least one flake of snow settles on a small plate, placed on a roof

5.12 THE CALCULUS AND REALITY

in London, at some time during the 24 hours of Christmas day. The idea could be adapted to cover "rain tomorrow", referring to a rain gauge at a specified location in Arizona. To be useful, several localities would be needed to distinguish the wetter parts from the drier ones. A lesson learned from this is that the statement, about which the belief is expressed in probability terms, should, in principle, be testable as to its truth and, in particular, be well defined. We say, in principle, because there are statements such as "the Earl of Oxford wrote *Hamlet*", which cannot be verified now, and maybe never, yet can be reasonably described as true or false, even if you do not see how to do the verification.

It was seen, with the discussion of scoring rules in §5.8, that when verification is possible, a penalty score can be constructed, which will provide assistance in assessing the meteorologist's ability, not only on one occasion, but over a period. This has been used with weather forecasts of rain at a specified place in the United States, where the professionals performed well, achieving a low penalty score. Even then there remains a problem with how detailed the forecast is. Compare the statement "rain tomorrow" with "at least two millimeters (2 mm) of rain tomorrow", both referring to the same, specified site. The latter must necessarily have the smaller probability, for the first event can happen in two, exclusive ways, "not more than 2 mm" and "at least 2 mm", so that by the addition rule, your probability of the first event equals the sum of the probabilities of the other two and, since probability is never negative, the first must exceed either of the other two. Generally, if the truth of one event implies the truth of a second, the first must have a smaller probability. Applying this to the forecast, the more specific the forecast, the smaller must be the probability, and it must be taken into account in assessing the meteorologist's ability.

The lesson to be learned from this study of weather forecasts applies generally and warns us to be careful in the specification of the uncertain events that are being referred to. This also applies to your knowledge base. It is not merely a question of calculating with probabilities but also that of relating the ingredients of your probability statements to reality. You do not need to think only about $p(E \mid K)$ but also about the precise nature of E and K. In §12.4 it will be important to consider how you came to know K.

5.13 CLOSURE

A point now arises that appears to be little more than a mathematician wanting to be tidy, or even fussy; but it can have practical relevance. When discussing a single event E, it is often necessary to include the complementary event E^c in the discussion (§3.7); indeed it is essential when using Bayes rule (§6.3), especially when expressed in the odds form (§6.5) with a likelihood ratio. If there are two events, E with F, they can combine in two ways (§5.1): the disjunction, E or F, and the conjunction, E and F, written as EF. Both combinations occur in the extended form of Bayes rule (§9.1), where there is a partition of E by exclusive and exhaustive events $F_1, F_2 \ldots F_n$, because we use both F_1 or F_2 and EF_1. Generally, when a situation involves several events E, F, G, and so on, it is mathematically convenient to suppose that *all* events that can be formed by complement, disjunction and conjunction from the original events are available. When this is true, the set of events is said to be *closed*; then none of the operations within the probability calculations can take one out of a closed system. There exists a small world (§11.7) from which you cannot escape, except by introducing another event, an immigrant event.

There are practical cases where closure is necessary. Suppose that you are in conversation with a stranger and are uncertain about her family. In the course of the conversation, she remarks that she has a daughter, thus changing your uncertainty about her family, and you will need to update your opinion about it using Bayes rule. You need to think about the complement of "I have a daughter". What is it? Perhaps "I have a son" or "I have no children". One possibility is that she made the remark because you had your daughter with you at the time of her statement. In §12.4, we shall meet a problem that does not permit a definitive solution until questions such as this are settled. We need to work in small worlds within which everything is possible.

CHAPTER 6

Bayes Rule

6.1 TRANSPOSED CONDITIONALS

This chapter is devoted to what is surely the most interesting rule in probability, with an overall importance that makes it fit to rank alongside the basic equations of Einstein or the fundamental rules of genetics. But first a few examples that are included to demonstrate the need for the rule.

No one who has absorbed the thesis of this book will confuse

$$p(E \mid F) \quad \text{with} \quad p(F \mid E).$$

The notation makes it apparent that they are different, reversing the orders of the two events, E and F. In the first probability, E is an uncertain event whose belief is being measured supposing, or knowing, that F is true (plus an unstated K). In the second probability, E, far from being uncertain, is supposed or known to be true; it is F that is uncertain and whose belief is being assessed. Despite the obvious differences, people are continually confusing one probability with the other. They are termed *transposed conditionals*, because E and F have been transposed in the two probabilities, each taking it in turn to be the conditional. They

---

Understanding Uncertainty, Revised Edition. Dennis V. Lindley.
© 2014 John Wiley & Sons, Inc. Published 2014 by John Wiley & Sons, Inc.

are sometimes referred to as Janus examples after the Roman god with two heads looking in opposite directions. Notice that the confusion occurs in ordinary logic, as in this example from a newspaper, the person's name being changed. "If it is true that you should never trust a man with a tidy desk, then you should have complete faith in Peter Brown, for his desk, indeed every surface in his room, is cluttered with papers and books." The first part of the sentence says "tidy" implies "untrustworthy", the second that "untidiness" implies "trustworthy", and the deduction of the second from the first ignores those who are both untidy and untrustworthy. The reversal here can be recognized by noting that E implies F is equivalent to F^c implies E^c, but not E^c implies F^c; here E is "tidy", F is "untrustworthy". Notice that the mathematical notation for probability makes the distinction, which is not easily apparent in the English language, very clearly, so much so that the language and notation have been advocated in legal cases, where the confusion is rife. The precision of the mathematical language should appeal to the legal mind. Following are some examples of the confusion:

EXAMPLE 1. ARMADILLOS

Armadillos frequently give birth to identical twins. A scientist took advantage of this fact to study the effects of environmental factors on the animals, confident that they would not be influenced by genetic differences, as there are none between such twins. One twin was enabled to live a sedentary life; the other was made to work in a treadmill for much of the time. It was observed that the worker developed much thicker and stronger legs than the sedentary animal. A puzzle in anthropology is the existence together of Neanderthal man and essentially modern man. The former had much thicker leg bones than us. On the basis of the observations on armadillos, the scientist concluded that Neanderthal man was more physically active than were our ancestors.

EXAMPLE 2. DISEASE SYMPTOMS

The first example was specific but the others are deliberately made more abstract to emphasize the generality of the situations. Doctors studying a disease D noticed that 90% of patients with the disease exhibited a symptom S. Later, another doctor sees a patient and notices that she exhibits symptom S. As a result, the doctor concludes that there is a 90% chance that the new patient has the disease D.

EXAMPLE 3. FORENSIC SCIENCE EVIDENCE

A crime has been committed and a forensic scientist reports that the perpetrator must have attribute P. For example, there may be DNA of type P at the scene that can only be accounted for as having come from the guilty party. The police find someone with P, who is subsequently arrested and brought to trial, charged with the crime. In court, the forensic scientist reports that attribute P only occurs in a proportion p of the population. Since p is very small, the court infers that the defendant is highly likely to be guilty, going on to assess the chance of guilt as $1-p$ since an innocent person would only have a chance p of having P.

EXAMPLE 4. SIGNIFICANCE TESTS

Scientists often set up an Aunt Sally and attempt to knock it down. (In America there is a sex change and Aunt Sally becomes a straw man.) Thus, they may suppose that a chemical has no effect on a reaction and then perform an experiment that, if the effect did not exist, would give numbers that are very small. If they obtain numbers that are large compared with expectation, they say that the straw man, usually called a *null hypothesis*, is rejected and that the effect does exist. By "large" here they mean numbers that would arise only a small proportion p of times, were the null hypothesis true. When they do arise, they speak of having confidence $1-p$ that the effect exists. The procedure summarized here is called a *significance test* and p is the *significance level* of the test. Scientific journals are unfortunately full of significance tests, often with $p=0.05$, and will be discussed in §§11.10 &14.4.

116   BAYES RULE

What do these examples have in common? They all turn things upside down, or put them back to front. The exercise affected the armadillo's legs but the inference for man was that the legs were indicative of exercise. The disease gave rise to the symptom but, with the new patient, the symptom was suggestive of the disease. If innocent, there is only a chance p of the evidence, leading to a statement, based on the evidence, that there is only a chance p that he is innocent. If the straw man was correct, the result would be small; the fact that it is not small is used as evidence that the straw man is false. In each case, the first statement has been turned around to provide the second. Now this is not entirely ridiculous; one can infer something about a disease from a symptom, but we need to do it with some care. It cannot be done in the naïve ways described in these examples. The proper inversion is accomplished by the probability rule that is the concern of this chapter.

6.2 LEARNING

Bayes rule has an even more important role than that of clarifying transposed conditionals; it tells us how we ought to learn. All of us learn as a result of new experiences. When we are children, we do it easily and almost without effort; at school, we make the activity more formal; in middle age, we get more set in our ways; in old age, learning becomes difficult and we engage but little in the activity. This is descriptive (see §2.5) in contrast to the prescriptive, or normative form; how ought we to learn? Let us see how we can answer this question by using the framework of probability so far developed. For any statement or event, F, you have a probability $p(F)$ on some knowledge base. Next suppose that you acquire new evidence, E, that bears on the truth of F, thus affecting your uncertainty of F; this will change your probability to $p(F|E)$ and learning is accomplished by your change from $p(F)$ to $p(F|E)$. The normative question is: How do you pass from the former, old uncertainty to the later, new one? In the formal learning process at school, with its emphasis on right and wrong, you learn that F is true, $p(F|E) = 1$, but the formulation to be presented here is a generalized, and a more realistic, model. It has already been explained, in connection with scoring rules in §5.8, that in teaching there can be too much emphasis on certainty and that a proper appreciation of uncertainty is to be encouraged. The rule that is the subject of this chapter tells you how to pass from your initial uncertainty of F to your revised uncertainty of F when you acquire new evidence E. The claim is made that a major aspect of learning is captured by the rule. It does not deal with the extraordinary inspiration of a genius when an epiphany is experienced, but it does explain how you routinely ought to learn by the acquisition of new evidence. And, of course, it describes how a computer might learn (§5.9).

Example 2 above of a disease and its symptom provides a simple illustration. $p(D)$ might be the doctor's initial probability that the new patient has disease D. The observation is then made that the patient exhibits symptom S and the doctor's opinion about D changes from $p(D)$ to $p(D\,|\,S)$. The problem to be discussed and solved here is how the change is to be made. As has been suggested in discussing this example above, a factor that the doctor will surely take into account is how probable it is that patients with the disease exhibit the symptom, in our terminology $p(S\,|\,D)$. It is here that transposed conditionals enter and become important, because the learning transition from $p(D)$ to $p(D\,|\,S)$ involves the Janus effect $p(S\,|\,D)$, looking in the opposite direction. The rule therefore simultaneously does two things: it provides a prescriptive account of learning and relates transposed conditionals. Therein lies its great importance.

Before leaving this "buildup" to the great rule, let us point out something else the rule does. Suppose that patients with D nearly always exhibit symptom S, so that $p(S\,|\,D)$ is very large, near 1. At first glance, this suggests that a patient with the symptom has high probability of suffering from the disease. But this is not necessarily so, for suppose patients without the disease also often exhibit the symptom, with $p(S\,|\,D^c)$ also near 1. Then it looks as if the symptom has little or no diagnostic power and the doctor cannot learn much from its presence because it often occurs, whether the disease is present or not. It will be seen that learning involves both $p(S\,|\,D)$ and $p(S\,|\,D^c)$ and leads us to a very important observation that is often forgotten. It is essential in learning not only to consider the evidence E (or S) on the basis that F (or D) is true but also on the basis that it is false. It is a good rule in life always to consider the alternatives—here not having the disease, as well as having it. Recall that $p(D\,|\,S)$ and $p(D^c\,|\,S)$ necessarily add to 1; this is not true of the transposed values, $p(S\,|\,D)$ and $p(S\,|\,D^c)$. Enough whetting of appetites, let us pass to this wonderful result about learning and the Janus effect.

6.3 BAYES RULE

To establish the rule, recall the multiplication rule of probability of §5.3. Omitting explicit reference to the knowledge base, it reads for two uncertain events E and F,

$$p(EF) = p(E)p(F \mid E).$$

In words, the probability that two events are both true is equal to the probability of the first, times the probability of the second, given that the first is true. Or the proportion of balls that are both red and spotted is the proportion that are red, times the proportion of spotted among the red. The result is still true if the two events are interchanged so, in the above result, write E wherever F occurs, and F wherever E, which, recall, are any two events. This gives

$$p(FE) = p(F)p(E \mid F).$$

But the event FE is exactly the same as the event EF, being only true when both events are true, and in particular has the same probability. In other words, the left-hand sides of the two equations just displayed are equal, so the same must be true of the right-hand sides. Consequently,

$$p(E)p(F \mid E) = p(F)p(E \mid F).$$

If $p(E)$ is not zero, both sides may be divided by it to obtain

$$p(F \mid E) = p(E \mid F)p(F)/p(E).$$

Here we have what we want: $p(F \mid E)$ on the left, in terms of the transposed conditional $p(E \mid F)$ on the right. We also have the learning process mentioned in the last section, with $p(F)$ on the right and $p(F \mid E)$ on the left, showing how you can learn about F on knowing that E is true. This is Bayes rule. Let us now state it formally and include the knowledge base, lest it be forgotten.

Bayes Rule. For any two events E and F and knowledge base K,

$$p(F \mid EK) = p(E \mid FK)p(F \mid K)/p(E \mid K),$$

provided that $p(E \mid K)$ is not zero.

The result is named after the Rev. Thomas Bayes, a nonconformist minister who lived in Tunbridge Wells, England. The strict rules of grammar demand the clumsy Bayes's rule, but we treat Bayes as an adjective. The result, or something near to it, is in a paper of his that appeared posthumously in 1763. It nowadays most commonly appears in a form given in §9.1, available with more than two events. In the next section, there follow several examples of its use.

The claim has often been correctly made that Einstein's equation $E = mc^2$ is of supreme importance because it underlies so much of physics. Here E is energy, m is mass, and c is the velocity of light in a vacuum. I would claim that Bayes equation, or rule, is equally important because it describes how we ought to react to the acquisition of new information; how the gaining of E (evidence, not energy), in addition to the earlier knowledge K should change your views about F from $p(F \mid K)$ to $p(F \mid EK)$. In one sense it is more important than the equation of physics because we all learn things new to us, whereas only theoretical physicists use Einstein, though we all benefit from their usage. My claim is that they are of comparable importance but that Bayes connects with our daily activities in a manner that is different from our involvement with Einstein.

6.4 MEDICAL DIAGNOSIS

Let us return to the medical example of diagnosis in §5.6. We were concerned with a diagnostic test for cancer that yielded either a positive or a negative result, and patients giving a positive result had to go for further tests. The test had the following probabilities for you:

$$p(-\mid C) = 0.01, \quad p(+\mid C^c) = 0.05,$$

where C denotes the event that the patient has cancer, C^c that they do not, and $+, -$ denote the two possible results of the test. The idea here is that a positive result is indicative of cancer and, in the language of §4.4, having cancer and having a positive test result are positively associated. The two probabilities describe uncertainties concerning the errors. The first is the error of failing to indicate cancer when it is present; the second of indicating cancer in a healthy patient. The first error is the more serious and, in this example, has the lower error probability. Using the fact that the probability of the complementary event is one minus the probability of the event, we have the probabilities of the correct indications to be

$$p(+|C) = 0.99, \quad p(-|C^c) = 0.95.$$

In §5.6, it was also supposed that

$$p(C) = 0.02,$$

or that 2% of patients taking the test had cancer. The rule of the extension of the conversation was used to evaluate

$$p(+) = 0.0688,$$

demonstrating that almost 7% of tested patients would give a positive result. It was pointed out that this value was much greater than the probability of cancer at 2%, an increase caused by the errors of which the test is capable.

Now let us look at the situation of a patient who has been given a test that yielded a positive result. What can be said about whether or not they have cancer? Here $+$ is known; the uncertain event is C. We therefore require $p(C|+)$. This immediately follows from Bayes rule, with C replacing F and $+$ replacing E in the statement of the rule above, giving

$$p(C|+) = p(+|C)p(C)/p(+).$$

All the numerical values on the right-hand side are available above and inserting them into the rule,

$$p(C|+) = 0.99 \times 0.02/0.0688 = 0.2878.$$

Similarly, for a patient with a negative result, the probability of cancer is

$$p(C|-) = p(-|C)p(C)/p(-) = 0.01 \times 0.02/0.9312 = 0.0002,$$

where the fact that $p(-) = 1 - p(+)$ has been used. It is an alarming experience for a person to take a test of this type and obtain a positive result, for the result is all too easily interpreted by them to imply that they have cancer. The truth is less alarming, for the probability is about 29%, a much higher figure than the initial 2%, but nothing like certainty. The result is a striking testimony to the advantage of attaching numbers to uncertainty, for no literary discussion could possibly convince one that cancer was still unlikely despite the positive test result. It would be better to have a test that gave more reliable indications with smaller errors, but this may be expensive or require visits to a hospital, whereas the one studied here may be given by nonmedical staff. They are often referred to as screening tests. Notice that a patient who records negative has almost no chance of having cancer, with only 1 in 5000 slipping through the net. A similar situation arises with roadside tests, used on the spot by the police, to assess the presence of alcohol in the blood of drivers.

The calculations just performed exhibit the normative learning phenomenon discussed in §6.2, the positive test result having changed your probability of cancer from 0.02 to 0.29, a negative one decreasing it to a negligible amount; either way you have learned something about whether or not you have cancer. Elaborating on this aspect of the test will be postponed until we have a more convenient form of Bayes rule in terms of odds in the next section.

When the cancer example was discussed earlier, the calculations for $p(+)$ were performed using the balls-in-urn approach, in addition to employing the rule of the extension of the conversation. The same arithmetical technique can be used here, avoiding Bayes rule. In §5.6, to which the readers may like to return to refresh themselves on the numerical results obtained there, it was shown that out of the 10,000 balls in the urn, 688 were positive and, of these, 198 were also with cancer. As usual, probability corresponds to a fraction of balls, and here the probability of cancer, given a positive test result, is the fraction of positive balls that are cancerous, namely 198 out of 688, giving a probability of $198/688 = 0.2878$ as before.

The analysis exhibited in this example is basic to many medical diagnoses where the test results, or symptoms, are not perfectly reliable, their performance being described by the error rates. The

122   BAYES RULE

value of the test depends on the two error probabilities and also on the incidence of the disease in the population presenting themselves for diagnosis, $p(C)$ in the example. With the three values available, all the uncertainties in the diagnosis can be found using the extension of the conversation and Bayes rule. The reader is advised to work through another numerical example to get the feel of the situation.

A common social reaction to tests such as the one just described is to criticize them for admitting errors and to demand certainty. Similar objections are heard elsewhere, for example, when errors occur in vaccination and society rejects the vaccine, or when a doctor makes a wrong diagnosis, a surgeon makes a faulty incision, or an innocent person is found guilty. The fact is that errors are an essential feature of the way we live and their elimination is often impossible and always costly. People who accuse a surgeon of making an error might ask themselves how often they make errors. Instead of reaching for the ideal, society would do better to recognize that some, hopefully small, uncertainty is inevitable and learn to live with it through a proper understanding of probability. How, except by Bayes rule, can one convince a lady who has experienced the trauma of having a positive test after a breast scan, that she is nevertheless more than twice as likely not to have, as to have, cancer. A probability of 0.2878 translates into odds of 2.4746 to 1, about 5 to 2 against.

It was emphasized in §4.2, with the example of inflation and unemployment, that all the uncertainties in a 2×2 contingency table could be found in terms of three probabilities. Our medical example can be written in the contingency form, with rows for the test result and columns for cancer, and uses three basic values, the two error probabilities, and the incidence probability: $p(-|C), p(+|C^c)$ and $p(C)$. From these we calculated $p(C|+)$, $p(C|-)$ and $p(+)$. Notice that in the statement of Bayes rule that yielded $p(C|+)$, only two of the basic probabilities appeared, $p(+|C)$ and $p(C)$, whereas the other probability needed, $p(+)$, had to be calculated. We now take a look at an alternative form of Bayes rule that avoids this last calculation, employing only the basic values, which exhibits the learning process more clearly.

6.5 ODDS FORM OF BAYES RULE

Recall that Bayes rule says that

$$p(F\mid E) = p(E\mid F)p(F)/p(E),$$

and that in the medical example we had to calculate $p(E)$, there $p(+)$, before it could be used. We now derive another form of the rule that does not involve this extra calculation. In the rule, replace F every time it appears by its complement F^c, with the result

$$p(F^c\mid E) = p(E\mid F^c)p(F^c)/p(E).$$

If we take the ratio of the two terms on the left-hand sides of these two equations, it must be equal to the ratio of the two terms on the right. But $p(E)$ appears in both of these and will disappear on taking the ratio, which is just what we want to happen. The result is

$$\frac{p(F\mid E)}{p(F^c\mid E)} = \frac{p(E\mid F)}{p(E\mid F^c)}\frac{p(F)}{p(F^c)}.$$

Here are three ratios. That on the far right is the ratio of the probability of F to that of its complement. This was encountered in §3.8, where it was called the odds on F and written $o(F)$. Similarly the ratio on the far left is the odds on F, given E, written $o(F\mid E)$. Using this notation for odds on and reinstating the knowledge base, the rule can be stated formally:

Bayes Rule. For any two events E and F considered with knowledge base K, for which $p(EF\mid K)$ is not zero,

$$o(F\mid EK) = \frac{p(E\mid FK)}{p(E\mid F^cK)} o(F\mid K).$$

(The qualification that your probability for the conjunction of the events is not zero is needed to avoid division by zero.) In words, this says that the odds on F, given E and K, are the odds on F, given K alone, multiplied by a ratio of two probabilities. Let us look at this ratio.

At first glance it looks like another odds with F in the numerator and its complement below. But this is not so, for the two probabilities do not concern F as the uncertain event, as with odds; they are both probabilities of E, not F nor its complement, and confusion with odds would involve a transposed conditional. The numerator and the denominator are both probabilities of the same event, but under different circumstances, the former when F is true, the latter when it is not, so that the ratio merits a different name from odds. It is called the *likelihood ratio* of F, given E. Bayes rule can now be stated in the form, the odds on F, given E, are equal to the product of the odds on F and the likelihood ratio of F, given E. In other words, a single multiplication is all that is required to pass from the original odds to the ones incorporating the extra information. Notice that with our convention of using odds on, the likelihood ratio also has the probability of the event in the numerator and of the complement in the denominator. Writers who use odds against, have to invert our likelihood ratio.

Before continuing with the discussion of the rule in this new form, a little must be said about the term, likelihood ratio. When mathematicians meet a new idea, which they often need to refer to, they give it a name. Where is this name to come from? Usual practice is to take a word from the English (or other) language and use it as the precise meaning of the term for the new idea. We have already seen this done once with the word "probability". Our mathematical usage is in the very precise form of comparison with the standard urn as a measure of belief, which does not include, for example, the perfectly proper English usage in the phrase "He could probably do it". The same thing is done here and the word "likelihood" is employed. In English, these two words are nearly synonymous, whereas the mathematical usages are for two very different things. In §7.8, another near-synonym "chance" will be used, as something different again. This habit of taking standard words and giving them very precise meanings is often found confusing to others, but experience shows that, with practice, it works very well. So while likelihood and probability may be near-synonyms in everyday English, they are totally different in our usage, which difference we now explain.

We have written $p(E \mid F)$ for your probability of E, given F. It is also referred to as your *likelihood* of F, given E. It may seem unnecessary to have two words, but the reason is that $p(E\,|\,F)$ depends on two things, E and F. In its dependence on the first, we think of it as probability; in its dependence on the second, as likelihood. This dependence is emphasized by the use of "given", $p(E\,|\,F)$ is your probability of E, given F, whereas it is your likelihood of F, given E. In the likelihood, the event E is to be thought of as fixed. Likelihood behaves differently from probability. We saw that the latter added in the addition law. Likelihood does not add; it is not true that

$$p(E\,|\,F) + p(E\,|\,G) = p(E\,|\,F \text{ or } G),$$

even when F and G are exclusive. To emphasize the distinction, different notations are sometimes used, and $\ell(F\,|\,E)$ is written for the likelihood of F, given E. Then Bayes rule may be written, omitting reference to the knowledge base, as

$$o(F\,|\,E) = \frac{\ell(F\,|\,E)}{\ell(F^c\,|\,E)} o(F).$$

An advantage of this form is that all the expressions therein have as main argument F, or its complement, with E in the conditions, though the earlier form, in terms of probability and odds, is often preferred. The value of the distinction between probability and likelihood will become apparent when the likelihood principle is treated in §14.1.

Notice that our object of removing $p(E)$ from the calculations has been achieved. All we need are your odds on F and your probabilities of E, both when F is true and when it is false (or the equivalent likelihoods). To appreciate the new form's importance, let us take an example that is important in its own right as it leads into the use of our ideas in legal trials in §10.14. It also emphasizes the role of Bayes in handling new evidence, beyond its use with transposed conditionals.

6.6 FORENSIC EVIDENCE

Suppose that you are a member of the jury in a criminal trial. The event that the defendant is truly guilty of the crime with which he has been charged is, for you, an uncertain event within our meaning of the phrase.

You will therefore, at any stage of the trial, have your probability of guilt, or equivalently, odds on guilt. Denote the event by G and the odds by $o(G)$. During the course of the trial, your odds will change as you listen to the evidence. Denote a particular piece of evidence by E. Then, on receipt of this evidence, your odds will change from $o(G)$ to $o(G\,|\,E)$ and Bayes rule will tell you how to effect this change.

Consider a specific form of evidence. Suppose the crime is one of breaking and entering and that the criminal has left DNA evidence at the scene, made when he broke the window to gain access. A forensic scientist has examined this evidence and found that the DNA is of a genotype that occurs in a proportion f of people. Furthermore, the defendant is of the same genotype. The evidence E thus consists of two parts: the match between the DNA of the defendant and the DNA at the scene, and the proportion of people with this genotype.

We are now ready to apply Bayes rule. Replacing F in the formulation in §6.5 by G here, and omitting reference to K, it reads

$$o(G\,|\,E) = \frac{p(E\,|\,G)}{p(E\,|\,G^c)} o(G),$$

the original odds on guilt being multiplied by the likelihood ratio for guilt, given the new evidence, to provide the final odds on guilt, given the new evidence. Consider the two likelihoods involved in the likelihood ratio. If the defendant is truly guilty, then there will be a match between the evidence at the scene and his own DNA because he would have left the DNA. It follows that the numerator, $p(E\,|\,G)$, equals 1. If the defendant is truly not guilty, then the true perpetrator of the crime is another member of the population within which the proportion is f. It therefore seems reasonable to take $p(E\,|\,G^c)$ to be f, leaving until §7.7 the consideration of whether this is correct, as it often is. Hence, the likelihood ratio is $1/f$ and Bayes rule gives

$$o(G\,|\,E) = 1/f \times o(G).$$

(The multiplication sign, usually omitted, is inserted for clarity.) In words, the original odds on guilt are multiplied by the reciprocal of the frequency of the genotype to obtain the new odds. If it is a common genotype that occurs in 20% of the population, then the odds are multiplied by 5. If it is a rare type that only occurs in 1%, then the odds are multiplied by 100. If you were contemplating breaking and entering, it would pay to be of a common genotype.

The analysis in this section is the correct treatment of Example 3 in §6.1, introduced to illustrate the dangers of transposed conditionals, a danger that was early recognized in legal circumstances and led to the error of confusing $p(E\,|\,G)$ with $p(G\,|\,E)$ being called the prosecutor's fallacy, which may be a little hard on the legal profession which is no worse than others in making the mistake. Experience has led to the useful suggestion that whenever you make a statement of uncertainty, you make it in a form where the uncertain event and the conditioning event are clearly stated and separated. Always state $p(A\,|\,B)$, making it transparent what A and B are.

6.7 LIKELIHOOD RATIO

The odds form of Bayes rule is more appropriate than the earlier probability form because it clarifies the learning process from $p(F)$ to $p(F\,|\,E)$ on receipt of evidence E. Remembering that odds are equivalent to probability, in which one can pass easily from one to another, the odds form shows that learning is accomplished by taking the odds $o(F)$ and multiplying it by the likelihood ratio of F, given E, to obtain the revised odds $o(F\,|\,E)$. The learning process is performed by a single multiplication. (Readers with an understanding of logarithms will appreciate that the process becomes even easier if they are used, the multiplication being replaced by the simpler addition. Actually log-odds are a better measure of uncertainty in some respects than either odds or probability.) The multiplying factor, the likelihood ratio, involves the probabilities of the evidence both on the supposition that F is true, and that it is false. It is often useful to quote the likelihood ratio separately from any consideration of the uncertainty of F. Values of the ratio near one have little learning effect; only very large, or very small, values give rise to substantial learning, the former favoring the truth of F, the latter its falsity. Its importance reminds us to emphasize again that you must consider how the evidence depends both on F and also on the alternative possibility F^c.

Significance tests were mentioned in §6.1 where evidence is said to be significant if it is improbable when F is true, that is, if $p(E|F)$ is small. But suppose that the evidence is equally improbable when F is false, then the likelihood ratio is 1 and the effect of the evidence on the odds on F is to leave them, and your opinion of F, unaltered by the evidence, despite significance. By contrast, if the evidence is highly probable when F is false, the odds are much diminished. Yet the scientific literature is full of significance tests that only take into account that $p(E|F)$ is small. The situation is not as bad as this might suggest because significance tests ordinarily are designed so that E not only is improbable on F but is highly probable on F^c. The likelihood ratio is then small and the odds on F, given E, are small, in agreement with ideas of significance. Nevertheless, the odds can differ substantially from the significance level, so that the latter can be misleading as will be seen in §14.4.

The example of the DNA evidence illustrates the general point of needing to look at alternatives. If guilty, the evidence is sure to match, but if innocent, the match is less certain, how certain depending on the frequency of the genotype. Thus, $p(E|F^c) = f$ is highly relevant. A similar legal illustration is provided by glass fragments scattered at the break-in. If guilty, the defendant will have matching fragments on his clothing and $p(E|F) = 1$ again. But if innocent, he may also be certain to have matching fragments if he is a builder who works with glass, with $p(E|F^c) = 1$ if K includes this knowledge.

Even when the likelihood ratio is large, it may still not convince you that F is true. The reason for this is that the ratio has to be multiplied by the odds, and if these are small, the final odds may also be small. Go back to the example of the DNA evidence in the last section, where the likelihood ratio was $1/f$. Suppose that any male in the town might have broken in, and that there are $n+1$ such men. (Recall the nuisance of the extra one when passing from probability to odds, see §3.8.) The initial odds on guilt are $1/n$. As a result of multiplying by the ratio $1/f$, the final odds are $1/fn$. Now fn is about equal to the number of people in the town whose genotype matches that at the scene of the crime, and may be quite large. Consequently, the DNA evidence does not, on its own, convince you of the defendant's guilt. Another way of looking at the same need to use both ratio and odds is to recall the point made in §4.2 that with two events, three probabilities are required to describe completely your uncertainty. In Bayes rule, three uncertainties are present, your probability of, or odds on, F, and your two probabilities for the evidence. The final odds depend on all three aspects of uncertainty and none of them should be forgotten. Admittedly, only the ratio of two of them is required, but it is rare to possess this ratio without knowing the individual values. The central lesson of this section is that you must consider the uncertainty of any evidence on the basis of all hypotheses that might explain it. Here only two, F and F^c, have been considered, but the point is quite general and will be studied further in §9.1.

6.8 CROMWELL'S RULE

Bayes rule in its original, probability, form says that

$$p(F \mid E) = p(E \mid F)p(F)/p(E),$$

provided that $p(E)$ is not zero, which is assumed throughout this section. Suppose that your probability for F were zero, then since multiplication of zero by any number always gives the same result, zero, the right-hand, and hence also the left-hand, side will always be zero, whatever be the evidence E. In other words, if you have probability zero for something, F, you will always have probability zero for it, whatever evidence E you receive. Since, if an event has probability zero, the complementary event always has probability one, it also follows that if you believe something so strongly that you give it probability one, then, whatever evidence you receive, you will continue to believe in it. No evidence can possibly shake your strongly held belief.

To many people, this last result seems unacceptable. Scientists often appear to have probability one for some hypothesis, but if you press them, they will admit that their probability is just a little bit less than one, enough for it to be diminished by very striking evidence, that is, evidence with a very small likelihood ratio. They accept this because the history of science shows them that theories do alter over time with additional evidence. Really striking evidence is usually agreed to damage seriously, if not destroy, a theory F.

This is a convenient point to remind the reader that there is almost nothing in this book to say what your beliefs should be, only how they should fit together, or cohere. Cromwell's rule is a slight exception, but all it does is to exclude values 0 and 1 in most circumstances because their use can lead to what many people consider unsatisfactory results. As an example of such a result, consider the case of a person who holds a view F with probability 1. Then coherence says that it is no use having a debate with them because nothing will change their mind. The discussion in §1.4 is relevant.

Almost all thinking people agree that you should not have probability 1 (or 0) for any event, other than one demonstrable by logic, for example $2 \times 2 = 4$. The rule that denies probabilities of 1 or 0 is called Cromwell's rule, named after Oliver Cromwell who said to the Church of Scotland, "think it possible you may be mistaken". Its acceptance means that the convexity rule of probability needs to be strengthened.

Convexity Rule. For any event E with knowledge base K, your probability of E, given K, $p(E \mid K)$ is a number between 0 and 1. Your probability is 1 if, and only if, K logically implies the truth of E.

This is the same as the original form of the rule in §5.4 with the addition of the words "and only if". Naming the rule after Cromwell is perhaps arbitrary, but recall Stigler's law mentioned in the Prologue. The same spirit of open mindedness occurs in the Jain philosophy where it has been encapsulated in the maxim "It is wrong to assert absolutely".

The adoption of Cromwell's rule means that you always admit the possibility that you might be wrong. Nothing, except logic, is incapable of being influenced by evidence. Much of the time you can admit probabilities of one, as in the legal case of §6.6, because the arithmetic would hardly be altered if you replaced it by "nearly one", yet occasionally, it will be necessary to admit that the one is really one less a very small amount. Mathematicians often refer to a very small quantity as *epsilon*, after the Greek letter commonly used for such a value. So let your beliefs have probability 1 minus ε; believe it possible, you may be mistaken. I have one minus epsilon for my belief that the thesis of this book is correct. The law should not treat the defendant as innocent until proved guilty but should admit a very small probability that he is guilty; for if not, no evidence could coherently lead to conviction, however strong that evidence.

The whole of the argument in this section depends on a fixed knowledge base. If that base changes, then the situation can be different from that described. Suppose that you learn that some part of K is false; something that you had supposed to be true is in fact not so. Then the knowledge base changes and you need to deal with a new one. In the new one, the probability that was zero need no longer be so and evidence can now affect your beliefs, whereas before it could not. If my knowledge base included the "fact" that viruses and bacteria are both killed by antibiotics, then my medical practice will change if I am persuaded that this is not so and only bacteria are affected.

6.9 A TALE OF TWO URNS

The inclusion of the following example has two purposes: to test your coherence and to show you Bayes rule as a learning tool in a simple case. Suppose that before you is an urn containing a large number of balls that are identical, except that some are colored red, the rest are white. An urn, in fact, of the type that was used as a standard, but unlike the standard, there are two, and only two, possibilities: either 2/3 of the balls are red, or 2/3 are white, and you do not know which. All that information constitutes your knowledge base. The first possibility, where the red balls predominate, will be called the red urn, and denoted R, while the second, with a majority of white balls, will be termed the white urn, W. Since there are only two possibilities, R is the complement of W. You are uncertain whether the urn you have before you is the white one or the red, so you will have odds on it being red, $o(R)$. For example, you might think it just as likely to be white as red, so that your odds are one, or evens. In the discussion that follows, let this be so, $o(R) = 1$. There is no obligation to take this value and you are welcome to try any other value, except probabilities of 1 or 0, in accordance with Cromwell's rule.

You would like to remove this uncertainty and one way would be to invert the urn, tip out the balls, and look at which color predominates. Suppose this is not available to you, as realistically happens in practical cases that the urn example models. In a consideration of whether the white-tipped or red-tipped beetle was more common, it would be impossible to look at all beetles. What you could do is look at some beetles and, in the urn case, you could take individual balls from the urn and look at their colors. Suppose that you do this in such a way that you think the selection of a ball is random, see §3.2. Let r denote the event that a withdrawn ball is red, and w that the ball is white. Thus, capital letters refer to the unknown constitution of the urn and lower-case letters to the color of the ball, which may be observed and hence known. You immediately have two probabilities that follow from your supposition of randomness: $p(r|R) = p(w|W) = 2/3$. These are your probabilities that the withdrawn ball will be of the same color as the purported color of the urn. It follows, since r and w are complementary events, that $p(w|R) = p(r|W) = 1/3$.

Before proceeding further, imagine that the number of balls in the urn is very large, so that the removal of a few hardly alters the proportion of the colors present, and ask yourself the question: if 12 balls have been withdrawn at random and 9 are found to be red, the other 3 white, what are your revised odds on it being the red urn? Answer the question intuitively, without the help of Bayes or any other probability rule. Answer it as you would if you were a member of a jury and the 12 balls were 12 pieces of evidence, of equal importance, which the court had produced, 9 by the prosecution and 3 by the defense.

Having given your intuitive response, let us do the calculations, starting with just one ball taken from the urn at random and found to be red. Recall Bayes rule in §6.5,

$$o(R|r) = \frac{p(r|R)}{p(r|W)} o(R),$$

where R replaces F and r replaces E in the earlier result. All the quantities on the right-hand side have been evaluated already. The odds were evens, $o(R) = 1$, and the two likelihoods for the ratio are

$p(r|R) = 2/3$ and $p(r|W) = 1/3$. The ratio is therefore 2, and the revised odds, as a result of withdrawing the red ball, are $o(R|r) = 2$. It is twice as probable to be the red urn as it is to be the white one, or $p(R|r) = 2/3$. It is easy to see, in the same way, that had the withdrawn ball been white, the odds would have been halved, instead of doubled, and $p(R|w) = 1/3$.

In summary, the withdrawal of a red ball doubles the odds on it being the red urn; a white one halves the odds. The same thing will happen for subsequent withdrawals, each red one results in a doubling, each white one in a halving of the odds on it being the red urn. To return to the numerical example where 12 balls were withdrawn and 9 found to be red, 3 white, there are 9 doublings and 3 halvings with the result that, since each doubling cancels out a halving, the total effect is of six doublings and the result is a multiplication of the odds by $2 \times 2 \times 2 \times 2 \times 2 \times 2 = 64$. Thus, starting from odds of 1, the 12 balls have resulted in your odds changing to 64. Your probability that the urn is indeed the red one has increased from 1/2 to $64/65 = 0.985$ to three decimal places. You have strong evidence that it is the red urn and, in the legal example, perhaps enough to pass a judgment of guilt.

Notice the strong use of coherence. The assumption that the balls were withdrawn at random, coupled with your initial belief that the urn could just as reasonably be the red one as the white, implies that after 9 red and 3 white have been seen, you must have a probability 0.985 that the urn is the red one. You may like to compare your intuitive answer, requested above, with this coherent value. For most people, the intuitive answer is much smaller. In other words, they are not as convinced by the evidence as coherence requires. This even applies to the withdrawal of a single ball; using common sense, the odds are not doubled but a factor less than 2 is used. I have even known people who use a factor less than 1; that is, the red ball indicates to them that it is less likely to be the red urn. One of them, in explanation, said, "life is always cussed". The claim being made here is not that evidence is always underrated, for there are cases where more is claimed for evidence than is reasonable, but only that reasonable use of evidence requires coherence and the calculus of probability.

Here is one of the most important lessons from this book: the probability calculus shows you how to interpret evidence sensibly.

It enables you to interpret the single ball, or the single beetle, in the context of the urn, or the population of beetles. It enables you to assess the evidence provided in the court of law. It enables you to assess the value of a medical test correctly. Generally, it shows how one set of beliefs inevitably leads to other beliefs.

There is an artificiality about the urn example, in that there were only two possibilities for the fraction of red balls. In reality, there would be many possibilities and any fraction might be possible. The argument already used extends to the general case. We do not include it here because the extension involves technical mathematics whose infliction on the reader would distract from the general point about coherence. The reader should be able to manage three possibilities, adding the possibility of equal numbers of red and white balls, by using the original form of Bayes rule in §6.3. A modified form of this example is studied in §7.5.

There is another lesson that can be learned from this little example. We said that 12 balls had been withdrawn and 9 were red. It did not matter what the order was, a consideration that will be important when we tackle exchangeability in §7.3. Furthermore, both the 12 and 9 did not concern us, for it was only the difference between the numbers of the two types of balls that entered into the final calculations. This was because each red ball made a doubling, each white a halving, so that a red ball canceled out the effect of a white one. All that mattered was the excess of one color, here red, over the other. What is happening here is that one has a lot of evidence, for example, *rrwrrrwrwrrr*, being the 12 balls in order, but most of it can be cast aside and all that matters is the excess of red over white, here six. Spotting what really matters in a mass of evidence is greatly helped by probability considerations. The full history of the 12 balls does not matter; the excess 6 is *sufficient*, which is the technical term used. It is useful in handling a lot of data, to see what is sufficient for the task in hand.

It was supposed that initially you thought the urn was as likely to be red as white, putting $o(R) = 1$. Suppose instead that you had a different value for the odds, then it would still be true that every red ball withdrawn randomly from the urn would double your odds and every white ball would halve them. If it were truly the red urn, there would be about twice as many doublings as halvings and your odds would increase, so that you would, after many balls had been withdrawn, become almost convinced (probability near one) that the urn was truly red. Similarly, were it the white urn, the halvings would occur twice as often as the doublings, and you would think it to be the white urn. In either case, truth will be revealed whatever you thought initially. When we discuss science in §11.6, it will be seen that this mechanism of Bayes rule is about how different views, here about the urn, are generally brought into agreement by evidence, here of balls withdrawn.

6.10 RAVENS

This section concerns an example that has been much discussed by philosophers, yet yields easily to a probability analysis. Although it superficially appears trivial, it does serve as a useful introduction to some aspects of scientific method, free of technical difficulties. Alternatively, the section can be omitted without damage to the appreciation of the remainder of the book.

People are often concerned with general statements; statements that are not confined to one, or a few, instances, but hold in many, if not all, cases. "All men are aggressive", "cheese is rich in calcium", "a body, when released, falls to the ground" are all general statements, as distinct from special cases such as "John is aggressive". We have called such statements, events, though the word "hypothesis" might be more apt here. Scientists are especially involved with hypotheses, which have been referred to as Aunt Sallies, or straw men, in §6.1. Evidence in support of such general statements can be obtained from special cases; as "John is aggressive" supports "All men are aggressive". People sometimes have difficulty with general statements, being more comfortable with special cases.

Consider the general statement, hypothesis, or event, "All ravens are black". You are uncertain about this because you have not seen all ravens, yet have never seen a raven that was not black. It is convenient to think in terms of a contingency table, as in §4.1, where the two rows refer to the type of creature, raven or nonraven, and the two columns to the color, black or nonblack. The entries in the body of the table are numbers in the four categories and lead to extra totals for rows and columns. The entries needed in the subsequent analysis have been inserted; thus the total number of ravens is n and the proportion of them that are black is f. If the general statement is true, $f = 1$. Similarly, the total number of nonblack creatures is N, of which a proportion g is not ravens. Again, if the general statement is true, $g = 1$. The general statement is also equivalent to saying the number in the top, right-hand cell of the table is zero, so that the statement "all ravens are black" (the first row) is the same as "all nonblack creatures are nonravens" (the second column).

	Black	Nonblack	
Ravens	fn		n
Nonravens		gN	
		N	

You are uncertain about the general statement, or hypothesis, but your uncertainty would be changed by seeing a raven and observing that it was not black, when the hypothesis is immediately seen to be false. This is obvious, but let us see how Bayes rule confirms this. Recall the rule in odds form

$$o(F \mid E) = \frac{p(E \mid F)}{p(E \mid F^c)} o(F),$$

where evidence E changes your odds on the hypothesis F. In the current usage, F is the hypothesis that all ravens are black, and the evidence E is that of a nonblack raven. But if F true, E is logically false and therefore has probability zero. So $p(E \mid F) = 0$ and inserting this into the equation just given, the right-hand side is zero, so the left-hand side must also be zero. Hence, $o(F \mid E) = 0$ and you have zero odds and so zero probability for the hypothesis. This is heavy going and does not need Bayes or even probability, only elementary logic, and is included here merely to show that Bayes works even in the extreme case.

In contrast, suppose you see a raven and note that it is black. Does this change your belief in the hypothesis? Now we do need Bayes rule. If the hypothesis is true, the raven is bound to be black, so $p(E \mid F) = 1$

and the rule gives

$$o(F\,|\,E) = \frac{1}{p(E\,|\,F^c)} o(F).$$

The analysis now depends on $p(E\,|\,F^c)$, the probability that a raven will be black, when "All ravens are black" is not true. This is the proportion of ravens that are black in the society in which not all ravens are black. (This point will be considered in more detail in Chapter 7 but is immediately appealing if you return to our original concept of belief in relation to balls in an urn. Here there are creatures considered both with respect to color and whether they are ravens.) From the table, this is f. Hence, Bayes rule gives

$$o(F\,|\,E) = 1/f \times o(F).$$

This is essentially the same result as the second displayed equation in §6.6. The likelihood ratio is $1/f$, which is greater than one, and the odds in favor of the hypothesis are increased by the observation of a black raven. How much they are increased depends on f, what you think the proportion of black ravens might be. You may well think f is nearly one, in which case the observation of a black raven will have little effect on your belief in the hypothesis.

The aspect of this situation that has puzzled philosophers is that since, as we have seen "All ravens are black" is logically the same as "All nonblack creatures are nonravens", the observation of a nonblack creature to be a nonraven should also affect your opinion of the original hypothesis. But this is not true, for the sight of a green creature and the observation that it is a snake does not affect your belief in the colors of ravens. Let us see what Bayes has to say. The evidence is that a nonblack creature is not a raven, denoted by E^* to distinguish it from the previous evidence. As before, if the hypothesis is true, that evidence has probability one. If it is not true, it has probability g, using exactly the same argument as before, referring to a column of the table, rather than to a row. Hence, Bayes rule gives us that

$$o(F\,|\,E^*) = 1/g \times o(F).$$

So the odds on the hypothesis have again increased, but now by the factor $1/g$ rather than $1/f$. But look at the table; g is the proportion of nonblack creatures that are nonravens. Ravens constitute a very small proportion of all creatures, and the same is true if we concentrate on nonblack ones. So g is indeed very close to one, and so is $1/g$. Hence, the change in your odds is negligible and the observation of the green snake has hardly any effect. Whereas f is not as close to one and the black raven has more effect.

The reader's understanding of what is happening here may be aided by changing the scenario from ravens to men, and the property of being black to that of being aggressive. The hypothesis is "All men are aggressive" and is equivalent to "All nonaggressive people are non-men, that is, women". Here f and g are of the same order of magnitude and the observation of a man behaving aggressively has almost as much weight as that of a peaceful person turning out to be a woman.

6.11 DIAGNOSIS AND RELATED MATTERS

In §§5.6 and 6.4, an example of medical diagnosis was discussed, and here we return to it for a third time because it has yet more features worthy of comment. Recall that patients either had cancer, event C, with probability $p(C)$, or not. They were also given a diagnostic test that could either yield a positive, $+$, or negative, $-$, result, the former being positively associated with cancer. The performance of the test was described by two error probabilities $p(-\,|\,C)$, false negatives, and $p(+\,|\,C^c)$, false positives. These three probabilities completely describe the uncertainties and from them, all other uncertainties, such as $p(+)$, can be found using the probability rules. In place of the error probabilities, practitioners often use the success ratios $p(+\,|\,C)$, termed the *sensitivity*, and $p(-\,|\,C^c)$, the *specificity*. In the numerical example, we had

$$p(-\,|\,C) = 0.01, \quad p(+\,|\,C^c) = 0.05, \quad \text{and} \quad p(C) = 0.02$$

and from these we calculated $p(+) = 0.0688$, about 0.07. As a result, it follows that while only 2% have cancer, 7% will respond positively, three and a half times as many. This increase from the true cancer rate to the apparent rate is typical of situations, where the rate is small and errors occur. As an extreme example, take a type of cancer that is very rare with $p(C)=0.001$ but with a test of the same sensitivity and specificity. The rule of the extension of the conversation establishes that

$$p(+) = p(+\,|\,C)p(C) + p(+\,|\,C^c)p(C^c)$$
$$= 0.99 \times 0.001 + 0.05 \times 0.999 = 0.0509.$$

Here the true cancer rate of 0.001 has yielded an apparent rate of 0.05, an increase by a factor of 50. The situation may have arisen in the United States where the National Rifle Association asked a sample of citizens whether they had used a gun in self-defense during the past year. Here C is replaced by true usage and $+$ by reported usage, recognizing that people do not always tell the truth. The error probabilities above are reasonable when questions of this type are posed. Yet if only one person in a thousand had truly used a gun in self-defense, it will appear that one in twenty did, providing grist to the Association.

The transposed conditional, or Janus effect, of Example 2 in §6.1 has already been mentioned in connection with the cancer figures. Here $p(+\,|\,C)$ at 0.99 is quite different from $p(C\,|\,+)$ at 0.29, see §6.4. This confusion has led to an error in cancer surgery where a predictor was used to classify younger women as at high $(+)$ or low $(-)$ risk of developing breast cancer later in life. Here $p(C)$ can be quite high at 0.1, or 10%, with good sensitivity at $p(+\,|\,C) = 0.92$ but poor specificity at $p(-\,|\,C^c) = 0.50$, the latter figure implying that among those women who do not develop breast cancer, high- and low-risk classifications are equally common. A surgeon who observed the large fraction, 92%, of high-risk patients among those with breast cancer, advocated removing the breasts of young women at high risk, so that they could not be affected later in life. This is absurd. What the surgeon is uncertain about is cancer, C; what is known is that the patient is at high risk, $+$; so what is required is $p(C\,|\,+)$, which here is evaluated by Bayes rule to be 0.17, a much lower figure than the 0.92, which the surgeon mistakenly used.

There is another type of error that rarely occurs in the medical context, where sensitivity and specificity are carefully distinguished, but has arisen in psychology and in law. A popular example concerns a town where the buses are either red or blue. An accident occurs at night in which the bus involved is driven away without the driver apparently being aware that anything untoward had happened. A witness says that it was a red bus and the lawyer acting for the company uses in defense the argument that the illumination at night was poor and the witness was mistaken. The law pondered the frequency of mistakes and asked a psychologist for their experience of errors of color identification in poor light and was quoted an error rate of 10%. What both experts failed to recognize is that two types of error are involved here, that of identifying a red bus as blue, and that of thinking that a blue bus is red. These could be different. The situation fits into the diagnostic schema used here, C corresponding to the bus being truly blue, C^c to it being red. A positive result, $+$, is replaced by the witness statement that it was blue, and negative, $-$, to their saying it was red. The two errors are the probabilities of thinking that a red bus was blue and vice versa. The other relevant uncertainty is $p(C)$, the proportion of blue buses in the town, a factor that can easily be forgotten.

A lesson to be learned from all the examples and discussions in this chapter concerning two associated events, here C and $+$, is to think of the basic probabilities, three in all, $p(-\mid C), (p+\mid C^c)$, and $p(C)$, and calculate all others from them. The notation can be enormously clarifying both in developing the concepts required and in calculating with them. The advantage of the notation becomes even more pronounced when considering three events in Chapter 8, but before doing this, it is needful to discuss a possible confusion between frequency, of say cancer in a population, with your uncertainty of cancer expressed as your probability. This is done in the next chapter.

6.12 INFORMATION

We ordinarily use data to provide us with information about something of which we are uncertain, as in the test for cancer in §6.4, where the data, the positive or negative result of the test, give information about the possibility of cancer in the patient. To see how this works, it is necessary to be more precise about what is meant by information.

6.12 INFORMATION

If you had probability near 1 for an event, you would feel that you had a fair amount of information about the event, feeling confident in its truth, and similarly, a probability near 0 would lead to some assurance that it was false. On the contrary, with probability of $\frac{1}{2}$, you have little information, feeling that the event is as likely to be true as to be false. Considerations such as these suggest that your information about an event depends on your probability p for the event, decreasing with p as p increases from 0, reaching a minimum at $p = \frac{1}{2}$ and then increasing to its original value at $p = 1$. Figure 6.1 illustrates the idea and a more detailed analysis would reveal the exact form of the curve and consequently the numerical values for information. It turns out that information has a unique, precise meaning, which is at the basis of what is now called the "information age". The analysis to derive this unique value is not performed here because it is rather technical. Instead, the concept is explored in a more qualitative form using the cancer diagnosis of §6.4 as an example.

Initially you had a probability $p(C) = 0.02$ that the patient had cancer, corresponding to a reasonable amount of information, since it is near 0. Suppose, seeking to increase your information about that patient, the test is performed with a negative result; then we saw that $p(C|-)$ was 0.0002, even closer to 0, so that, referring to the figure, information has been gained as one might have anticipated. But suppose, on the contrary, the test had yielded a positive result, which we saw gave

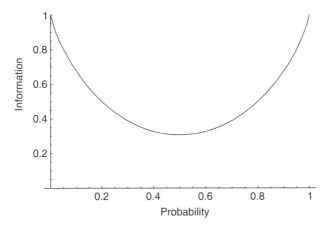

FIGURE 6.1 Information about an event as a function of your probability for that event.

$p(C\,|+) = 0.2878$, then information would be lost as your probability had increased from near 0. The example shows what is, in fact, a general phenomenon, that data can both increase and decrease information, in apparent conflict with the idea that data are collected, or evidence presented in court, with the hope of acquiring more information. To resolve this, notice that the information was increased with a negative result and we saw that this had $p(-) = 0.9312$, whereas it only decreased with a positive result, $p(+) = 0.0688$. Since the first probability is vastly larger than the second, the test would nearly always (perhaps 93% of the time) increase information and rarely (7%) decrease. The example illustrates the general phenomenon:

Data may increase or decrease your information, but you always expect it to increase your information. (The term "expect" will be encountered in §9.3 and given a precise meaning; for the moment treat it in its usual linguistic sense.)

Rather loosely, the result displayed above says that data are always expected to be of value. That is one reason why we need a Freedom of Information Act. The result can be extended even further and used to justify the public dissemination of data that, at the moment, we consider private. For example, why are not tax returns public, for their general availability would seriously hinder tax evasion? A legal application will be found in §10.14. The ideas do not find acceptance because of a limitation of all the methods in this book, that is, they only apply to an individual; they are of less relevance when two people are involved, especially if there is antagonism between them.

CHAPTER 7

Measuring Uncertainty

The study of the rules of probability is interrupted in order to deal with an important, outstanding issue: the measurement of uncertainty. The method of comparison with a standard, which was used to obtain the rules, is rarely satisfactory and other methods need to be developed.

7.1 CLASSICAL FORM

With any event E is associated the complementary event E^c that is true whenever E is false, and false whenever E is true. It was shown in §3.7 that your two probabilities for these events necessarily add to one: $p(E) + p(E^c) = 1$. It follows that the measurement of the uncertainty of any event may be replaced by that of its complement because one probability can be calculated from that of the other. We saw in §5.5, an example involving birth dates where this was advantageous. Here we study the special case where your beliefs in the event and its complement are the same; $p(E) = p(E^c)$. In that case, since they add to one, both probabilities must equal one half; $p(E) = p(E^c) = \frac{1}{2}$. An example is provided by the genuine toss of what appears to you to be a coin from a reputable mint, where your belief that it will land heads equals that

---

Understanding Uncertainty, Revised Edition. Dennis V. Lindley.
© 2014 John Wiley & Sons, Inc. Published 2014 by John Wiley & Sons, Inc.

for tails; hence both events, "heads" and "tails", have probability one half. Notice that there is no obligation on you to have the same beliefs in the two outcomes, only that if you do, your probabilities are both one half. The idea extends to the throw of a cubical die; if you have the same beliefs for each of the six faces falling uppermost, then each must have probability one sixth, for six equal numbers, adding to one makes each of them equal to one sixth. Strictly, we have yet to prove an addition rule for probabilities with more than two events; it will be done in §8.1.

Generally, if an uncertain outcome has N possibilities, only one of which can occur, and if your beliefs in each possibility are the same, then your belief in each is $1/N$. The coin has $N=2$, the die $N=6$, and roulette $N=37$ or 38. This is the *classical* definition of probability and has essentially been used in §3.2 when considering an urn containing N balls numbered consecutively from 1 to N, for if all numbers are equally uncertain when a single ball is withdrawn, we said the ball was drawn at random and each value had probability $1/N$. The classical definition is fine in a limited context but is deficient in that for most cases such a split into equally uncertain possibilities does not exist; for example, contemplating uncertainty about tomorrow's weather has no such split. The real importance is as a standard with which other events may be compared. Notice that a tangible account of equal beliefs was provided by your attitude to a reward, in that if you are indifferent between a prize if ball 7 is withdrawn, or one contingent on ball 37, and this for any pair of different numbers, then your beliefs, and hence your probabilities, are all the same. The classical definition is therefore operational.

It is perhaps worth repeating the point, illustrated with the toss of a coin above, that there is no obligation on you to have a probability of one half for heads, since you may judge the coin to be biased and take 0.55 or any other value between zero and one. Similarly, you may judge the roulette wheel to be biased. These are not illogical values, merely unusual ones; indeed, they may be sensible if you have reason to suspect the casino. Some people argue that if there are N possibilities, only one of which can arise, and if you assign probability $1/N$ to each, then you are ignorant of the outcome using this as a definition of ignorance. This is unsound, for it is a strong statement to judge all possibilities equally uncertain, surely not one of ignorance. Why, with $N=2$, is the probability of 0.50 ignorance but one of 0.55 knowledge? Our attitude is that judgments of uncertainty are always made against a knowledge base and that ignorance, or an empty base, is not a sensible position. As soon as you understand the meanings of "toss" and "coin", you are not ignorant of coin tosses. Ignorance has no place here, but this does not mean that the assignments of $1/N$ should be avoided; on the contrary, they often provide a convenient default position. Thus, suppose geneticists are attempting to isolate a gene in a species having N chromosomes, then their knowledge base may not guide them as to which chromosome it lies on, and, in default of more information, they may assign the same probability to each. Similarly, at the commencement of a police investigation with N suspects, the police might reasonably regard all equally probable of being guilty. In neither case is it ignorance, but merely a sensible position describing uncertainty.

The concept of equal beliefs, basic to the classical form, can often be used to advantage in other situations; we illustrate this with the example in §4.1 of inflation next year. It may be convenient to think of an inflation figure that you think is equally likely to be exceeded, or not be attained. If you settle on 3%, then your probability of it being less than 3% is $1/2$ and the same value of $1/2$ holds for values greater than 3%. The idea can be extended to find a value, like 2%, such that you think it is as likely to be less than 2% as between 2% and 3%. Similarly, 5% might be a value such that you feel it is as likely that the inflation might exceed it as be between 3 and 5%. Now you have four ranges of inflation, all equally probable. The idea can be extended to provide a probability distribution, as will be shown in §9.8. There remain many phenomena where the classical definition cannot be used, so we pass to a more powerful device based on frequency.

7.2 FREQUENCY DATA

Earlier we met the idea that if you, as a doctor, had seen many patients with a disease and noted that a proportion p of them had exhibited a symptom, then your probability that a further patient with the disease would show the symptom was also p. That is, you pass from a frequency among the patients seen, to a probability or belief about a further patient. This passage is so common that there has grown up a confusion between frequency, which refers to data, and probability, which is belief, so that people speak of the frequency interpretation of probability. There is a connection between the two concepts, but it is wrong to identify them, so let us investigate the situation carefully, starting with a simple example.

Suppose that you have before you a drawing-pin; the American term is a thumbtack. Such a pin has the property that, when tossed, it can either land with the point down on the table, D, or sticking up in the air, U. You are uncertain about the event that the pin will fall with the point up and will express this by your probability $p(U)$. We assume some knowledge base that remains fixed throughout the discussion. Now let the pin be tossed a number of times under conditions that remain stable; you do not, for instance, alter the tossing procedure. To be specific, let the results of 10 tosses in order be $UUDUDUUUDD$ and denote this result by x. Notice that 6 times the pin fell uppermost and 4 times it fell with the point down, giving a frequency of Us of 0.6. You are about to toss an eleventh time and are uncertain about the event U on that occasion. What is your probability $p(U|x)$? A natural response is 0.6, the frequency in the series of 10 tosses, and this is the procedure used by the doctor in the example. Is it sound? Can you pass from a frequency to a belief in this way? Is it coherent to do so? There are three reasons for thinking that the passage from frequency to belief is not so straightforward.

First, suppose that you had only tossed the pin once instead of 10 times. Looking back at the series of 10 results given above, you see that the first toss gave U and that the frequency of Us is therefore 100%. Is your probability that the second toss will result in the pin falling point uppermost the same as the frequency, that is, 1? Surely not, it might have increased a little from the original value $p(U)$, as with the red and white urns in §6.9, but not so far as to make the event certain for you, thereby violating Cromwell's rule (see §6.8). So you cannot make the identification of frequency and belief when the former is based on little information, here a single toss. If 10 is enough to allow the identification, but 1 is not, where do you draw the borderline; is 7 enough?

A second reason for doubting the identification is demonstrated by shifting the example. Suppose that the pin and the tosses are replaced by observations of the weather on successive days. Each day you observe if it is dry D or unsettled U, meaning "not dry". Suppose you record the weather on 10 successive days and obtain the same sequence x as before. What is your probability that the eleventh day will be unsettled? I suggest that the frequency of unsettled days in the last 10 days, here 0.6, is not a reasonable answer, at least under my knowledge base, because successive days of weather tend to be alike. Indeed, the forecast that tomorrow's weather will be the same as today's is often better than the one based solely on the frequency of weather. The last two days in x were both dry, so we are in a dry spell and your probability that tomorrow will be unsettled may be less than the frequency 0.6. This example is based on weather in Western Europe, and readers in other parts of the world may need to adapt it to their own conditions, using their knowledge base. The key point is that the order of the Us and Ds may matter with weather, but usually not with drawing-pins.

Those are two reasons for doubting the identification of frequency with belief. Here is another of a different character. Suppose, after having tossed the pin with the results given, you are now provided with a different type of drawing-pin and told that the next, eleventh, toss is to be made with this pin. It would not be sensible to ignore the 10 tosses already made, since they do provide you with some information about pins in general, but on the other hand, it is a different pin and the direct use of the frequency is dubious. You may, for example, look at the new pin and see that the head is heavier than the one used in the tossing, so perhaps this one is more likely to fall point upward than the other. You might therefore express your belief with a value greater than the frequency of 0.6.

So while the idea of identifying belief with frequency is attractive, it cannot be used in all circumstances. Nevertheless, frequencies surely do influence beliefs and what has to be done is to understand the relationship between the two ideas. This we proceed to do.

7.3 EXCHANGEABILITY

Consider again the drawing-pin and the result of the 10 tosses $UUDUDUUUDD$ that was abbreviated to x. Each toss could result in one of two outcomes, so there are $2 \times 2 \times \ldots \times 2$ (with 10 twos), or

1,024 possible results for the 10 tosses. Before you perform the tossing, you are uncertain about the outcomes and therefore, by the general thesis, ideally have probabilities for each of the 1,024 possibilities, the assessment of which is a formidable task if only because of the number involved. An assumption is now introduced, whose adoption will make this task much easier. It needs to be emphasized that the assumption is not always appropriate.

Suppose that when you think about the possible results of the 10 tosses, you feel that your probability for any series depends only on the number of times the pin falls with the point upward and not on the arrangement of the Us and Ds in the series. Thus, in the case cited, your probability of the result depends only on the fact that there are 6 Us (and therefore 4 Ds), so that $UUUUDDUDUD$, still with 6 Us and 4 Ds but in a different order, is, for you, just as probable as what you actually observed. If this were so, you would have a much easier assessment task, for there would now be only 11 possibilities, not 1,024, that is, from 0 to 10 Us. One way of expressing this is to say that any one toss, with its resulting outcome, may be exchanged for any other with the same outcome, in the sense that the exchange will not alter your belief, expressing the idea that the tosses were done under conditions that you feel were identical. Here is the formal definition:

A series of results, each of which can be one of the same two types, is *exchangeable* for you under knowledge base K if your probability for the series under K depends only on the numbers of the two types and not on their positions in the series. It will be called the assumption of *exchangeability*. In the example, the two types are U and D. Your probability, assuming exchangeability, for the series x with 6 Us out of 10, may be written $p(6|10)$. Given 10 tosses, this is your probability for 6 Us. Series that are exchangeable are of special importance because there are many series that almost all people agree are exchangeable, and because of the simplicity that they introduce into the structure of your beliefs. The concept is related to that of sufficiency mentioned toward the end of §6.9, the number of Us, rather than their order, being sufficient.

The assumption of exchangeability implies that the series of outcomes $UUUUUUDDDD$, in which the 6 Us and 4 Ds each occur together, is just as probable for you as the original series in which the

Us and Ds were mixed up. People are often unhappy with this, but its resolution is to notice that the series in the last sentence has a pattern to it, whereas the other is chaotic, and there are vastly more chaotic series than there are those with a pattern. There are 210 possible arrangements of 6 Us and 4 Ds, very few of which exhibit a pattern. It is the pattern that singles out that series, not its uncertainty; it is a coincidence that the Us and Ds form clumps. Coincidences are hard to discuss because the striking pattern, which owes nothing to uncertainty, is easily confused with your uncertainty. And there is the question of what constitutes a pattern; does $UDDUUUDDUU$ have a pattern because the last 5 tosses are identical to the first 5?

Exchangeability implies that your probability of U at any place in the series is the same as U at any other place. To see this, consider the first two terms in the series with the $2 \times 2 = 4$ possibilities

$$UU \quad UD \quad DU \quad DD.$$

Exchangeability implies that UD and DU have the same probability. U occurs at the first place if either UU or UD occurs, whereas it occurs at the second with UU or DU, and in either case, your probability of a U is the sum of your probabilities for the two possibilities. Now UU is common to both and UD has the same probability as DU, so the two sums are equal and U is just as probable at the first place as at the second. Generalizing this argument, it is apparent that any arrangement, say UDU, is just as probable at any place in an exchangeable series as at any other. An exchangeable series is stationary; its uncertainties do not change with place. In thinking about these results, you need to distinguish between your probability of U in the second place, on knowledge base K, and the same uncertainty when you have already observed U in the first place. The notation makes this clear, comparing $p(U_2)$ and $p(U_2|U_1)$, where a subscript refers to the place in the series and K is understood. Notice that $p(U_2|U_1)$ is easily calculated in terms of the basic, exchangeable values, as $p(U_1U_2)/p(U_1)$ by the multiplication rule (§5.3). Since, as was seen above, $p(U_1) = p(U_2)$, it follows from this last result that $p(U_1|U_2) = p(U_2|U_1)$, so that looking backward, on the left-hand side of the equation, is the same as looking forward, on the right.

Although in general for an exchangeable series, U_1 and U_2 are not independent, $p(U_2|U_1) \neq p(U_2)$, in §7.5 it will be seen that any exchangeable series can always be built up from series in which independence does obtain.

Most people would consider the series of tosses of a drawing-pin to be exchangeable. They would not think it true of the series of weather on successive days, because consecutive days tend to be more alike than widely separated days, so that *UUUDDD* is more probable for them than *UDUDUD*, despite the frequency being 0.5 in both cases. The records of the doctor observing the presence or absence of a symptom with a disease, you might think exchangeable, though if you knew the sexes of the patients and thought the disease was sex related, you might not. This example also serves to illustrate an important point, that since the definition of exchangeable depends on your probabilities, it depends on your knowledge base, and a series exchangeable under one base, without knowledge of sex, may fail to be under another, with knowledge of sex.

Let us return to the question of the connection, if any, between frequency and probability. In the case of the pin, you wanted to pass from the results of the 10 tosses to an eleventh toss about to be made. The only aspect of the 10 tosses that matters under exchangeability is the 6 *U*s. One possibility is for you to consider the 11 tosses, the 10 already seen and the new one, exchangeable. If so, we say that the new toss is *exchangeable* with the others. This would be reasonable with the single pin, but not when the eleventh toss was to be performed with a different pin. It might be fine with the medical example, but perhaps not if you were the next patient and you thought yourself different in some relevant way from those patients the doctor had already seen. For example, the study may have been made in one country and you are the resident of another.

Three examples were mentioned in §7.2: tosses of a pin, weather on successive days, and tosses of one pin aiding beliefs about another. The second series is not exchangeable, and in the third, the further toss is not exchangeable with the previous ones. In both these cases, we ruled out the possible identification of frequency with belief. It is only in the first example with exchangeability in the series and extended to a further toss that the identification might be reasonable, and it is this case that we study further, beginning with a special type of exchangeable series.

7.4 BERNOULLI SERIES

To illustrate the series, let us return to our basic urn with balls, all indistinguishable from each other except for color, some being red, the rest being white, the numbers of both types being known to you and from which you think a ball is to be drawn at random. Denote the proportion of red balls by the Greek letter θ, pronounced "theta" with a long e. (There is an important reason for going outside of the Roman alphabet that will appear in § 11.4.) Remember that you know the value of θ. Under these circumstances, your probability that a withdrawn ball will be red is θ. Suppose the number of balls in the urn is vast, so that the withdrawal of even a few balls will not affect the constitution of the urn and, in particular, will not change θ. Then your probability that a second ball will be red is still θ and it remains θ however many balls are withdrawn. Furthermore, your probability is not affected by the results of all previous withdrawals; even if 10 withdrawals have each produced a white ball, your probability of drawing a red ball on the eleventh withdrawal remains θ unless you discard the premise of randomness. In the terminology of §4.3, the withdrawals are independent, though there only two events were considered; the extension to many will be introduced in §8.8. With this example in mind, we make the definition:

A series, each member of which can have one of only two outcomes, is for you, a *Bernoulli series* if your probability of one outcome is the same for every member of the series and is independent of any earlier outcomes in the series. It is named after a Swiss mathematician

A Bernoulli series is somewhat artificial because you never learn from it, in the sense that your probability remains fixed at θ, whatever happens. In the artificiality of such a series, even 100 withdrawals, all of which resulted in a white ball, would not change your belief that withdrawal 101 would be red. Despite this, Bernoulli series is most important for a reason to be explained now. It is easy with a Bernoulli series to calculate your probability of any result. For example, take the series we had before with the toss of a drawing-pin, $UUDUDUUUDD$ with 6 Us and 4 Ds. If it is Bernoulli, the probability for each U is θ, for each D, $1-\theta$, and since the outcome of any one toss is judged by you to be independent of previous tosses, these probabilities may be multiplied (see §4.3). Hence, your probability for that series is $\theta^6(1-\theta)^4$, depending only on the number of Us, here 6, out of the 10 tosses. It follows that a Bernoulli series is exchangeable, since the dependence solely on the numbers of Us was our criterion for exchangeability. It is a special type of exchangeable series in which you judge the individual events to be independent. With independence, it is easy to write down your probabilities for any series by multiplication. With series that are exchangeable, but not Bernoulli, we do not, at the moment, know how to do this. For an exchangeable series of length 10, we saw there were 11 probabilities to think about, whereas in the Bernoulli case, there is only one, namely, θ, your probability for any U, and once you know that, you can find all the others by multiplying the appropriate numbers of θ and $1-\theta$ together. However, there is a link between exchangeable and Bernoulli series that enables the exchangeable calculation to be made in terms of the Bernoulli.

7.5 DE FINETTI'S RESULT

We return to the familiar urn with a large number of balls that are identical except for their colors, some red, and the rest white. Suppose that, unlike the case in the last section, you do not know the proportion of red balls but are told truthfully that it is one of two values θ_1 or θ_2. In §6.9, the case where θ_1 was 1/3 and θ_2 was 2/3 was considered, the former being referred to as the white urn, the latter, the red urn. Now the values of θ_1 and θ_2 are not restricted but, merely for identification, it is supposed that θ_1 is the smaller, so having a lesser proportion of red balls, it is referred to as the white urn. You do not know whether it is the red or the white urn that is before you and since you are uncertain, you will have a probability that it is the white one, p say and $1-p$ that it is the red urn.

Were you to know the proportion of red balls, you would, on withdrawing balls at random, have a Bernoulli series. Suppose that the balls are drawn at random from the urn without knowing whether it is the white or the red one, and that the result of 10 such drawings is RWRWRRRWW, abbreviated to x, essentially the same as with the tosses, though for ease of relating the result to the urns the notation has been changed, U to R, D to W, retaining x for the data. In the original urn treatment, lower-case letters were used for the data and upper-case for the true constitution. Here θ replaces the latter, freeing the capital letters. Complete consistency of notation is rarely possible. What is your probability, $p(x)$, for this result? Before the drawings were made, what is your belief that this result would be obtained? It is not easy to see directly but recall from §5.6 the rule of the extension of the conversation and extend your discussion of the series to include the value of θ. You require $p(x)$ which, by the rule is

$$p(x) = p(x|\theta_1)p(\theta_1) + p(x|\theta_2)p(\theta_2).$$

Now all the terms on the right-hand side of this equation are known, since once you know the proportion of red balls, the series is Bernoulli with $p(x|\theta_1) = \theta_1^6(1-\theta_1)^4$ and similarly for the other possibility, θ_2. Your probability of it being the white urn, corresponding to θ_1, was written p, so substituting these values into the right-hand side, you have

$$p(x) = \theta_1^6(1-\theta_1)^4 p + \theta_2^6(1-\theta_2)^4(1-p), \quad (7.1)$$

and the calculation is complete. The argument has been presented here for the case where there were only two values of θ. If there were three, the same procedure of extending the conversation would be available, except that now there would be three terms on the right-hand side of (7.1). Generally, any number of values of θ can be included, resulting in that number of terms on the right-hand side.

It is obvious that this series, with two possible values of θ, is exchangeable because the result just obtained depends only on 6, the number of red balls, out of 10, and not on their order of withdrawal. So we have established that the withdrawal of balls at random from an urn of unknown composition generates an exchangeable series. De Finetti showed that *every* series with two possibilities, here R and W, that you judge to be exchangeable can be represented as random withdrawals from an urn of unknown composition. In other words, the procedure just described is available for *every* exchangeable series and exchangeable series necessarily reduce to combinations of Bernoulli series. The result is of considerable importance because it enables you to think about your beliefs for an exchangeable series in a simple way. In the case where θ can take only two values, all you need to think about is the probability of θ_1, denoted by p above. Generally, each possible value of θ has to be assigned a probability and the extension of the conversation is then used to perform the evaluation. The last stage can be left to the mathematician or the computer and need not concern us here, but the assignment of probabilities, such as p, needs further thought.

An objection might be raised. Suppose you were to think about θ to two places of decimals; that is, you admitted values 0.01, 0.02, . . . , 0.99 so that there were 99 possibilities in all. (The extreme, special values of 0.00 and 1.00 being omitted.) Then there are 99 values of p for you to think about, whereas for the series of Rs and Ws of length 10, there were only 11 to be considered, and all this fuss has only made your task harder. This is perfectly sound, but once the 99 have been settled upon, the calculation will work for any length of series of Rs and Ws, not just 10; so that the 99 will replace the 1,000 needed for a series of length 999 and there is a real simplification. It will be seen in §9.8 that there are compact ways of studying the values of p that are not available for the raw series.

7.6 LARGE NUMBERS

In order to think about a series with two possible outcomes that you judge to be exchangeable, by de Finetti's result you need to think only about the values of θ underlying a Bernoulli series. In the case of the urn, θ had a concrete interpretation, as the proportion of red balls, but in other cases, such as patients with a disease, some exhibiting a symptom, it is not clear what meaning to attach to θ under exchangeability, so that before de Finetti's result can be used, we need to be able to escape from the tyranny of the Greek alphabet and think in medical terms. To do this, we need a mathematical result called a *law of large numbers*, which says that for any series, each member of which has two possible outcomes and that you consider exchangeable, you have probability 1, that is, you are sure, that the frequency of one of these outcomes tends to a fixed value as the length of the series increases, rather than wobbling about all over the place. The fixed value to which the frequency tends is an interpretation for θ. (Probability 1 may appear to violate Cromwell's rule in §6.8, but the law is the result of logic in the form of mathematics and is therefore exempted from the rule.) Consequently, to think about an exchangeable series of two outcomes, you need, apart from the Bernoulli calculations, only to think about your beliefs about the frequency of outcomes in a long series. This value is termed the *limit* of the observable frequencies.

Let all the threads be put together to produce your probabilistic description of a series with two outcomes that you judge to be exchangeable.

1. By exchangeability, you admit that the frequency of an outcome in the series will tend to a limit. Denote this limit by θ.
2. Assess your probabilities $p(\theta)$ for the various values of θ.
3. Combine this with the Bernoulli probabilities, $\theta^r(1-\theta)^{n-r}$, giving a term $\theta^r(1-\theta)^{n-r}p(\theta)$, and take the sum of these over the various values of θ. This is your probability for r outcomes of one type in an exchangeable series of length n.

Consider the case of a drawing-pin and, to illustrate, take the nine possible values of θ, 0.1, 0.2, . . . , 0.9 corresponding to the outcome that the pin falls with point upward, U. You need 9 numbers, adding to one, to describe your beliefs about the pin. If you feel that the frequency of it falling with point up will probably be less than its falling with point down, then the larger probabilities would be assigned to the smaller values. For example, you might assign probabilities

$$0.03, 0.10, 0.20, 0.25, 0.17, 0.12, 0.07, 0.04, 0.02$$

to the 9 values; thus, $p(\theta = 0.1) = 0.03$. Contrast this with the case of a coin, where you might attach high probability to heads and tails occurring with equal frequency in the long run but, with due attention to Cromwell's rule, would not rule out bias. A possible set of probabilities, spread over the same 9 values as before, might be

$$0.01, 0.01, 0.01, 0.01, 0.92, 0.01, 0.01, 0.01, 0.01.$$

This means that your probability that the coin is fair, and being tossed fairly, is 0.92, but admit that other values are possible with small probabilities.

We began in §7.2 by considering how frequency and belief were related; how the doctor's observation that a symptom, occurring with frequency p in the patients already seen, related to the belief that the next patient would exhibit the symptom. With the 10 tosses of the pin giving the result x, we sought $p(U|x)$, your probability that the next toss, judged exchangeable with the other tosses, would result in it falling with point up, event U. (To avoid fussy notation, U now refers to the eleventh toss.) Next, we show how this probability can be calculated using the three-step procedure just described. By the multiplication rule (compare the case of U_1 and U_2 in §7.3),

$$p(U|x) = p(Ux)/p(x).$$

The denominator $p(x)$ was calculated in §7.5, the last displayed equation therein, for two values of the limiting frequency θ, with its obvious generalization. The numerator $p(Ux)$ follows in exactly the same way since Ux has one extra U, giving 7 Us and still 4 Ds. Hence, the required result, $p(U|x)$. This method is available for every exchangeable series and a future outcome judged exchangeable with it.

There is another way of arranging the calculation, which makes use of Bayes rule in learning about θ from the observed data, and which is illustrated using the example of the red and white urns in §6.9. Here θ_1 as in §7.5, is the proportion of red balls with $\theta_1 = \frac{1}{3}$ in the white urn W and $\theta_2 = \frac{2}{3}$ in the red R. Suppose that some balls are withdrawn at random and let the result be denoted x. (In §6.9, 12 balls were withdrawn and 9 found to be red, 3 white, but the exact nature of the data, x, need not concern us here.) In analogy with the doctor, uncertain about the next patient, let us consider your probability that another random ball will be red, an event denoted r. Extend the conversation to include θ, the true but unknown constitution of the urn, with the result

$$p(r|x) = p(r|Rx)p(R|x) + p(r|Wx)p(W|x).$$

Now, if you know it is the red urn R, the data x tells you nothing about the next ball, so $p(r|Rx) = p(r|R) = 2/3$ and similarly, $p(r|Wx) = 1/3$. (In the terminology of §4.3, r and x are independent, given R.) The other probability $p(R|x)$ was found in §6.9, by Bayes rule, to be 64/65 and naturally, the complement $p(W|x) = 1/65$. Inserting these numerical values into the result just displayed yields

$$p(r|x) = 2/3 \times 64/65 + 1/3 \times 1/65 = 129/195 = 0.6615$$

to four decimal places. This is a little less than $2/3 = 0.6667$ to the same accuracy, the slight reduction being caused by the fact that, although you are almost sure it is the red urn, just a little doubt, expressed through your probability 1/65, that it is the white one remains.

There remains a general problem, that of summing the various terms and performing the calculations in Equation (7.1) of §7.5 above. This is a technical matter and has been attended to by mathematicians. My best advice to you is to consult a statistician, just as you would consult a plumber if the repairing of your plumbing system was outside your capabilities. However, it is possible to describe one of the results that have been obtained in a form that is of immediate use without technical skills.

7.7 BELIEF AND FREQUENCY

Take a series with two outcomes, U and D, of length n that you judge to be exchangeable and suppose that you have just observed r Us and therefore $(n - r)$ Ds. By exchangeability, it does not matter to you where the Us and Ds appeared in the series. Now consider your probability that the next term, judged exchangeable with the series, will be U. This is $p(U|r, n)$, your probability of U, given the result (r, n). Although it is tempting to equate this with r/n, the frequency of Us in the series, we saw in §7.2 that it would not be realistic to do this for short series with small n. The methods of the last section tell us how to proceed but they involve technicalities. It is now shown how they may be overcome if an assumption is made about your opinion, $p(\theta)$, of the hidden value of θ, the limiting frequency of Us.

Denote the observed frequency in the series by $f = r/n$, which is firmly based on data and has no element of uncertainty. There is another frequency, the limiting one, θ, that is conceptual and not data based, about which you are uncertain and have beliefs. Let g be your best guess as to the value of θ before you have any data on the series. Exactly what is meant by "best guess" will be explained in a moment. Now you have two pieces of information about the frequency with which U, rather than D, will arise: f, which is based on data, and g, which is based on initial beliefs about the series. It surely seems natural, in assessing the probability of a further U, to incorporate both these pieces of information, combining them in some way. The simplest way to do this is to take a bit of one and add it to a bit of the other; addition being the simplest arithmetic operation, So consider the expression $(nf + mg)/(n + m)$, where m is a positive number. If $m = n$, the expression gives equal prominence to f and g, being the average $(f + g)/2$. If n is much larger than m, little attention is paid to g and the expression is near to f; similarly, if m is by far greater, the emphasis is on g. Generally, the expression lies between f and g, exactly where depends on the balance between m and n. Technical analysis shows that it is often appropriate to equate the result $(nf + mg)/(n + m)$ to the required probability $p(U|r, n)$. Leaving the discussion of m for the moment, the final result is

$$p(U|r, n) = \frac{nf + mg}{n + m}. \qquad (7.2)$$

Consider an example. Suppose with the drawing-pin, you believed initially that D might be little more probable than U and that your best guess at the limiting frequency of Us was 0.4. This is g. Now you have data of 6 Us in 10 tosses, $r = 6$, $n = 10$, $f = 0.6$, and the formula gives

$$p(U|6, 10) = \frac{10 \times 0.6 + m \times 0.4}{10 + m}. \qquad (7.3)$$

This is a simple combination of the two frequencies, which necessarily lies between them, greater than what you initially believed, because of the observations, but less than you observed, because of your lower, initial belief. It is now possible to see what g, your best guess of θ, means, for if the general result (7.2) is used with $n=0$, that is, before any observations have been made, $p(U|r=n=0)$ reduces to g when $n=0$. Hence, your best guess is your belief that the first member of the series will be U, rather than D. There remains the value of m to consider.

A clue to m can be found by reflecting that so far you have not inserted any indication of how strongly you felt g reflected your initial opinion. Thus, with the pin, you may not have much strength of conviction about 0.4, whereas had it been a coin that was being tossed, you would have had a firm opinion that the frequency in the limit would be 0.5 and these feelings were reflected in the two sets of nine probabilities chosen in the last section. m measures this conviction, being small in the case of the pin and high in the case of the coin. But what of an exact value? There are several ways to assess this. One of them is to assess $p(U|r, n)$ directly and then equate it to the above expression, so obtaining m since all the other quantities are known. For example, suppose with the pin, you felt 0.55 was your probability after the 6 Us and 4 Ds, then arithmetic shows that $m = 10/3$, a little more than three. (Put $m = 10/3$ in (7.3) and you will obtain the result 0.55.) Let us take 3, rather than the more precise value. Then what you are saying in using the formula is that you are taking 10 parts of the data to 3 parts of your initial belief, out of 13 parts in all. Roughly, $m = 3$ says that your initial belief is worth about three observations in the series. Had m been 10, you would have given equal weight to the two frequencies. With the coin and $g = 0.5$, you might have had a large value, say $m = 100$. Equation (7.2) then gives a probability for U on the next toss of 0.509 and the observed frequency of 0.6 has only slightly affected your belief that the coin is being tossed fairly. Notice how the fact that m measures your strength of conviction about g goes some way to answering those who feel that a single probability is inadequate, instead preferring upper and lower probabilities in order to incorporate this conviction (see §3.5). The analysis demonstrates that when the conviction is relevant, it can be included within our simpler framework by introducing m. Furthermore, the introduction of m is balanced by your conviction about the data f, naturally expressed by the number n of observations. Here our simplicity has paid off and the additional complexity is unnecessary. The expression above requires your best guess g about θ, in the sense of your probability that the first toss will result in U, and also the strength of your conviction about θ measured by m in comparison with n, the length of the series.

As remarked above, if n is large, the formula weighs f, the frequency, very high and the effect of g is small, so the formula says that it is sensible to identify frequency and belief, provided the exchangeable data are numerous. Thus, if the doctor had seen a lot of patients whom he judged exchangeable, with a proportion f exhibiting the symptoms, a patient judged exchangeable with them would, for him, have a probability effectively f of exhibiting the symptom. This is the justification for a procedure, adopted in many cases, of equating the probability of an aspect of the future with a frequency observed in the past. Notice that it requires three conditions: an exchangeable series, a long series, and a case exchangeable with the series. The first condition rules out the weather; the last excludes a different pin.

There is one extremely important point to be made about (7.2), a point that will repeatedly arise in probability calculations and is not confined to exchangeable series. Once you have chosen the two values, g and m, to reflect your initial opinion and the strength of that opinion, you are committed to $p(U|r, n)$ for *all* values of r and n, and not just those that you originally contemplated. Thus, in the case of the pin with $g = 0.4$ and $m = 3$, a series of five tosses all of which resulted in U and hence $f = 1$, would give your probability for another U on the sixth toss to be 0.78. When considering the values of g and m, you need to bear in mind that all these probabilities can be affected, and it is often useful to consider several hypothetical values of r and n.

A consequence of the rules of probability and the coherence they reflect is that while a few probabilities can be chosen at will, many others are automatically determined from the few by the rules. This is a general principle and affects all calculations of beliefs. In the exchangeable case, there are many implications from your choice of g and m, one for every possible series of data, and for every possible combination of f and n. If you find that there are no values of g and m that can accommodate your beliefs for all combinations, then you have two alternatives. You can retain exchangeability, but go back to the original $p(\theta)$, which will give you more flexibility. If this is still not enough, then your only resource is to abandon your view that the series is exchangeable. Here is an example.

There are many people who believe that if you have a long series almost entirely of Us, then there is a greater probability for a D next time than if you had experienced fewer Us. The idea being that compensation is needed to make up the appropriate frequency of Ds that has so far been too low. One can easily see that this view conflicts with (7.2) since the bigger the r is, the larger is the probability of U next time. It follows that if you have belief in compensation, then you cannot simultaneously have beliefs that (7.2) accommodates. More can be said, for the compensation concept and exchangeability do not even cohere and you cannot believe both. Mathematically, $p(U|r, n) = (nf + mg)/(n + m)$ increases with r and the more Us you see, the greater is your belief in U next time.

It may be felt that excessive attention has been paid to the notion of exchangeability and that we have labored unduly over a rather narrow, specialized concept. The reason for our labors is that the notion is used throughout the analysis of data, where many series, not just of two but of any number of outcomes, are generally accepted, not only by you, but by nearly everyone, to be exchangeable. Even series, such as weather, that are not exchangeable, have been studied by connecting them with other series that are exchangeable, though the technicalities are beyond us here. So exchangeability arises all over the place and our hope is that by studying it in a simple case of two outcomes, you will gain an appreciation of its value elsewhere, even though the technicalities are understandably beyond you. The quantity θ that was introduced above is called a *parameter* and it will be seen in Chapter 11 how parameters play a central role in science. Next, we take a closer look at the Bernoulli parameter θ.

7.8 CHANCE

It was seen in §4.3, with the discussion of two events, that there was some simplification if the two events were independent; in particular, the product rule was simplified. Also, instead of three probabilities needed for a complete description of the uncertainty surrounding two events, A and B, for example, $p(A)$, $p(B|A)$, and $p(B|A^c)$, independence required only two, $p(A)$ and $p(B)$. (As usual, a fixed knowledge base is assumed, for independence can be destroyed or created by changes in the base, as we will see in a moment.) The simplification produced by independence is even greater with more than two events, considered in the next chapter. It would therefore be most desirable if you could create independence in your beliefs in some way; that is what the quantity we have denoted by θ does with exchangeable series. To see this, consider the first two tosses of the pin and the result UD. These are not independent for you, since your probability for the D on the second toss is influenced by the occurrence of U on the first. But now introduce θ and you have independence since $p(UD|\theta) = p(U|\theta)p(D|\theta) = \theta(1-\theta)$ by the Bernoulli nature of the series, given θ. Generally, for any length of an exchangeable series, you have independence, given θ, but not without θ.

It is not a topic that will arise much in this book, but there are many uncertain situations that are most profitably studied by introducing a new, and perhaps a little artificial, quantity such as θ, to create independence. For example, in agricultural experiments, two varieties will behave similarly, and therefore not independently, because they experience similar weather conditions; so a quantity representing weather is introduced to create independence, given the weather, and thereby simplifying the analysis, without weather necessarily being described in terms of sunshine, temperature, humidity, and so on. Readers who are familiar with even the simplest statistical literature will have encountered the mantra "independent and identically distributed", which occurs so frequently that it has acquired an acronym, iid. Yet the authors hardly ever mean what they say. What they intend is iid given some quantity such as θ.

Returning to the exchangeable series of two possible outcomes, U and D, let us look at θ in more detail. First notice that it behaves like a probability; indeed, within the Bernoulli series it is a probability, namely your probability of U were you to know its value, $p(U|\theta) = \theta$. Also it obeys the probability rules, for example, in calculating the result $\theta^r(1-\theta)^{n-r}$ for your probability of r Us. Does θ therefore correspond to your belief in something? You already have beliefs about its value expressed by a probability $p(\theta)$, yet according to the attitude adopted in this book, it is nonsense for you to have a belief about your belief, if only because doing so leads to an infinite regress of beliefs about beliefs about beliefs . . . Another feature of the Bernoulli θ is that it has a degree of objectivity in the sense that if Peter and Mary both judge a series to be exchangeable, then the value of θ, as a limiting frequency, will be common to them both, though unknown to them both. The objectivity is limited though because if Paul disagrees with exchangeability, θ may not have a meaning for him. Experience shows that there is a massive agreement about some series being exchangeable, so that objectivity can be at least a convenient approximation.

The upshot of these considerations is that θ, while it obeys the rules of the probability calculus, is not a probability in the sense of a belief. As a result, we prefer to give it a different name and it is often referred to as a *chance*. Thus, de Finetti's basic result in §7.5 is that an exchangeable series of two outcomes is always a mixture of Bernoulli series with different chances. Notice that there are now three words that are almost synonymous in the English language but to which we have assigned special, different meanings. *Probability* always refers to your belief, *likelihood* to your uncertainty of a single event under different circumstances, and *chance* is a concept pertaining to a Bernoulli series. It may appear pedantic to fuss in this way, but experience has shown that the separation of the ideas is essential for a proper appreciation of uncertainty. It has the minor misfortune that we cannot vary the language, as modern writers like (see §2.8) switching between probability, likelihood, and chance, for if probability is meant, then probability it has to remain. It also helps to understand why mathematical modes of thought differ from those of poets. Poets like to invest words with many shades of meaning and encourage ambiguity, while mathematicians are precise and a word has a single, unambiguous meaning. Poets make simple things complicated; scientists try to make complicated things simple.

The relationship between probability and chance is profitably explored a little further using the pin as an example. First, $p(U)$ expresses your belief that the first toss will result in the pin falling with point up, U. Strictly, it should be $p(U|K)$, referring to your knowledge base but, as usual, K will be kept constant and conveniently omitted. On the contrary, $p(U|\theta)$ is θ, your belief concerning the first toss, were you to know the value of θ. The relationship between $p(U)$ and $p(U|\theta)$ is, for the case where two values of θ, θ_1, and θ_2, are being considered, as in §7.5, obtained by using the extension of the conversation from U to include θ, and introducing $p(\theta)$ (§7.6)

$$p(U) = p(U|\theta_1)p(\theta_1) + p(U|\theta_2)p(\theta_2)$$
$$= \theta_1 p(\theta_1) + \theta_2 p(\theta_2).$$

Generally, if there are many values of the chance that you consider possible, there will be a term equal to the value of the chance θ, times your probability for that value $p(\theta)$, the terms being added to provide your probability for U. The expression on the right-hand side plays an important, general role that is encountered in §9.3.

The "probabilities" that are basic to quantum mechanics are really chances, in our usage of the terms. Those who accept quantum mechanics accept exchangeability as part of that acceptance and therefore have chances. In statistical mechanics, there are two forms of exchangeability, Fermi–Dirac and Bose–Einstein. The same situation is observed in genetics, which is based on chances, not on probabilities. Furthermore, since the "probabilities" that physicists and geneticists recognize are really chances, the chances are associated in their minds with frequency, so that probability is thought of in terms of frequency.

We now have two methods of assessing probabilities: using the concept of cases, which have equal uncertainties, the classical method, and that based on frequency allied with the concept of exchangeability. The former applies only to a limited class of situations, such as games that use cards or dice. The second is of such wide use that probability is often confused with frequency. There remain situations where neither of these methods apply, for example, when you attempt to assess your probability that the political party you support will win the next democratic election. Here there are no equally probable cases and the frequency with which your party has won the previous elections is no guide, only because you do not make the judgment that those elections are, for you, exchangeable. We, therefore, need a further method. This is based on coherence and is treated in Chapter 13 when we have examined the phenomena that can arise when you contemplate three events.

CHAPTER 8

Three Events

8.1 THE RULES OF PROBABILITY

So far in this book we have almost entirely been concerned with studies involving only two events. The ideas developed there are now extended to situations with three or more events. The rules of probability that were developed in Chapter 5 are perfectly adequate to deal with the extension, and no new rules are required, but they do lead to some surprising results when more events are contemplated. We begin by looking again at the three rules.

The convexity rule in §5.4 deals with a single event and requires no elaboration. The addition rule, in the simpler form of Equation (5.2) of §5.2, says that if two events, E and F, are exclusive (that is, cannot both be true) then

$$p(E \text{ or } F) = p(E) + p(F). \tag{8.1}$$

The extension to three is immediate. Suppose E, F, and G are three events that are exclusive, in the sense that no two of them can both be true, then

$$p(E \text{ or } F \text{ or } G) = p(E) + p(F) + p(G), \tag{8.2}$$

Understanding Uncertainty, Revised Edition. Dennis V. Lindley.
© 2014 John Wiley & Sons, Inc. Published 2014 by John Wiley & Sons, Inc.

where E or F or G means the event that is true if, and only if, one of them is true. There are several ways to see that this is correct. One is to return to the urn and suppose that some balls are emerald, E, some fawn, F, some green, G, and the remainder without color. A colored ball has only one color, so the colors are exclusive and the total number of colored balls is the sum of the numbers that are emerald, fawn, and green. Dividing by 100, the total number of balls, to obtain proportions or probabilities, the probability of being colored is seen to be the sum of the three probabilities of the individual colors. Exactly the same argument can be used for any number of different colors, not just three, and correspondingly for any number of exclusive events.

The mathematical way of seeing that the result is true is to take the original two-event form, (8.1) above, and replace F by F or G giving, since E is exclusive of F or G,

$$p(E \text{ or } F \text{ or } G) = p(E) + p(F \text{ or } G). \tag{8.3}$$

We then use the two-event form again to yield

$$p(F \text{ or } G) = p(F) + p(G).$$

Combining these two results gives the three-event form of Equation (8.2).

It is worth stopping for a moment to look at the mathematical argument in the last paragraph because it demonstrates the power of mathematical notation, which, as remarked in §2.7, is really another language. Since the two-event form of the addition rule applies to any two events, the notation reflecting this, the flexibility can be used to advantage and, in particular, we can use an event that combines two others, as in Equation (8.3). Repeatedly doing this, we obtain general statements, not tied to special concepts such as balls in an urn. Furthermore, the method applies to any finite number of exclusive events by repeated use of the method.

The addition rule says that, to obtain your probability of one of a number of exclusive events happening, you add your probabilities of the individual events, a result used in discussing the classical theory in §7.1. Notice that it is essential that the events be exclusive. There is an extension of the general form of §5.4 without the restriction to exclusive events but, since we shall not need it, and it is somewhat complicated, it is not given here. In all forms, recall that a fixed knowledge base is assumed.

The multiplication rule of §5.3 also extends to three events in the form

$$p(EFG) = p(E) \times p(F \mid E) \times p(G \mid EF). \tag{8.4}$$

In words, your probability that all three events occur is your probability of the first, multiplied by your probability of the second, conditional on the first, and then multiplied by your probability of the third, conditional on both the first and second. Incidentally, this provides a good illustration of the simplicity and clarity of the mathematics in (8.4) compared with ordinary English in the last sentence. Recall that EF is the event that is true if, and only if, *both* E and F are true; whereas E or F is true provided *either* one of E and F is true. The reader might like to refer again to the truth table in §5.1.

As with the addition rule, the multiplication rule may be demonstrated with balls in an urn. There would be red balls corresponding to E, the others being white. Some balls would, in addition, be spotted, corresponding to F, the others being plain. Finally, some balls would be plastic, corresponding to G, the others being wood. Thus there would be eight types of balls, for example, some balls would be plastic, painted red, with spots, corresponding to EFG. The proportion of balls in the urn that are simultaneously red, plain, and plastic is equal to the proportion of red, times the proportion of plain among the red, times the proportion of plastic among the red, plain ones. Mathematically, the two-event form with the events E and FG gives

$$p(EFG) = p(E) \times p(FG \mid E).$$

Applying it again to F and G with everything conditional on E (as well as the implicit knowledge base) we have

$$p(FG \mid E) = p(F \mid E) \times p(G \mid EF).$$

Combining these last two results gives the form stated in (8.4). Notice that unlike the addition rule, there is no need for the events in the multiplication rule to be restricted in any way.

168   THREE EVENTS

So there is nothing really new in passing from two to three events, insofar as the rules of probability are concerned, and we do not need new rules. We derived modified forms, by the mathematical arguments just given, from the old. Since all the properties of probability follow from the three basic rules (see §5.4), it means that the other rules, such as that of the extension of the conversation (§5.6), and most importantly Bayes rule, (§6.3) to be discussed in §9.1, similarly extend to three events. Although the rules are adequate for any number of events, it turns out that they lead to surprising results when passing from two to three events, while beyond three, nothing surprising happens, life just gets more complicated. To investigate the surprise, we start with an example, which is extreme but has been chosen to emphasize a point. The phenomenon it displays ordinarily occurs in a less extreme form, where it is very common.

8.2 SIMPSON'S PARADOX

Below are some data in the form of a contingency table (see §4.1). The context is medicine, where 80 patients with a disease took part in a clinical trial. 40 of them were given a treatment, in the form of an experimental drug, and the remaining 40 were provided with a placebo, none of them knowing which they had received. At the end of the trial, each patient was classified as recovered or not. The outcome of the trial is given in the table, with the obvious notation of T for treated and R for recovered. As before, the raised letter c denotes complement: T^c for the placebo and R^c for a patient who had not recovered by the end of the trial. In addition to the raw data, the recovery rates, calculated from them, have been included in the last column.

	R	R^c	Total	Rate
T	20	20	40	50%
T^c	16	24	40	40%
Total	36	44	80	

The treatment by the drug would appear to be beneficial since the recovery rate for the treated patients is 10% higher than for those untreated, and medical opinion might be that the treatment is a "good thing". It may be objected that the trial is too small for reliable conclusions to be drawn. If you feel this, add a couple of zeros to each of the raw figures, with 8000 in all and 3600 recovered. The analysis that follows will not be affected.

It was seen in §7.7 that under exchangeable conditions, and with sufficient data, you would assert $p(R\,|\,T) = 0.5$ and $p(R\,|\,T^c) = 0.4$. Here the first probability refers to the event of your recovery were you to receive the treatment, the second to your recovery with the placebo, and on this basis you might decide to have the treatment, thereby increasing your probability of recovery by 0.1. Notice the use of the subjunctive here, since these are assessments before your treatment regime is decided.

Both men and women took part in the trial and the results were available for each sex separately. The results for the 40 males, presented in tabular form as above, were

Males	R	R^c	Total	Rate
T	18	12	30	60%
T^c	7	3	10	70%
Total	25	15	40	

Now the position is reversed and instead of the treatment increasing the recovery rate by 10%, it has decreased it by the same amount. The treatment would appear not just to be ineffectual, but to be positively harmful. Using exchangeability again, a male might argue that $p(R\,|\,TM) = 0.6$, whereas $p(R\,|\,T^cM) = 0.7$, where M denotes male. The interpretation is as before, that were he to receive the treatment, his probability of recovery would be 0.6, whereas without it, it would be 0.7 and the treatment is to be avoided. Notice that the man is making a different exchangeable judgment from that made with all the data. There he was supposing himself to be exchangeable with all the data; now, with more information, he is restricting himself to being exchangeable with the males only. Exchangeability is always conditional, just like probability.

It might be thought that a treatment that is good overall, but is bad for the males, must be good for the females, if only to compensate. So let us look at the data for the 40 females who took part in the trial. These can be obtained by subtracting the numbers in the table for the males from those in the complete table and no new information is needed. The result, as the reader may easily verify, is

Females	R	R^c	Total	Rate
T	2	8	10	20%
T^c	9	21	30	30%
Total	11	29	40	

The result is not that anticipated because the treatment is just as bad for the females as it is for the males, namely, a reduction in the recovery rate by 10%. As before, a female might argue that for her $p(R \mid TF) = 0.2$, whereas $p(R \mid T^cF) = 0.3$. Here F denotes female or M^c. (Lest feminists object, $F^c = M$.) Thus, a woman might decide not to use the treatment.

The situation is that a treatment that appears to be good for all of us (the first table) is bad for the men and bad for the women. This is the paradox. It is usually known as Simpson's paradox, after a UK civil servant who came across it in his salad days, though it had occurred earlier in the literature, recalling Stigler's law of eponymy in the prologue. In his form, the paradox says that the overall behavior may be contrary to the behavior in each of a number of subgroups, here male and female. People's first reaction is to disbelieve the paradox and to think that there has been a mistake in the arithmetic, but careful perusal of the figures shows that this is not so. Recalling that probabilities are equivalent to proportions of balls in an urn, you could envisage the paradox in terms of balls, colored red or white (for T), spotted or plain (for R), and plastic or wood (for M). The paradox, as we now try to show, is of considerable practical importance even in its most modest form.

8.3 SOURCE OF THE PARADOX

How has the paradox arisen? First notice that the disease the treatment was designed to cure is more serious for the women than it is for the men. Confining ourselves to the data for the placebo, where the disease effectively remained untreated, apart from the possible psychological encouragement from participation in the trial, we see that only 30% of the women recovered, whereas 70% of the men did, so it is a disease that is more serious for women than for men. Second, observe that in the case of the men, 75% (30 out of the 40 men in the trial) were treated, whereas with the women, only 25% (10 out of the 40 women) received the treatment. Thus, the treatment went predominately to the men, who were more likely to recover anyhow, and kept from the women who were the main sufferers. Consequently, the treatment looked good, not because of real merit, but because it was mainly applied to the men with their higher recovery rate. Perhaps the person in charge of the trial was distrustful of the treatment, feeling that it might do harm, so gave it predominately to the men, who were likely to recover anyhow, and kept it from the principal sufferers, the women. Whatever the reason, it is the confusion between sex and treatment that has given rise to the paradox. We leave it to the reader to do the arithmetic to convince themselves that had the sexes been handled equally with respect to the allocation of the treatment, in the sense that the proportion of men receiving the treatment equalled the proportion of treated women, then the original table, that did not refer to sex, would have exhibited the same 10% reduction in the recovery rate as exhibited in the other two tables, that did record sex. For this calculation, assume the same recovery rates as in the last two tables.

Simpson's paradox therefore arises because the allocation of treatments depended on another quantity, sex, that itself had an effect on the recovery rate. This type of dependence is called *confounding* and the two quantities, treatment and sex, are said to be *confounded*. Because of the confounding, it is not possible to be sure, in the original table, that the apparent treatment effect is real and not due to the confounded quantity, sex. What is therefore required is an allocation of treatments to the patients that is not confounded with any other quantity that might have an effect. This is a tall order and before we see how it can be achieved, let us draw some lessons from the paradox.

8.4 EXPERIMENTATION

An immediate consequence of the paradox is that one cannot believe the message that a simple contingency table appears to deliver without more investigation. Recovery appears to be helped by the treatment in the example but the effect is an illusion due to sex. Even the tables that include sex cannot be guaranteed to send a correct message since there may exist another quantity that reverses the effect of treatment again. For example, it might happen that breaking up the last two tables according to whether the patients came from a rural or urban community, thus producing four tables, rural males, rural females, and so forth, would exhibit a different effect.

The paradox has repeatedly arisen in practice. The early work on the relationship between smoking and lung cancer revealed a strong positive association (§4.4) between smoking and the occurrence of the disease, just as our first table showed one between treatment and recovery. An eminent statistician suggested that there might be a genetic factor that encouraged smoking and also made the person prone to lung cancer, playing a similar role to that of sex in our example; if so, the causal relation between consumption of cigarettes and lung cancer might be spurious. The suggestion was sensible and much further work was required to eliminate this possibility and establish the causal link. It has now been demonstrated that the original table did not lie and smoking is a cause of lung cancer. Here are examples where the table did misrepresent the situation. It was once claimed that the consumption of yogurt increased one's life span, again on the basis of a contingency table. Here, unlike smoking, there was a confounding, genetic effect because consumption of yogurt was greatest in Bulgaria, where there appears to be a gene for longevity, so that longevity was confounded with a gene. A trial of the effect of giving milk to schoolchildren in Scotland appeared to show that milk was harmful, because the teachers had given milk to those they felt most in need of it and kept it from the healthier ones, so confounding the consumption of milk with health. We will meet a sociological example in §8.7.

A general lesson from the analysis of the paradox, and the examples, is that one should always be suspicious of a claimed association between two factors, because there may be confounding with other factors. Most scientists today are fully cognizant of the difficulties and try to eliminate the confounding by methods about to be described, but this may be hard in fields where observational data, rather than experimental data, are all that are available. It is often found that people without training in numeracy fail to appreciate the difficulties the paradox exemplifies. Neither arts nor science has a monopoly of truth, so that better understanding and control of the world will surely come through combining both standpoints.

How are the difficulties revealed by the paradox to be overcome; how can contingency tables be presented that really mean what they appear to mean? One way is to think of all the quantities that might affect the feature of interest; in our example, all the quantities that might affect recovery. These are termed *factors*. Thus, treatment, sex and urban environment, are all factors. An experiment is then performed with all factors fixed (e.g., placebo, male, rural) and a second with all the factors the same except for the one of primary interest (e.g., treatment, male, rural) giving a measure of the treatment for rural males. This is repeated for all combinations of the other factors; then if every factor has been included, any differences between the two experiments must be due to the single factor that changed (e.g., treatment). Experiments with all factors fixed are said to be *controlled*. It may happen with a controlled experiment that the effect is seen to be present only for certain values of the other factors; thus the treatment could work only for the men. In that case, we talk of an *interaction* between the treatment and the factor. Physics and chemistry abound with examples of controlled experiments. Unfortunately, it is usually possible to perform such experiments only within the confines of a laboratory, where conditions can be controlled, while in subjects such as medicine or agriculture, where the experiments cannot be confined to a laboratory, such control is rarely possible. The situation is even worse in sociology, where even modest amounts of control are difficult to arrange. Some physicists can be contemptuous of attempts by sociologists to be scientific, but often the contempt is unjustified because of a failure to recognize the difficulties of experimentation in the latter field compared with the precision attainable in their own, often at considerable expense. We will return to this point when we discuss the scientific method in §11.1.

8.5 RANDOMIZATION

If the doctor is not able to know about or to control all the possible factors that might reasonably effect the disease under investigation, how is a medical trial to be performed? How is a dietitian to determine whether some modest amount of alcohol is good for one? How is a criminologist able to assess the main causes of crime or how they might be treated? How is a farmer to find out the best fertilizer for his crops? In none of these cases is complete control possible. Nevertheless, there is a possible answer that does not demand the complete control beloved by the laboratory scientist.

Let us return to the example of the medical trial and Simpson's paradox in §8.2, where we saw that it was important to avoid factors that were confounded with the treatment under investigation, the factor in the example being sex. We wanted sex and the treatment to be independent (see §8.3), or that the treatment should be assigned to a patient irrespective of their sex. How is this to be done? There are two ways.

The first is obvious and is related to the controlled method often used in the laboratory. It is to recognize the factor and make the assignment of treatment accordingly. Thus, in the medical trial, since it was known that sex was influential, the disease being more serious in women than men, sex should have been recognized from the outset and the same proportion of men should have been allocated to the treatment as women, and similarly for the placebo. One difficulty with this method is that there may be many factors that are recognized as possibly influencing the result. Suppose there are 8 factors, each of which can exist, like sex, in two forms, then there are $2 \times 2 \times \ldots \times 2$ (with 8 2s) or 256 in all, possible groups of patients. Even with only one patient of each type allocated to the treatment and one to the placebo, the experiment will involve 512 people and will likely be too big. There are ways of reducing the size by ignoring some interactions (§8.4) between factors, but these will not be considered here. A second difficulty is that one can never be sure that every factor has been thought about. Maybe the blood type of the patient could affect the result, so another factor and an even larger experiment would be needed. It is all too easy to think of possible factors and criticize an experiment because they have not been included.

There is an ingenious way out of these troubles that uses randomization. We met the idea of randomness when drawing balls from the standard urn in §3.2. You would think the balls to be drawn at random if any ball was judged by you to be as likely as any other to be taken; or if you were indifferent between a prize contingent on a specified ball being drawn, and the same prize contingent on a different specified ball. Let the same device be used with the medical trial and suppose the balls in the urn are either labeled treatment T or placebo T^c. With a list of the patients who have agreed to participate in the trial available, a ball is withdrawn at random and the first patient is assigned according to what is on the ball. Proceed in this way with subsequent withdrawals and patients. By this device, the patients are said to be allocated to treatment *at random*. If there are the same numbers of balls of each type, then there will be equal numbers of patients receiving treatment as placebo. What is more important is that the proportions of men, and of women, receiving the treatment will be about the same because sex had nothing to do with the allocation. In practice, the balls in urns are not used and there are tables of random numbers that operate in a similar way. The important point is that the method of withdrawal of the balls, or the use of the tables, is not confounded with sex. Indeed, as far as you are concerned, the allocation of treatments by these methods is not confounded with anything because random means that the withdrawal of the balls is not affected by anything. By this device you can be reasonably sure that the final results from the trial will really mean what they appear to say and that no factor can disturb your conclusion. Actually you cannot be quite sure because it could happen, just by chance, that all the men got the treatment and all the women the placebo, just as all tosses of a coin could fall heads, but it is unlikely and, in any case, if you were aware of sex as a source of concern, you could check on this before carrying out the trial.

Experience shows that the following is the best procedure to use in designing an experiment such as the medical trial. First think of factors, such as sex, that might be influential and, if there are not too many of these, allocate treatment to patients so that no confounding with them takes place. Having made sure that the important factors are not confounded, allocate the treatments purposefully as regards these, as in a controlled experiment, but otherwise at random. Having done all this, check that the randomization has not produced something odd; for example, if all the treated patients are rhesus negative and all the placebos positive, it would be better to do the randomization again, since otherwise the claim could legitimately be made that treatment is confounded with blood type. It is not essential to use a table of random numbers and it would suffice to allocate the treatments in a haphazard way, checking afterward for any possible confounding. But random numbers are often a convenient way of getting a haphazard result. It is always important to check the result of the haphazard or random selection, preferably before carrying out the experiment, to check for any possible confounding. Sometimes it is not necessary to perform any randomization; all controls can be implemented, as in the laboratory, but the general form is a suitable method of experimentation that permits reliable conclusions to be drawn. Unfortunately, there are cases where even the haphazard allocation is not feasible and an example will be encountered in §8.7.

As an example of applying these methods, consider the original medical trial. With 80 patients available, 40 of each sex, the men could have been allocated the treatment at random or haphazardly, and similarly the women. Before beginning the trial, the allocation should be carefully inspected to make as sure as one can be that the randomness has not thrown up some factor that could influence the conclusions. For example, the random allocation might have resulted in the treated men coming predominately from a city; in which case, to forestall possible criticism on the basis of town versus country, something more haphazard might be attempted. Of course, one can never cover all possibilities; the best one can hope is to reduce the uncertainty surrounding any conclusion.

8.6 EXCHANGEABILITY

In the case of the medical trial of §8.2, which illustrated the paradox, it was pointed out that, in order for you to use the information the trial provided, you would ordinarily make some assumption of exchangeability of yourself with the patients who took part in the trial. For example, presented with only the first table, with no reference to sex, you might feel that, were you to receive the treatment, you would be exchangeable with the 40 patients in the trial who also had T and that your probability of recovery would therefore be 50%, $p(R\,|\,T)=0.5$. Similarly, were the treatment not taken, exchangeability would be with the other 40, $p(R\,|\,T^c)=0.4$, and as a result, you would accept the treatment. Recall that it is being assumed that the numbers in the trial are large so that the more delicate considerations of §7.6 are not needed.

If you now received the additional results that included information about sex, you would alter your assumption of exchangeability. Thus a woman might judge herself to be exchangeable with the women who took part in the trial and, using an argument similar to that advanced in the last paragraph, would take her personal probabilities to be $p(R\,|\,TF)=0.2$ and $p(R\,|\,T^cF)=0.3$, so would refuse the treatment. In this section, we point out how careful you have to be about these judgments of exchangeability, using an example from agriculture.

The medical example has patients affected by three factors: treatment, recovery, and sex. In the agricultural example, plants replace patients and the three factors are variety (black or white), yield (high or low), and height (tall or short). For convenience, the two sets are listed and you may find it helpful in what follows to make repeated reference to this list.

Medical Trial	Agricultural Trial
Treatment (or placebo)	Black (or white) variety
Recovery (or not)	High (or low) yield
Sex (male or female)	Height (tall or short)

Black corresponds to treatment, high yield to recovery, and tall to male. In both trials, interest lies in the association between the first two factors, the third factor being there because it might influence the conclusions. The farmer wants to know whether to plant the black or white variety with the aim of getting a high yield. The results in the agricultural trial could be written out in the form of contingency tables exactly as in the medical case. Suppose that in doing so, the numbers in the agricultural case are the same as in the medical one. Thus corresponding to the entry 18 for *TRM* in the medical table for the males, the number of black plants that both grew tall and had high yield was 18. For the reader's convenience, the new tables are given here, with the obvious notation: *H*, high; *L*, low; *B*, black; *W*, white.

	H	L	Total	Rate
B	20	20	40	50%
W	16	24	40	40%
Total	36	44	80	

Tall	H	L	Total	Rate
B	18	12	30	60%
W	7	3	10	70%
Total	25	15	40	

Short	H	L	Total	Rate
B	2	8	10	20%
W	9	21	30	30%
Total	11	29	40	

We have two entirely different sets of data, the numbers happen to be the same. The conclusion reached in the medical trial was that the treatment was not to be used, since it is harmful for both the men and the women, and the placebo was to be preferred. Before reading beyond this sentence, ask yourself, and answer this question: do you think the white variety is better; white, recall, corresponds to placebo in the list above?

Most people answer the question by preferring the black variety as the one giving the higher yield. That is, they take their conclusion from the first agricultural table, rather than from the two that incorporate the breakdown by height (replacing sex). The conclusion is correct and despite the fact that the numbers are exactly the same in the two trials, the conclusions are different. Why is this? We argue that the difference lies in the use of exchangeability with the data.

In the medical trial, as we have just seen, a woman would consider herself exchangeable with the women in the trial and her personal probability of recovery, were she to have the treatment, would be $p(R \mid TF) = 0.2$. Consider a farmer having to decide, after seeing the data, whether to plant the white variety or the black. What exchangeability judgment is reasonable for him to make between the new planting and the plantings in the trial? If the same judgment were made as with the medical experience, the farmer would reach, in analogy with $p(R\,|\,TF)$, the probability of high yield, were the black variety planted and grew short. But this probability is of no relevance to the farmer because when the black variety is planted, he does not know whether it will grow tall or short, unlike the woman who knows her sex. A relevant probability is $p(R\,|\,T)$, the probability of high yield when the black variety is planted, which is obtainable from the first of the tables and, with a judgment of exchangeability between the new planting of that variety and those in the trial, has the value $p(R\,|\,T) = 0.5$. Similarly, with the white variety, $p(R\,|\,T^c) = 0.4$, a smaller value, so that the black variety is to be preferred. The difference is that the doctor could know the patient's sex, whereas the farmer could not know the plant's height.

It is surprising that although the numbers are the same in the medical and agricultural trials, the conclusions are exactly opposite, placebo (white variety) in the medical case and black variety (treatment) in agriculture. The example demonstrates the need to think carefully about the appreciation of data, as well as the data themselves. It is possible today to purchase computer packages that purport to analyze data. As a result, people put their data into the computer, together with the package, expecting to obtain sensible results. They may, but they may not, for although the computer is a wonderful tool for computing, it is not, at the time of writing, a substitute for thought. Our example, simple though it is, demonstrates the necessity for more than just calculation; what to calculate is also relevant. A good computer package would ask the user to make exchangeability and other assumptions, as well as perform the calculations. Statistical textbooks can also be misleading in presenting analyses for contingency tables without adequate attention to the practical circumstances surrounding the numbers.

Since the farmer's conclusion depended only on the results of the first table, where height was ignored, it might be felt that the additional data with height included are irrelevant for him, but this is not so. For example, were the black variety to be planted, corresponding to B in the tables, more plants would grow tall (male) than short (female). In the trial there were 30 of the former and only 10 of the latter. From this it might be deduced that $p(\text{tall} \mid B) = 3/4$ and it is perhaps because of this tendency of the black variety to grow tall that it provides a higher yield. Notice that in the medical trial, sex was controlled. For example, the doctor chose to give the treatment predominately to the men. In the agricultural trial, height (corresponding to sex) was not controlled but was influenced by the variety selected. This affects the judgments of exchangeability subsequently made.

Many people, in discussing these two examples, would speak of causation, saying that giving someone the treatment does not cause them to be male, whereas planting the black variety is the cause of the plants growing tall. They would claim that it is this causal difference that distinguishes the two cases, medical and agricultural. This is surely sound but there are difficulties, that need not concern us here, in providing a precise definition of causation. It fits better with our approach to uncertainty, to use the concept of exchangeability. It is possible to come very close to the concept of causation by using the concepts of "doing" and "seeing" mentioned in §4.7, where seeing a quantity to have a value can have entirely different consequences from doing something to make the quantity have that same value. In the medical example of this chapter, the doctor can do something to control the sex of the patient receiving treatment, not by changing someone's sex, but by selecting for treatment or placebo, according to their sex. In contrast, the farmer cannot control the height of an individual plant but can merely see how tall it grows. So, in a distortion of the English language but fitting within our specialized use of it, the doctor can "do" sex, whereas the farmer can only "see" height, and it is this difference that, perhaps better than exchangeability, explains the distinction between the two experiments. There is still a need to pass from the data to the conclusions about which is better for you, as a patient, or you, as a farmer, where the correct judgment of exchangeability is essential.

The intimate connection between "doing" and "seeing" on the one hand, and causation on the other may be clarified by the consideration of the following famous illustration. The arrival of low atmospheric pressure in an area causes the barometer to fall and later rain to arrive, so that seeing the barometer fall leads you to anticipate rain, whereas making the barometer fall by artificial means does not make it rain. Low pressure causes rain but a low barometer does not.

8.7 SPURIOUS ASSOCIATION

The police in Britain recently produced data showing association between the crime of mugging and ethnic group, which we present in our familiar form of a contingency table, taking the liberty of changing the numbers slightly to simplify the arithmetic. The reason for the change is that our concern is with illustrating the phenomenon, rather than making judgments about crime and ethnicity. It is supposed that the police took 64 people at random, in accordance with the ideas expounded in §8.5, from the Blacks in a population; and similarly, 64 at random from the Whites. We return to this point after the data have been analyzed.

	C	C^c	Total	Rate
B	26	38	64	41%
W	11	53	64	17%
Total	37	91	128	

In the table, C denotes those arrested on the charge of mugging, C^c is, as usual, the complement, not arrested, and the factor will be referred to as crime, hence the letter C. B means that the person was Black, W that they were White, again the complement. Naïve use of the table would say that since the crime rate is 41% among the Blacks but only 17% among the Whites, race was a cause of crime. The argument can be compared with that applied to the table for the medical trial, with ethnic group replacing treatment and crime substituting for recovery.

As with the medical trial, it is instructive to include a third factor, which is here, not sex, but unemployment. The breakdown follows.

Unemployed	C	C^c	Total	Rate
B	24	24	48	50%
W	4	4	8	50%
Total	28	28	56	

Employed	C	C^c	Total	Rate
B	2	14	16	$12\frac{1}{2}\%$
W	7	49	56	$12\frac{1}{2}\%$
Total	9	63	72	

The situation has now changed dramatically because, both within the unemployed and within the employed, there is no difference between the crime rate for the Blacks and that for the Whites; the marked difference that the police saw in the original table has quite disappeared. In the language that was used in §4.3, for both those with and without work, the crime rate is independent of race. The reason for the apparent association between race and crime that was suggested by the first table is that the unemployment rate among the Blacks is 75% (48 out of 64), whereas among the Whites it is only $12\frac{1}{2}\%$ (8 out of 64). A black person is six times more likely to be out of work than a white person. Do not forget, the original figures have been massaged as an aid to clarity.

This example is of the same form as the medical one but the influence is not reversed by the inclusion of the additional factor, as it was there. Instead, it is eliminated and replaced by a new one. Here we have one factor, race, being blamed for crime, when the culprit is really unemployment. Of course, we have to be careful, there may be some other factor that has so far not been considered, which could change the situation yet again. The Blacks in the study might be younger than the Whites. Also it would be pertinent to ask how the 64 Whites and 64 Blacks were selected; were they randomly selected from a some larger group, or were they from those 64 people who were stopped by the police, some of whom were charged, others not? A sound sociological study would need some clear thinking and experimentation in the style of §8.5 may be difficult, if not impossible.

The lesson to be drawn is again that a naïve analysis of a contingency table can be dangerous and the fact that a rate is high in one group and low in another does not establish that the factor defining the groups is responsible for the variation in the rate. Only a carefully designed experiment that eliminated confounding can provide a reliable assessment of the reason for the variation in the rate.

This is one reason, as we have said before, why sociological data, such as ours on crime and race, are so difficult to interpret and why sociology is, in some ways, a harder subject than physics. We will return to this point when the scientific method is treated in Chapter 11.

False association may affect government policy. In Britain, a proposal to charge university students for their education has been defended on the grounds that graduates earn more than nongraduates. The apparent association between graduation and earnings may be explained by the factor of intelligence. Universities may select students on the basis of their intelligence, and the real connection is between earnings and intelligence.

It was remarked above that crime and race appeared to be independent when employment was taken into account. Independence has only been systematically studied in connection with two events, so in the next section we look at the concept for three events, stimulated by this example.

8.8 INDEPENDENCE

Independence for two events was studied in §4.3 and association in §4.4. Two events, E and F, are said to be independent if your probability of their both occurring is the product of their separate probabilities,

$$p(EF) = p(E)p(F), \qquad (8.5)$$

for a fixed knowledge base. It was seen that this was equivalent to saying that your probability of one event being true did not depend on the truth of the other,

$$p(E \mid F) = p(E). \qquad (8.6)$$

Either of these definitions leads to a variety of other statements, such as E^c and F^c being independent, or $p(F \mid E) = p(F)$, reversing the roles of the events in (8.6). It is not easy to go wrong with independence when only two events are under consideration, but with three or more, the analysis becomes subtler. When presented with a new book on probability or statistics, the first thing I do is to turn to the definition of

184 THREE EVENTS

independence for three events and see if the author has got it right; often it is wrong, so I must be careful here.

Two events are independent if, whatever you learn about one, does not affect your uncertainty about the other. It is this form of definition that extends to three, or more, events. Events are *independent*, for a fixed knowledge base, if information about the truth or falsity of any set of them does not affect your uncertainty about the remainder. Thus, the independence of three events, E, F, and G, implies that, being told that both F is false and G is true, does not alter your uncertainty for E. In symbols,

$$p(E \mid F^c G) = p(E). \tag{8.7}$$

The multiplication rule (§5.3) says that $p(EF^c G) = p(F^c G)p(E \mid F^c G)$. Using (8.7), we have $p(EF^c G) = p(E)p(F^c G)$. The definition of independence just given also says that $p(F^c \mid G) = p(F^c)$ and a second use of the product rule enables this to be written as $p(F^c G) = p(F^c)p(G)$. Putting these together gives

$$p(EF^c G) = p(E)p(F^c)p(G). \tag{8.8}$$

There are many statements such as (8.7) and (8.8), all of which follow from the definition of independence. With two events, we saw from the contingency table that they all stemmed from one statement, Equation (8.5), but this is no longer true with more than two events; for example, neither (8.7) nor (8.8) on their own is enough for independence for three events. It is not even enough that the events be independent in pairs. In words, it can happen that E and F are independent, so are F and G, also G and E, yet the three are not. Here is an example with no suggestion that the numbers about to be given correspond to any actual case. It is derived from that of the previous section with the minor change that C means criminal and I innocent, its complement. Black or White, B or W, and unemployed or employed, U or E, remain unaltered. A population may be divided into four groups as follows:

White and employed, WE. Suppose these are all criminals, C. They fiddle their income tax or defraud their employers.

Black and employed, BE. These are all innocent, *I*. They are so pleased to be in a job that they are careful never to give an excuse for dismissal.

White and unemployed, WU. These are all innocent, *I*. They pay no tax and have no one to defraud.

Black and unemployed, BU. These are all criminals, being bored with the hopelessness of their situation.

This accounts for everyone. Next suppose there are 25 people in each of the groups, 100 in all. Let us write out a list of the events and their numbers:

$$WEC\ 25;\quad BEI\ 25;\quad WUI\ 25;\quad BUC\ 25.$$

If you like, think of an urn with 100 balls, 25 of each type. The probability of each single event is $1/2$. For example, 50 out of the 100 are White, so $p(W) = 1/2$, and similarly, $p(U) = p(C) = 1/2$. Next take any pair of events, U and C say; an inspection shows that they only occur together in one of the four groups displayed above. In the case of U and C, only in the group BUC, from which it follows that $p(UC) = 1/4 = p(U)p(C)$ so that U and C are independent. The same argument works for any pair of events. It follows that the three events are pairwise independent but in any reasonable meaning that we might attach to independence, the three events are not independent. For example, as soon as it is known that a person is both Black and employed, you know for sure, with probability 1, that they are innocent. Your probability of innocence has increased from $1/2$ to 1 as a result of the knowledge that they are Black and employed, in contradiction to the definition of independence. While such extremes do not happen in practice, it is not unusual for the association between pairs to be weak, near independence, yet there exist strong connections within the triplet.

With this interpretation of independence, probabilities may be calculated by the product rule without introducing conditions. Equation (8.8) gives an example for three events. Again, we repeat, this is for a fixed knowledge base; if that changes, then independence may arise or disappear. Thus, in the example just presented, W and E are independent, but if the knowledge base changes by learning that C is true, they are highly dependent; an employed criminal is necessarily White. True independence is a concept that introduces considerable simplification into a problem because if it obtains one need only think about the probabilities of the individual events. Provided that you have described your uncertainties for E, F, and G through $p(E)$, $p(F)$, and $p(G)$, then all your uncertainties about the three events together are described. Thus,

$$p(EF^cG) = p(E)p(F^c)p(G) = p(E)[1 - p(F)]p(G).$$

In contrast, without any independence, seven probabilities are required before you have a complete description of your uncertainty surrounding three events, for example,

$$p(E), p(F|E), p(F|E^c), p(G|FE), p(G|F^cE), p(G|FE^c), p(G|F^cE^c),$$

obtained by taking the events E, F, and G in that order and displaying how the uncertainty of one event depends on all the possibilities for the previous events. The reduction from seven statements of probability to three, due to independence, results in considerable simplification.

8.9 CONCLUSIONS

The main lesson to be learnt from the material in this chapter is that the relationship between two uncertain events is not always what it appears to be because there may be a third event that influences them both and distorts the apparent connection. Thus the connection between treatment and recovery of the patients may be completely changed by consideration of the patient's sex or the apparent dependence of crime on race can be destroyed by the inclusion of unemployment. Although we have not explored cases where there are four or more events, the reader will appreciate that the more events that are taken into consideration, the more complicated is the process of trying to understand what is truly happening. No new concepts are involved, only increased complexity.

8.9 CONCLUSIONS

In this chapter and the preceding ones, attention has been confined to events that can assume only two values, true or false, but the ideas extend, as explained in the next chapter, to quantities that can assume a range of values beyond two, and that are uncertain for you. Thus we can pass from simple recovery to degrees of recovery; or the yield, instead of being high or low, may be measured in kilograms per hectare. Yet more complexity arises as a result and there are real difficulties in appreciating the connections between several quantities. Scientists have developed methods for handling large numbers of quantities, but they are necessarily complicated and need the utmost care in interpretation. I was once asked to investigate allergies in a data set with a large number of factors that might possibly trigger a smaller number of types of allergy. I declined because of the complexity of the problem and the consequent small probability of reaching a reasonably firm conclusion. The proper understanding of allergies is likely to come through an attempt to understand the allergic process, rather than through massive contingency tables, just as progress in cancer therapy appears to be coming through a study of how cells become cancerous, rather than from data on cancer patients. A good theory is better than a lot of data without a theory.

Alternative medicines, that have become so popular recently, are full of associations that are of doubtful validity. A friend recently told me that bananas produce mucus when eaten, an association that may well be true, but how could it have been established without very careful experimentation, an activity that practitioners of alternative medicine do not engage in as often as regular doctors, or by understanding the physiological process of mucus production? When anyone asserts that "A is associated with B", a good riposte is "how do you know?" When I tried this on my friend, she replied that she experienced increased mucus whenever she ate a banana, ignoring the fact that one experience is not enough to establish a relation, just as one throw of the pin landing uppermost does not convince you that the next will also land in the same fashion (see §7.2).

My friend's reaction to her personal experience is understandable because we all find it easier to pay attention to what we directly encounter than to careful and numerous studies performed elsewhere. Today a newspaper has a story of a mother and father whose two children were vaccinated and subsequently developed autism, from which the parents concluded that the vaccine causes autism, despite the fact that the analysis of thousands of vaccinated children has revealed no evidence of a link with autism. In effect, the parents are saying that their two children count more than thousands of others; of course they do to them, but for the rest of us they are just two out of thousands. One of the hopes I have for this book is that it will enable you to assess beliefs more sensibly than these parents and you will appreciate the value of proper, scientific methods.

CHAPTER 9

Variation

9.1 VARIATION AND UNCERTAINTY

Variation often gives rise to uncertainty. Though we can recognize a group of objects called teapots, the variation present from one teapot to another results in our being uncertain whether we shall spill some of the tea when first pouring from a strange pot, for some pots are good pourers, some are not. More seriously, all biological material exhibits variation; even the simple influenza virus varies, with the consequence that we are uncertain what vaccine to use against it. Human beings show variation that we rightly cherish, yet it gives rise to uncertainty, whether in what size of trousers a retailer should stock, or in a stranger's reaction to a request.

There are only a few topics where variation is not present. Precision engineering is capable of making objects, like the balls in the urn, that are, for practical purposes, indistinguishable and portray no obvious variation. One atom of an isotope of hydrogen is regarded as the same as any other atom and the behavior of the isotope in the presence of oxygen can be predicted perfectly. We can say, in the spirit of §7.3, that one atom can be exchanged for another. Physics and chemistry are both founded on this lack of variation that partly explains

Understanding Uncertainty, Revised Edition. Dennis V. Lindley.
© 2014 John Wiley & Sons, Inc. Published 2014 by John Wiley & Sons, Inc.

why physical scientists were so uncomfortable with quantum physics and its unpredictability. It also helps to explain why those two subjects have advanced more than others, like biology, because they are not hampered by variation and so have less uncertainty. Biology is now advancing more quickly since some aspects of it have been reduced to the chemistry of amino acids. However, the laws of genetics contain randomness and the resulting variation is basic to the concept of evolution, where variants more suited to their environment stand a better chance of producing offspring that survive and breed.

Variation produces uncertainty because you cannot be sure what the variable material will do. Uncertainty inevitably necessitates description in terms of probability, hence probability is an essential tool in the handling of variation. This chapter is devoted to a study of variation and probability, beginning with a simple example, the familiar balls in an urn. Before doing so, we need to look at the rule of the extension of the conversation again because it plays a key role in the analysis and provides another form of Bayes rule. In equation (5.7) of §5.6, the rule was presented as extending from one event E to include another event F, with its complement F^c; and in §8.1 it was mentioned that, just as the basic rules applied to any number of events, not just two, so would the extension rule. It is the precise form of this that needs to be discussed. Consider events $F_1, F_2, \ldots, F_n$, which are *exclusive* (§5.2) in that at most one of them can be true, and also *exhaustive*, in that one of them must be true, or they exhaust the possibilities; they are said to form a *partition* of the events. Clearly the original pair, F and F^c, form a partition. Consider the events $EF_1, EF_2, \ldots, EF_n$. They are exclusive but do not exhaust the possibilities since E^c might be true. What they do exhaust is E since they describe all the ways that E might occur. By the addition rule, Equation (8.2) of §8.1, for n events,

$$p(E) = p(EF_1) + p(EF_2) + \cdots + p(EF_n). \qquad (9.1)$$

In words, if the events F_i form a partition, $p(E)$ is equal to the sum of n terms of which a typical one is $p(EF_i)$. By the product rule, $p(EF_i) = p(E \mid F_i)p(F_i)$, so the general form for the rule of the extension of the conversation is as follows:

If, on some knowledge base, events $F_1, F_2, \ldots, F_n$ form a partition and E is another event, $p(E)$ is equal to the sum of n terms, of which a typical one is $p(E\,|\,F_i)p(F_i)$. It is this form that will be repeatedly used in the rest of the book.

Bayes rule, applied separately to each member of the partition, and omitting reference to the knowledge base, says

$$p(F_i\,|\,E) = p(E\,|\,F_i)p(F_i)/p(E).$$

In words, your prior $p(F_i)$ is multiplied by your likelihood $p(E\,|\,F_i)$ and divided by $p(E)$, your probability of obtaining evidence E. This last term is difficult to think about but can be obtained from the prior and likelihood by use of the extension of the conversation from E to include the partition (9.1). It is this form of Bayes rule that is commonly used in statistics (Chapter 14). There is an equivalent way of approaching this form. For each F_i, $p(F_i\,|\,E)$ is the product of the prior and the likelihood, divided by the same term, $p(E)$ for each i. But we know the $p(F_i\,|\,E)$ must add to one since the F_i form a partition. So if all the products are added, the result may be divided by some value to make the total one. Clearly this total is just that given by the extension of the conversation (9.1). Bayes rule can therefore be expressed by saying

$$p(F_i\,|\,E) \propto p(E\,|\,F_i)p(F_i),$$

where the symbol $\propto$ means "is proportional to". This form is especially attractive because it clearly demonstrates that your posterior probabilities depend on your likelihood and your prior, and on nothing else.

9.2 BINOMIAL DISTRIBUTION

Take our usual urn containing a vast number of balls, identical except that a known proportion θ are colored red, the rest white, and suppose you take one ball at random, then your probability that it will be red is θ. Generally, if you take n balls from the urn at random, you will have a Bernoulli series (§7.4) in which the withdrawals are exchangeable and your probability for any outcome of the n drawings depends, not on the order of red and white balls, but only on the number, say r, of red balls out of the n. Even with θ and n fixed, the number r of red balls will be variable and uncertain. We now calculate your probability of r red balls, given n and θ, which, in the standard notation, is $p(r\mid n,\theta)$, the presence of θ after the vertical line reminding us that θ is supposed known, as well as n. There is also an unstated knowledge base, which incorporates things like the withdrawals being random. Take an example of drawing 6 balls from the urn and determine your probability of just one red, and therefore 5 white balls; $n=6, r=1$. One way this can happen is to have the red ball appear first, followed by the 5 white, with probability $\theta(1-\theta)^5$ by the multiplication rule for the random (and therefore independent) drawings, as in §7.4. There are 6 such possibilities, for the single red can appear in any of 6 positions, and each has the same probability, so $p(r=1\mid n=6,\theta)=6\theta(1-\theta)^5$ by the addition rule. This method works for any values of r and n. The product $\theta^r(1-\theta)^{n-r}$ is obvious, but the number of ways of obtaining r red in n drawings is a little tricky, so will be omitted.

Here is a numerical example with $\theta=1/3$ and $n=6$. Your probabilities are given to two significant figures; that is, two digits after the 0's, if any, that follow the decimal point (§2.9).

Number of red balls	0	1	2	3	4	5	6
Probability	0.088	0.26	0.33	0.22	0.082	0.016	0.0014

Thus your probability of 1 red ball, when 6 are drawn from an urn with one third of the balls red, is 0.26, as the reader can verify by putting $\theta=1/3$ in the expression $6\theta(1-\theta)^5$. The table shows that 1, 2, or 3 red balls are each quite probable but 0, 4, or 5 are somewhat unusual and having every ball red, $r=6$, is most surprising. The numbers here reflect your uncertainty when 6 balls are removed randomly from the urn, but if you were repeatedly to remove 6 and recognize the connection between these ideas and frequency expressed in the law of large numbers (§7.6), then, in the long run, you would have exactly 2 red out of the 6 in 33%, or one third, of the time. Here we have variation, one drawing of 6 balls typically differing from the result of another 6, so that the variation and the uncertainty are intimately related.

9.2 BINOMIAL DISTRIBUTION

A useful way of looking at this situation, which generalizes to many others, is to note that there are 7 exclusive events in the table, which exhaust the possibilities and so form a partition. They may be written E_0, E_1 up to E_6, where E_r corresponds to r red balls; thus, from the table $p(E_3 \mid n=6, \theta=1/3) = 0.22$. The 7 probabilities add to 1 (apart from rounding errors) and we say the total probability of 1 is *distributed* over the 7 values that form the partition. Generally, if there are a finite number of events, which are exclusive (only one can be true) and exhaustive (one must be true), then the corresponding set of probabilities is said to form a *probability distribution*. Since this is a book about probability, we shall typically drop the adjective and refer to a *distribution*. The distribution tabulated above is an example of a *binomial* distribution that applies whenever there is a fixed number, here 6, of observations, each of which can result in an event being true or false (red or white). The chance (§7.8) of truth is the same for each observation (here $\theta = 1/3$) and the events are independent. The binomial distribution relates to a Bernoulli series (§7.4) where the exchangeable property reduces consideration to the number of true events, not their order. The number of observations, n, is termed the *index* of the binomial and the chance of truth (red), θ, is called the *parameter*. The distribution tabulated above has index 6 and parameter 1/3. The variation in the number of red balls, when a fixed number of balls is withdrawn, is described by the binomial distribution.

The number r of red balls can take any value from 0 to n inclusive, with probabilities $p(r)$ for you, omitting the conditions from the notation. The idea generalizes to every quantity that can take a finite number of possible values with probabilities assigned by you to each value. Such a quantity is called an *uncertain quantity*. Thus the number of red balls is an uncertain quantity and the probabilities form your distribution of that quantity. (The term *random variable* often replaces uncertain quantity.) An event can be considered as an uncertain quantity taking two values, 1, true and 0, false, so that an uncertain quantity is a generalization of an uncertain event and its distribution generalizes the probability of an event. Most of the examples considered in Chapter 1 concern uncertain quantities or can easily be extended to do so. Thus the uncertain event of "rain tomorrow" can be extended to "millimeters of rain tomorrow" (Example 1 of §1.2) or the event of "ace" to the number on the card (Example 7). The amount of inflation (Example 10) or the proportion of HIV (Example 11) are examples of quantities which are uncertain.

The binomial distribution is relevant to many practical situations. If you observe a number n of people taken at random, called a *sample* of people, then r, the number of women, will have a binomial distribution with index n and parameter $\theta = \frac{1}{2}$, or slightly less than $\frac{1}{2}$ if the sample is of babies, or more than $\frac{1}{2}$ if it is of persons more than 80 years of age. If, for the same people, the gene with alleles A or a, with A dominant, were investigated, then the number r of double recessives aa, often those with a defect, will be binomial with parameter θ^2, where θ is the proportion of alleles a in the population from which the sample was taken. If n observations are made of the fall of a ball in roulette, played in a reputable casino, then r, the number of balls falling in slot 22, is binomial with index n and parameter $\theta = 1/37$ if there are 37 slots. Notice that in these examples three requirements for the binomial are satisfied: the number n is fixed, the individual occurrences are random, and you have the same, known probability θ of the outcome under consideration (sex, defective or 22) for each occurrence. A more common situation is where these conditions obtain except that θ is unknown to you and is a chance, about which you have a probability distribution. As an example, consider the case at the beginning of this paragraph when sex is replaced by voting intent, with only two candidates and the "don't knows" omitted. Then θ is unknown and you can apply Bayes rule to modify your opinion of it after hearing the intentions of the voters. Notice that the samples must be taken at random to preserve exchangeability. It would not be correct to ask all members of a household since there exists a tendency for members to be in agreement within households.

Although we shall not explore the point in any detail, it is worth noting why n is fixed in this, and other, examples. Suppose the balls were taken from the urn until you had two red balls in succession, and then stopped. There would be a distribution for the total number of red balls withdrawn but it would be different from the binomial obtained when the number of withdrawn balls was fixed. For example, your probability of finishing with one red ball is zero, compared with 0.26 in our binomial example with $n = 6$. It is not unknown for people to sample until they reach a situation that is favorable to them, two consecutive reds in our example. It is then incorrect to treat the sample as binomial. On the other hand, a useful practice in medical trials is to stop when the evidence is thought enough to establish the merit, or inadequacy, of the drug under test. This is permissible provided the correct probability is used. The point is treated in more detail under the term "optional stopping" in §14.3.

9.3 EXPECTATION

A distribution for an uncertain quantity is a rather complicated affair, even in the binomial case with $n=6$ it consists of 7 numbers, adding to one, and it would be desirable to encapsulate the main features of a distribution in far fewer numbers. In doing this, some knowledge of the distribution will be lost, but there will be an increase in understanding. In this section the most important feature of an uncertain quantity and your distribution for it will be developed. Again we resort to our familiar urn, but it will be used somewhat differently from the last section and to emphasize the difference, and hopefully prevent confusion, a slight change in notation will be employed.

Consider an urn containing a known number m of balls, identical except for the fact that s of them are scarlet, the rest white. If s is unknown to you, it is an uncertain quantity and you will have a distribution for it, $p(s)$, being your probability that the number of scarlet balls is s. Suppose that one ball is to be drawn at random from the urn and denote by S the event that it is scarlet. What is your probability for this when s is uncertain? (There is an unstated knowledge base that includes m, the total number of balls in the urn.) If the number of scarlet balls in the urn were known to you, the answer would be simple from the basic definition of probability in §3.3, $p(S|s) = s/m$. This suggests that it might be worthwhile extending the conversation from S to include s. Using the general form at the end of §9.1 to calculate $p(S)$ it is necessary to evaluate the products $p(S|s)p(s)$ for each value of s and add over all values of s from 0 to m. Since $p(S|s) = s/m$, the products to be added reduce to $sp(s)$ and their sum has to be divided by m. The sum of the products $sp(s)$ is called your *expectation* of the uncertain quantity s and will be denoted by E and not confused with the use of E for an event, or for evidence. Often E is called the *expected value* of s. Generally, for any distribution of an unknown quantity, the result of taking the probability of any value of the quantity, multiplying it by the value, and adding all the products, is called the expectation of the uncertain quantity. Since every distribution can be conceptually associated with the random withdrawal of a ball from an urn, in the manner employed here, the idea is of wide applicability. Part of its importance lies in the fact that if the quantity, s, is known, your probability for the scarlet ball is s/m, whereas when unknown, it is E/m, replacing the unknown value s by the known expectation E. As far as the random withdrawal of one ball is concerned, the uncertain state of the urn can be replaced by an urn with a known number E of scarlet balls. When we discuss decision analysis in §10.4, we encounter another case where uncertainty can be replaced by expectation without any loss of power. Of course, some features of a distribution are lost if only expectation is employed, but it is far and away the most important feature of a distribution, or of the quantity to which it relates. In many cases, as with the urn, it provides all the information you need. So important is it that other names are in use. It is sometimes called your *prevision* of the uncertain quantity, your vision of it before determining its true value. When referring to a distribution, without having any particular quantity in mind, it is often called the *mean* of the distribution. The same term is frequently used for the quantity, thus we talk about the mean income or the mean size of family.

The connection between probability and expectation is even closer than the development just given suggests. We saw in §9.2 that one could associate any event A with a quantity taking the value zero if A is false, and one if true, these being the appropriate limits of your probability for A. What is your expectation of this quantity? Recall we have to take each value of the quantity, multiply by its probability, and add the resulting products. Here

$$E = 0 \times [1 - p(A)] + 1 \times p(A) = p(A),$$

so that your expectation and your probability are identical. Some writers have based their whole treatment of uncertainty on expectation, rather than on probability. This is entirely satisfactory, but we have chosen not to adopt that approach for three reasons:

1. It can happen that the quantity can take so many values that the sum of all the products becomes unwieldy. This is essentially a mathematical reason and, in that language, the sum diverges.
2. We have seen that it is often hard to assess probability (§§3.5 and 5.6). It is even harder to assess expectation since a quantity can assume so many values, whereas probability is just expectation for a quantity that can only assume two, 0 and 1.
3. Expectation can be more easily misunderstood than probability. Suppose a standard die is sensibly rolled, then you will ordinarily associate probability 1/6 with each of the possible values 1, 2, 3, 4, 5, 6 for the number of spots that might appear uppermost when the die comes to rest, and hence have expectation $(1+2+3+4+5+6)/6 = 3\frac{1}{2}$. Yet in the ordinary use of the English language, you will never "expect" to see $3\frac{1}{2}$ spots because it is impossible. However, if you were to receive \$1 for every spot you would reasonably expect to receive $\$3\frac{1}{2}$. I once experienced communication problems with an official because I had said $2\frac{1}{2}$ defectives were expected in a batch of 100 components. He was never convinced and went around his department joking about the statistician who was half defective.

There is an alternative interpretation of expectation but this is left until another distribution has been discussed in §9.4. Notice that the concept of expectation, as presented here, is not just a convenient quantity but arises naturally from a probability rule, namely the extension of the conversation. Also its derivation has nothing to do with frequency, the ball being withdrawn only once. Compare the comments in the final paragraph of §3.4.

9.4 POISSON DISTRIBUTION

Suppose you are a telephone operator who handles calls for an emergency service and are beginning a tour of duty of 2 hours. You will be uncertain about the number of calls you will have to deal with during the

TABLE 9.1 Poisson distribution with expectation 4

Number of calls	0	1	2	3	4	5	6	7	8	9	10	>10
Probability	0.018	0.073	0.15	0.20	0.20	0.16	0.10	0.060	0.030	0.013	0.0053	0.0028

tour and will therefore have a probability distribution for that number as an uncertain quantity. Table 9.1 gives a possible distribution.

For example, your probability of just 4 calls is 0.20, which can also be interpreted by considerations of frequency in §7.6 as meaning that, over a long sequence of tours, when conditions remain stable, you can anticipate 4 calls on about 20% of tours. Notice that more than 10 calls (>10) is thought to be a very rare event and all values above 10 have been lumped together. Again the probabilities add to 1 and they can be partially added, for example, your probability of eight or more calls, a busy tour, is $0.030 + 0.013 + 0.0053 + 0.0028 = 0.051$, or roughly 1 in 20 tours are anticipated to be busy.

The tabulated distribution has been derived from two assumptions:

1. For any small period of time, like 5 minutes the chance of a call is the same, irrespective of which 5 minutes in the 2 hours is being considered.
2. This chance is independent of all experiences of calls before the 5 minute period.

The first assumption says roughly that the demands for the emergency service are constant, and the second that what has happened so far in your tour does not affect the future. Notice that the assumptions are similar to those for the binomial distribution, the constancy of θ and the random withdrawals. In practice, neither of the assumptions may be exactly true, but experience has shown that small departures do not seriously affect the conclusions and that larger departures can be handled by building on cases where they do, rather as exchangeable series can be built on the Bernoulli form (§7.5). As a result, the ideas presented here are basic to many processes occurring naturally.

A distribution resulting from the assumptions is called a *Poisson distribution*, after a French mathematician of that name, and depends on only one value, the chance mentioned in the first assumption, called the *parameter* of the Poisson. The tabulation above is for the case where the chance is about 1/6. Notice that, in the description of the parameter, the unit, here 5 minutes, is vital, for 1 minute the parameter would be about 1/30, a fifth of the previous value.

There is an alternative parametric description of the Poisson distribution that is often more convenient and uses the expectation, or mean, of the distribution. For that just tabulated, the expectation is

$$0 \times 0.018 + 1 \times 0.073 + 2 \times 0.15 + \cdots + 10 \times 0.0053 + 11 \times 0.0028,$$

where the dots signify that the values from 3 to 9 calls have to be included and where values in excess of 10 have been replaced by 11. A simple exercise on a calculator shows that the sum is 4.03. Because the probabilities have been given only to two significant figures and that all values in excess of 10 have been put together, this result is not exact and the correct value is exactly 4. The tabulation above is for a mean of four calls in a 2 hour period. It is intuitively obvious and can be rigorously proved that if you expect 4 calls in 2 hours, you expect 1 in $\frac{1}{2}$ an hour and 1/6 in the 5 minute period above. Recall comment 3 in §9.3.

The Poisson distribution, or a close approximation to it, occurs very frequently in practice. It is a good approximation whenever there is a very small chance of an event occurring, but lots of opportunities when it might occur, and where one happening does not interfere with another. There are lots of 1 minute periods when a call might be received but a very small chance of one in any such period. In the example, 120 such periods each with chance about $4/120 = 1/30$. There is little chance of your falling ill but there are lots of people who could fall ill, so illnesses in a population often satisfy a Poisson distribution. An example of this appears in §9.10. Historically, an early instance was deaths from the kick of a horse in the Prussian cavalry, where there were lots of soldiers interacting with horses, providing many opportunities for, but few casualties from, horse kicks. Indeed, the Poisson distribution is so ubiquitous that any departure from it gives rise to suspicions that something is amiss. Childhood deaths from leukemia near nuclear power plants provides an example, clusters of cases suggesting departure from the Poisson assumptions.

There is another way of thinking about the Poisson distribution that sheds further light on what is happening. To see this, suppose you are the operator on your shift of 2 hours, expecting 4 calls, and suppose, instead of fixing the duration and seeing how many calls arise, you think about the next call and wait to see how much time elapses before it occurs. You might query whether there is time for a cup of coffee before the phone rings. The second of the two assumptions above means that at any time, say 3.45, what has happened before then does not affect your uncertainty about the future, so forget the past and at 3.45 wait until the next call comes. Will you have to wait 1 minute, 2 minutes, or more? The number of minutes is, for you, an uncertain quantity and you will have a distribution for it. What can be said about this distribution? Common sense suggests that if you expect to receive 4 calls in 2 hours, you expect to wait half an hour for that one call. Here is a case where common sense is correct and generally if you expect C calls in an hour, you expect to wait $1/C$ hours for the first one. But now look at the situation in a different way and ask what is the most probable time, to the nearest minute, that you will need to wait for that call? Will it be 1 minute, 2 minutes, or perhaps 30 minutes, the expected time? The answer surprises most people for it is 1 minute. As the time increases, the probability of your having to wait until then decreases, so that, in particular, the expected time has small probability. Here is an example where, for most people, common sense fails and our basic idea of coherence provides a different answer, an answer that stands up to rigorous scrutiny. The incorrect, common sense has led to the belief that the calls should be spread out somewhat uniformly, rather than occurring in clusters. In fact, even in a Poisson distribution, clusters often arise simply because small intervals between calls are more probable than large ones. This clustering has led to a popular tradition that events occur in threes; a tradition that comes about because of your large probabilities for small intervals. Clusters are natural and it does not require a special explanation to appreciate them. This is why it is hard to separate real clusters from the ones that occur solely from the Poisson distribution, as with leukemia mentioned in the last paragraph.

There is another interpretation for expectation that deserves notice. In your role as an operator, coming on duty at 16.00, you can expect 4 calls before you have a break at 18.00. Instead of just one specific tour of duty, suppose you are employed over a long period and accumulate 1000 tours. You expect 4000 calls in all and, in an extension of the law of large numbers in §7.6, you will actually experience something very close to 4000. In other words, the expectation of 4 is a long-term average. This interpretation is not as useful as the earlier one, referring to a specific tour, because it requires not just stability over 2, but over 2000 hours. It also confuses probability as belief with frequency, a confusion, which, as we saw in §7.2, is often misleading.

9.5 SPREAD

Your expectation of an uncertain quantity says something about what you anticipate or, in the frequency interpretation, tells you what might happen on the average. But there is another important feature of an uncertain quantity and that is its variation, referring to the departure, or spread, of individual results from your expectation. A simple way of appreciating the variation is to suppose the uncertain quantity is observed twice; for example, take the 6 balls from the urn as in §9.2 and observe the number of red balls; then repeat with a further 6. It will be rare that you obtain the same number of red balls on both occasions, the difference providing a measure of the variation, or spread. The operator experiencing two tours of duty will rarely have the same number of calls in the first as in the second. Exactly how the difference is turned into a measure of spread, or how it is employed when there are several observations, not just two, is an issue that is too technical for us to pursue here. The measure of spread ordinarily used is called the *standard deviation*. It is discussed further in §9.9. Instead we concentrate on a result that requires no technical skill beyond the appreciation of a square root. Recall (§2.9) that the square root of a number m is that number, written $\sqrt{m}$ or $m^{1/2}$, which, when multiplied by itself, $\sqrt{m} \times \sqrt{m}$ is equal to m. Thus the square root of 9 is 3 since $3 \times 3 = 9$. Of course, typical square roots are not integers or even simple fractions, a result that caused much distress in classical Greece, so that $\sqrt{2}$, for example, is about 1.41.

Let us return to making several observations on an uncertain quantity, in the last paragraph we took just two. Throughout the treatment that follows it is supposed that the observations obey two conditions:

1. Your distribution of the uncertain quantity remains fixed.
2. You regard the observations as independent (on a given knowledge base).

These are similar to the conditions that, in a different context, lead to the Poisson distribution. In the case of 6 balls drawn from the urn, condition (1) means the constitution of the urn remains fixed and your selection continues to be random. (2) demands that you do not allow the result of the first draw to affect the second. In the Poisson case with the emergency service, the second tour of duty is not influenced by the first, as might happen were there to be a serious fire extending over both tours.

Under these conditions, let x and y be two observations of the uncertain quantity. Then, as already suggested, the difference $x - y$ tells us something about the spread, whereas $x + y$ reflects the total behavior. Clearly the latter has more spread than x or y. Thus, with the urn, the total number of red balls from two sets of 6 can vary between 0 and 12, rather than 0 to 6 for a single observation. The key question is how much does the spread increase in passing from x to $x + y$? The answer is that for any reasonable measure, including the one hinted at above, but not developed for technical reasons, the spread is multiplied by $\sqrt{2}$. This is a special case of the *square-root* rule, which says that if m observations are made, under conditions (1) and (2), then the spread of the total of those observations is $\sqrt{m}$ times that of each individual observation. The example had $m = 2$. The important feature here is that the variability of the total of m observations is not m times that of any one, but only $\sqrt{m}$ times. $\sqrt{m}$ is much smaller than m; for example, $\sqrt{25}$ is only 5.

The square-root rule is often presented in a slightly different way which agrees more with intuition. When we study science in §11.11, we will see that a basic tenet of the scientific method is the ability to repeat experiments. If the experiments obey the conditions above as they often do, then scientists will sensibly take the average of the observations in each of the m experiments, in preference to a single one. The average is the total divided by m, and since the spread of the total is $\sqrt{m}$ times that of a single observation, the spread of the average must be $\sqrt{m}$, divided by m, or that of one observation divided by $\sqrt{m}$. ($\sqrt{m}/m = 1/\sqrt{m}$ since $m = \sqrt{m} \times \sqrt{m}$). Thus the square-root rule says that the variation of the average is that of one observation divided by $\sqrt{m}$, so the scientists' use of repetition is effective in reducing variation, dividing it by $\sqrt{m}$. In this form, the square-root rule was, for many years, regarded by some experimental scientists as almost the only thing they needed to know about uncertainty. Although this is no longer true, it remains central to an understanding of variability. If 16 observations of the same quantity are made, the variability, or spread of the average is only one quarter that of a single observation.

The occurrence of the square root explains a phenomenon that we all experience when repetitions of an activity can be less interesting than doing it for the first time and ultimately can sometimes become of no interest at all. To divide the variation by 2, we need 4 repetitions; to divide it by 2 again, dividing by 4 in all, we need 16 repetitions so that the second halving in variation requires $12 = 16 - 4$ repetitions rather than the 4 required first time. It expresses a law of diminishing returns, observation 16 having much less effect than observation 2. The square-root rule is not universal for, as we have emphasized, it requires independent and identical repetitions; but it does occur frequently and is very useful.

Although the spread of the average decreases as the number of repetitions increases, according to the rule, the expectation of the average remains the expectation of any single observation, as is intuitively obvious. Let us see how these ideas work, first for the binomial distribution (§9.2) where θ is the parameter and n the index, as when randomly removing n balls from an urn in which the proportion red is θ. It was seen in §9.3 that, for a single ball, the probability of being red, θ, and the expectation were the same. The expectation of the total number of red balls is therefore $n\theta$. Calculation shows that the spread of the number of red balls from n drawings is the square root of $n\theta(1-\theta)$, in accordance with the square-root rule. (Readers who want to know where the $\theta(1-\theta)$ comes from will find an explanation at the end of this section.) In particular, there is no spread when $\theta = 1$ or $\theta = 0$, with all balls of the same color, red or white respectively, for the two extreme cases. The Poisson distribution is even simpler, for if the expected number in a fixed period of say 1 hour is E, then over m hours the expected number is mE. Calculation shows that the spread for the 1 hour is $\sqrt{E}$, so that over m hours, is $\sqrt{(mE)}$, again in accord with the square-root rule. In §9.10 the use of these results will be discussed. In the meantime here is an example of how variation can be handled with profit, but before presenting it, we promised to show the origin of $\theta(1-\theta)$ above. The demonstration can be omitted without disturbing subsequent understanding.

Your probability of drawing a red ball from the urn is θ, and it was shown in §9.3 that if a quantity is defined as 1 if the ball is red, and 0 if white, your expectation of the quantity is also θ. More abstract language concerns a quantity, which is 1 if an event is true and 0 if false, when your probability and your expectation are the same. How far does the quantity depart, or spread, from its expectation? Clearly $1-\theta$ if the event is true and $0-\theta$ if false. Interest centers on the amount of the departure, not its sign, so we square the departures, getting $(1-\theta)^2$ with probability θ and θ^2 with probability $(1-\theta)$. The expected spread is, on multiplying the values by your probabilities and adding, $(1-\theta)^2\theta + \theta^2(1-\theta)$. The first term is $\theta(1-\theta)$ times $(1-\theta)$; the second is $\theta(1-\theta)$ times θ, so that on addition the total multiple of $\theta(1-\theta)$ is $1-\theta+\theta=1$, leaving the final expectation as $\theta(1-\theta)$. Having used squares, the units will be wrong, so take the square root, obtaining $[\theta(1-\theta)]^{1/2}$ as promised. Notice that the role of the square here has nothing to do with the square-root rule; it is introduced because we are interested in the magnitude of the departure from expectation, and not in its sign.

9.6 VARIABILITY AS AN EXPERIMENTAL TOOL

Although in many ways variability, and the uncertainty it produces, is a nuisance, it can be exploited to provide valuable insights into matters of importance. Here is a very simple example of a procedure that is widely used in scientific experiments. An agricultural field station wishes to compare the yields of two varieties of wheat and, to this end, sows one variety in one half of a field and the second in the other half. As far as possible the two halves are treated identically, applying the same fertilizers and the same herbicides at the same times, ensuring that the two conditions of identical and independent repetitions are satisfied, except for the varietal difference. Suppose the yield is 132 tonnes for one variety and 154 for the other, then is the second variety better, or is the difference of 22 tonnes attributable to natural variation that is present in the growing of wheat? One way to investigate this is to divide each half of the field devoted to a single variety into two equal parts, each a quarter of the total, and to harvest the parts separately. Suppose the results are 64 and 68 for the first variety, totaling 132, and 74 and 80 for the second. The two differences, of 4 and 6, give an indication of the natural variation since the same varieties are being compared. The original difference of 22 between varieties is much greater than these, suggesting there is a real difference between the varieties, not attributable to natural variation. But stay, there is a slip there, this last difference of 22 is based on half fields, the others on quarter fields, so a correction is needed. Each yield based on half the field is the sum of two yields from the two quarters that make up the half, and therefore, by the square-root rule, has $\sqrt{2}$ times the spread of a yield based on a quarter. Therefore the varietal difference of 22, based on halves, has $\sqrt{2}$ times the spread associated with the natural differences, of 4 and 6, within the varieties. Dividing 22 by $\sqrt{2}$ gives about 15, a figure which is comparable with the 4 and 6. Being much larger than either of these, the suggestion is that there probably is a real difference between the two varieties because of the inflation from 4 and 6 to 15.

The discussion in the last paragraph is a very simple example of a technique called *analysis of variance*. (Variance is just a special measure of variation; it is the square of the standard deviation mentioned in §9.5.) Here the variation present in a body of data, the yields in the four quarters, is split up, or analyzed, into portions that can each be attributed to different facets, natural variation and variation between varieties, that may be compared with one another. A century ago it used to be common, when examining how different factors affected a quantity, to vary one factor at a time. Modern work has shown that this is inefficient and that it is better to vary all the factors simultaneously in a systematic pattern, and then split up the variability in such a way that the effects of the factors may be separated into meaningful parts. Another advantage of this method over that in which the factors are viewed separately is that the scientists can see how factors interact, one with another. For example, it is possible that neither factor on its own has any influence but both together can be beneficial. In §8.9 mention was made of a claim that eating a banana caused mucus. To test this one could vary the factor, banana, and measure the variation in mucus, yet, remembering Simpson in §8.2, it would be sensible to think of other factors that might be relevant, such as time of day, other foods consumed besides banana, and variation between individuals, and then devise an experiment that explored all factors and analyzed the variation. Determining the connections between bananas and mucus is not easy, and the same is true of many claims of an association that are made. As we have said before, a useful riposte to a claim is "how do you know?" Both Simpson's paradox and variation can make it hard to acquire sound knowledge.

9.7 PROBABILITY AND CHANCE

It was seen in Chapter 7 that if there is a series that you judge exchangeable, the individual terms of which assume only two values, 1 or 0, true or false, success or failure, red or white, then you can regard the series as a Bernoulli series, with chance θ of red, about which you have a probability distribution. This result of de Finetti is now applied more generally. (Readers may like to refer to §7.8 in order to clarify their understanding of the distinction between the chance, and your probability, of an event). Take an uncertain quantity, which can assume any integer value, not just 1 or 0, and suppose you repeatedly observe it in a series that you consider exchangeable. An example is provided by a scientist who repeats the same experiment. Now concentrate on a particular value of the quantity, say 5, and observe whether, for each observation you get 5 or not; counting the former as a "success" and the latter as a "failure". Imagine playing roulette and always betting on 5. You now have a series of successes and failures, which you judge exchangeable, because the complete observations of the quantity were. De Finetti's result may be applied to demonstrate that there is a chance such that your series of successes or failures is Bernoulli with that chance. Denote this chance by θ_5 including the subscript 5 to remind us that success is obtaining 5. There is nothing special about 5, so that you have a whole slew of θ's, one for each value of the quantity. Recalling from §7.8 that the chances correspond to limiting frequencies, they will all be positive and add to 1. In other words, they form a *chance distribution*.

It has therefore been established that if you have an exchangeable series, not simply of 0's and 1's, but of a quantity capable of assuming many integer values, then there is a chance distribution such that knowing it, you can regard the observations in the series as independent with your probabilities given by the chances. For example, your probability that the first observation is 2 and the second is 5 is $\theta_2 \theta_5$ by the product rule. This supposes that the chances are known. The analysis when the chances are uncertain for you is more complicated. Recall from §7.8 that chances are not expressions of belief but rather you have beliefs about them. So here you will have beliefs about θ_2 and θ_5. To analyze your beliefs about the observations, it will be necessary to extend the conversation from the observations to include the chances, in generalization of the method used in §7.8 when the observations were only 0 or 1. The details are not pursued here.

The situation described in the last paragraph has found widespread use but, as presented there, it has a difficulty that there are lots of chances to think about, one for each value the quantity could possibly take. It is hard to contemplate so many and make uncertainty statements about them. It is often adequate to suppose all the chances are known functions of a few other values. We illustrate this with the Poisson distribution. Suppose that the operator experiences several tours of duty that are thought of as exchangeable. Then there will be a chance distribution of the numbers of calls per tour. But our operator in §9.4 made two additional assumptions, numbered (1) and (2), about independence and constancy within a tour. Adding these assumptions to that of exchangeability, the chances become severely constrained so that they are all functions of one value, the expectation E of the number of calls in any tour. It is unfortunate that the description, let alone the derivation, of these functions lies outside the modest mathematical level of this book. E is called the *parameter* of the Poisson distribution. Recall the tabulation in §9.4 for the Poisson distribution when $E=4$. Generally with an exchangeable series, the usual practice is to suppose the chances are all functions of a small number of parameters. The

Poisson has only one, E. The binomial has two, the index and what has been denoted by θ, though usually the index is known, so that θ is the sole parameter. In the example of §9.2 with $n=6$, θ_1, the chance of 1 success, is $6\theta(1-\theta)^5$ and, if θ is known, is your probability of 1 success. Many chance distributions depend on two parameters, one corresponding to the expectation, the other to the spread. The Poisson is exceptional in that the spread $\sqrt{E}$ is itself a function of the expectation E.

To recapitulate, a commonly used procedure is to have a series of observations that you judge exchangeable, such as repetitions of a scientific experiment, or a sample of households, with which, by de Finetti's result, you associate a chance distribution. By adding extra assumptions, as with the Poisson, or just for convenience or simplicity, you suppose these chances all depend on a small number, often two, of parameters. The parameters are uncertain for you and accordingly you have a probability distribution for them. Your complete probability specification consists of this parametric distribution and the chance distribution. With this convenient and popular model, you can update your opinion of the parameters as members of the series are observed. Thus with the Poisson parameter E, $p(E)$ can be updated by Bayes rule, on experiencing r calls in a tour, to give

$$p(E \mid r) = p(r \mid E)p(E)/p(r),$$

where $p(r \mid E)$ is the Poisson chance. Thus for $E=4$, $r=7$, it has the value 0.060 from Table 9.1. Notice the difference between $p(r \mid E)$ and $p(r)$. The latter is your probability for r calls when E is uncertain, and is calculated by extending the conversation from r to include E, as in §7.5. $p(r)$ would be relevant when you were starting a tour with uncertain expectation and wished to express your uncertainty about the number of calls you might experience in the tour.

9.8 PICTORIAL REPRESENTATION

There are quantities that do not take only integer values. We met one in §9.5 when considering the uncertainty about the time to the next phone

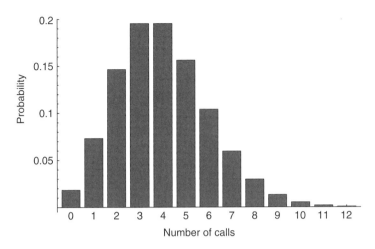

FIGURE 9.1 Poisson distribution with $E=4$ from Table 9.1.

call. At 3.45 this time can take any value, not just an integer. In practice we measure it to the precision of a convenient unit, like a minute, but in some situations more precision may be needed and recording to the nearest second might be used. Such a quantity is said to be *continuous*, whereas the integer-valued ones are *discrete*. To see how the associated uncertainties can be handled, it is convenient to use a pictorial representation. Figure 9.1 describes the Poisson distribution with expectation 4 in Table 9.1. The horizontal axis refers to the number of calls and upon this rectangles are erected, each with base length of 1 and each centered on a possible number of calls, 0, 1, 2, and so on. The height of the rectangle is the probability, according to the Poisson with expectation 4, of the number of calls included in the base. Thus that centered on $r=2$ has height 0.15. The vertical axis thus refers to probability. The important feature of this manner of representing any distribution of an integer-valued, uncertain quantity is that the area of the rectangles provides probabilities, since the base of the rectangle is 1. The key element in the interpretation of such figures is area.

This style of representation is now extended to continuous, uncertain quantities, beginning with the time to wait for a call as experienced by the operator. Table 9.2 provides your probabilities for the 12, 10 minute intervals within the 2 hours of the tour. Thus your

TABLE 9.2 Probabilities of the time to wait for the first call, divided into 10 minute intervals with the upper limit of each interval given $E = 4$ for a tour of 120 minutes

Time	10	20	30	40	50	60	70	80	90	100	110	120
p	0.284	0.203	0.146	0.104	0.074	0.055	0.038	0.028	0.020	0.014	0.010	0.007

(in addition there is a probability of 0.018 of no calls in the tour.)

probability that you will have to wait between 40 and 50 minutes for the first call is 0.074. In addition, your probability of having to wait more than 120 minutes is 0.018. This event corresponds to having no calls in the tour, agreeing with the corresponding entry in Table 9.1. Since the tour ends at 2 hours, this value will be omitted from future calculations. Figure 9.2 gives a pictorial representation along the lines of Figure 9.1. Thus on the first interval from 0 to 10 minutes is erected a rectangle of height 0.0284, so that its area, in terms of minutes, is 0.284, your probability of waiting between 0 and 10 minutes for the first call. Remember that the important feature here is the area of the rectangle, not its height. (The reason why the vertical axis is labeled "density" is explained later.) Notice confirmation of the surprising fact pointed out in §9.4 that the areas, and therefore the probabilities, diminish as time increases. Thus there is a chance of more than a quarter that the wait will be less than 10 minutes, despite the fact that only 4 calls are expected in 120 minutes.

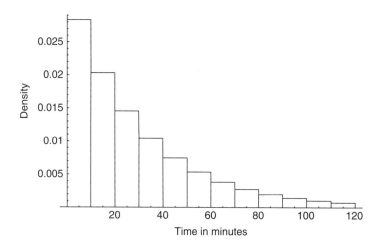

FIGURE 9.2 Pictorial representation of Table 9.2.

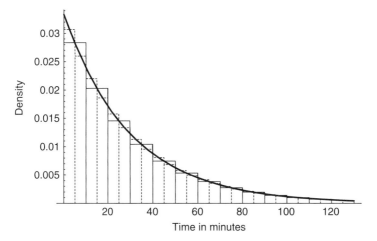

FIGURE 9.3 Pictorial representation of Table 9.2 with further division into 5 minute intervals and also the continuous density.

Figure 9.3 repeats Figure 9.2 with, superimposed upon it, the same rectangular representation when intervals of 5, rather than 10, minutes are used. Thus between 0 and 5 the rectangle has height 0.0307 and therefore area 0.154, your probability of getting a call almost before you have time to settle in. Between 5 and 10, the height is 0.0260, while the area and probability is 0.130. The two probabilities add to give 0.284, agreeing with Figure 9.2. Thus the two thinner rectangles, base 5, have total area to match that of the thicker rectangle, base 10, and the three heights are all about the same, around 0.03. Similar remarks apply to the other pairs. Now imagine these procedures for 5 minute and 10 minute intervals repeated for intervals of 1 minute, then 1 second, continually getting smaller. The rectangles will get thinner but their heights will remain about the same, so that if we concentrate on the tops of them they will eventually be indistinguishable from a smooth curve. This curve is also shown in Figure 9.3. It starts at height 0.033, or exactly $4/120$, corresponding to an expectation of 4 in 120 minutes, and descends steadily. Although it has only been shown up to the end of the tour, it continues beyond, as would be needed if the tour were longer.

It is this curve that is important. Its basic property is that the area under the curve between any two values, say between 40 and 50 minutes, is your probability of the quantity lying between those values, of waiting

212 VARIATION

more than 40 but less than 50 minutes. It is sometimes described as a curve of probability but it is not probability, it is the area under it that yields our uncertainty measure. Here it will be referred to as a probability density curve, or since this is a book about probability, simply as a *density*. (The familiar density is mass per unit volume; ours is probability per unit of base.) It is often referred to as a *frequency* curve because if you were to observe the quantity on a series of occasions that you judged exchangeable, the areas would agree with the frequencies with which the quantity lay between the boundaries of the areas.

9.9 PROBABILITY DENSITIES

There are three matters that need attention before we leave densities. The first is to remark that the ideas of this section easily extend to two uncertain quantities in a manner similar to the extension to two events in Chapter 4. Denote the two quantities by X and Y and divide both their ranges into intervals of equal, small lengths as in Figure 9.2. Next consider the event that both X lies in a selected interval and Y in another. Then we can think of X and Y lying in a square, bordered by the two intervals, and can erect a "tower" on that square of a height such that the volume of that tower is your probability that both X and Y lie within the square base of that tower. This is the same procedure used to construct Figure 9.2 with length of an interval replaced by area of a square, and area of the rectangle erected on the interval by volume of the tower erected on the square base. In this manner the height of the tower is the density $p(x, y)$ where (x, y) is any point in the base area. The pictorial representation is not normally useful because volumes are hard to picture, but the concept of density is most important. The procedure adopted here extends to three, or more, quantities.

The second point is that the results developed in this book for probabilities extend to densities. The point will now be demonstrated for Bayes rule but it applies generally. Consider Figure 9.3 and denote by w the width of the interval into which the range of the uncertain quantity, X, the waiting time, has been divided. The figure shows the cases $w = 10$, solid lines, and $w = 5$, dotted. If $p(x)$ is the density at value x of X, then $p(x)w$ is your probability that X lies in the interval that contains x. In other words, $p(x)w$ is a probability and obeys all the rules of probability. Recall Bayes rule (§6.3), omitting reference to the knowledge base,

$$p(F\,|\,E) = p(E\,|\,F)p(F)/p(E),$$

and let E be the event that X lies in the interval of width w that contains x. Then $p(E\,|\,F) = p(x\,|\,F)w$ and $p(E) = p(x)w$ where $p(x\,|\,F)$ and $p(x)$ are densities. Notice there are two densities for X here, one given F, the other on the knowledge base, and the range has been divided into intervals of the same width w. Inserting these values into the formula, the ws cancel and

$$p(E\,|\,F)/p(E) = p(x\,|\,F)/p(x),$$

exactly the same but with densities in place of probabilities. If F similarly corresponds to a continuous quantity Y, the interval there will occur on both sides of the equation and therefore again cancel. Hence Bayes rule reads

$$p(y\,|\,x) = p(x\,|\,y)p(y)/p(x).$$

The third matter is a variant of the addition rule that says that if F_i form a partition, then $p(E)$ is equal to the sum of the terms $p(EF_i)$ over all values of i (§9.1). For a density $p(x, y)$ of two uncertain quantities, X and Y, we may sum over all values of y (the technical term is integrate) to obtain the density of X alone, $p(x)$. This is often called the marginal density. The concept is frequently used in statistics, Chapter 14.

9.10 THE NORMAL DISTRIBUTION

This representation through a density is most useful, both to the mathematician and to lay persons, for describing their uncertainties. Figure 9.4 presents a typical density for incomes in a population. We have deliberately refrained from giving the units since these will differ from country to country. Recall the essential aspect is the area under the curve between any two values, so that keeping the distance between

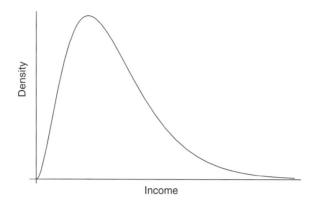

FIGURE 9.4 Density of income distribution.

these values constant, it is the height of the density that matters. Starting from the left, the density begins with low values, showing that few people have very small incomes. It rapidly ascends to a point where there are many people with these incomes. The further descent from the maximum is much slower than the ascent, showing that incomes somewhat above the common value do occur fairly frequently. The curve continues for a very long way, showing that a few people receive very high incomes. This type of income density is common in market economies and large spread thought undesirable.

There is one type of density that is very important, both because it has many simple, useful properties that make manipulation with it rather easy, and because it does arise, at least approximately, in practice. Two examples are shown in Figure 9.5. Features common to both are symmetry about the maximum, in a shape reminiscent of a

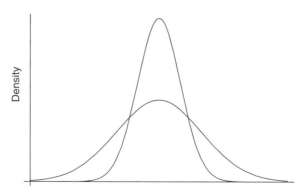

FIGURE 9.5 Normal densities with identical means, different spreads.

bell, and continuing at very small values for a long way. The maximum occurs at the mean, or expectation, so that the two in Figure 9.5 have the same mean. They differ in their spread. As the density flattens out, the value at the maximum necessarily decreases to keep the total area at one. These are examples of a *normal* density, the name being somewhat unfortunate because a density that is not normal, like that for income, is not abnormal. An alternative name is *Gaussian*. Each normal density is completely described by two parameters, its expectation or mean and its spread. The latter can be described nonmathematically in terms of the following property expressed in terms of a measure of spread called the *standard deviation*, abbreviated to s.d. (This was the measure used with the binomial, $[n\theta(1-\theta)]^{1/2}$ and the Poisson $\sqrt{E}$, and also mentioned in §9.5.)

For any normal density the probability of being within 1 s.d. of the mean is about 2/3, within 2 s.d. 19/20 and within 3 s.d. 997/1000.

Thus two-thirds of the total area under the curve is contained within 1 s.d. Values outside 2 s.d. only occur with frequency 1/20, or 5%. This latter figure has been unduly popular with statisticians. Values outside 3 s.d. are extremely rare, rather less than 3 in a thousand occur there. Figure 9.6 illustrates this property. One important property of the normal is that if X is a quantity with a normal distribution, then rescaling it by multiplying by a constant a and relocating it by adding a

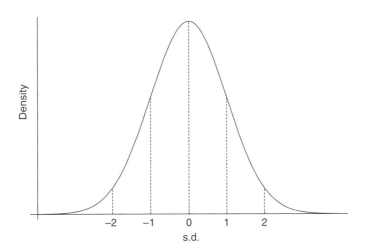

FIGURE 9.6 Normal density with zero mean and unit s.d.

constant b, results in another normal quantity whose expectation is similarly rescaled and relocated and whose s.d. is multiplied by a, the relocation having no effect on the spread.

Here are some reasons for the popularity of the normal distribution. The binomial (§9.2) with index n and parameter θ is, for large n, approximately normal, the approximation being best around $\theta = \frac{1}{2}$ and worst near 0 and 1. Similarly the Poisson is approximately normal for large expectation E. This latter result can be illustrated with the Poisson in Table 9.1 with mean, or expectation, 4. The s.d. is $\sqrt{E} = 2$, so the values 2, 3, 4, 5, and 6 are within 2 s.d.'s of the mean. The total probability of these is

$$0.15 + 0.20 + 0.20 + 0.16 + 0.10 = 0.81,$$

rather larger than the value 2/3, about 0.67, quoted above. But recall we are approximating a discrete quantity, number of calls with the Poisson, by a continuous one, the normal. Looking at Figure 9.1, it will be seen that the rectangle at $r = 2$ for the Poisson has only half its area within 1 s.d. of 4. Therefore the probability at $r = 2$ of 0.15 should be halved, as should the probability of 0.10 at $r = 6$. This reduces the total probability of 0.81 above by $0.07 + 0.05 = 0.12$, yielding 0.69 in excellent agreement with the normal value of 0.67. The halving here may appear suspect but it is genuinely sound.

Suppose you take (almost) any quantity, make a number n of observations of it that you judge exchangeable and then form their average, their total divided by n; then this average will have, to a good approximation, a normal distribution. Your expectation of the normal will be the same as that of the original quantity, the s.d. will be that of the original quantity divided by $\sqrt{n}$ in accordance with the square-root rule in §9.5. Since so many quantities are, in effect, averages, the normal distribution occurs reasonably often, though there is a tendency to use it even where inappropriate because of its attractive properties. Doubtless this tendency will diminish now that our computing power has increased. The result stated in the first sentence can be applied to both the binomial and Poisson distributions to justify the assertions in the previous paragraph about their approximations by the normal. Thus the binomial is based on a quantity taking values 0 and 1 whose values are totaled to give the binomial. The average is just this total divided by n, so the normal distribution for the average will translate, by the result above, into a normal distribution for the total. The s.d. of the 0–1 quantity, we saw in §9.5, was $[\theta(1-\theta)]^{1/2}$. That for the average will be $[\theta(1-\theta)/n]^{1/2}$ by the square-root rule, and that for the total $[n\theta(1-\theta)]^{1/2}$.

A classic example of a normal distribution is provided by the heights of men in a population. The same remark applies to women but since, in respect of height, men and women are not exchangeable, the expectations are different, women being slightly shorter. The s.d.'s are about the same. A similar normal property holds for most measurements of lengths on people, like those of leg lengths. This fact is of use to clothing manufacturers since they know, for example, that only about 1 in 20 of the population will lie outside 2 s.d.'s of the mean.

9.11 VARIATION AS A NATURAL PHENOMENON

In §1.3, it was mentioned that people do not like uncertainty and often invent concepts that appear to explain it. One instance of this is the introduction of gods who control variable phenomena like the weather, but we do not need to be as drastic as this, for people are prepared to see real cause and effect where nothing but natural variation is present. Here is an example that occurred recently and provoked action to remove discrepancies, which was unnecessary because only natural variation was present and the discrepancies explained in terms of it. The original figures have not been used because to do so would involve subtleties that might hide the key point to be put across. Effectively the figures have been rounded to present equalities that were not there originally but the conclusions are unaffected. Before entering into the discussion recall several facts learnt earlier in this chapter. First, the Poisson distribution is present when there are a lot of independent occasions when something might happen but the chance of the happening is small. In our example, there are a lot of people but each has a small chance of dying from the disease being considered. Second, the spread about the Poisson mean, expressed through the s.d., is equal to the square root of that mean. Finally, for expectations that are not too small, the Poisson distribution is well-approximated by the normal, for which about 2 out of 3 of the observations lie within 1 s.d., and 19 out of 20, within 2 s.d. of the mean, or expectation. With all these facts at our disposal, facts it might be pointed out that were likely unknown to the participants in the study, we can proceed with the example.

A disease had a death rate per year throughout a region of 125 per 100,000 people older than 30 years. The region was divided into 42 health authorities, each responsible for 100,000 such persons, and each recorded the number of deaths in a year from the disease among people older than 30 years in their area. There were therefore 42 instances of variation about an expectation of 125 and it is reasonable to approximate the situation with 42 examples of a Poisson with mean 125. Applying the square-root rule, the square root of 125 is about 11, so that it would be anticipated that about two-thirds of the authorities would have rates between 114 and 136, while only 1 in 20, just 2, would have rates outside an interval of twice this width, from 103 to 147. In fact there were three, at 97, 148, and 150. This is in good agreement with the Poisson proposal. Seeing these figures, the administrators in the health service were worried that two authorities had death rates 50% greater than that of the best authority. The media looks with horror at this, scents a story, and both groups try to find reasons for the discrepancy. The administrators punished the apparent errant authorities and praised the successful. In fact they were inventing causes for a variation that is natural to the patterns of death. Randomness is enough explanation and it is doubtful if anything can be done about that. Basically, the square-root rule was not appreciated.

Indeed, we can go further and say that if all, rather than two-thirds, of the authorities had death rates within 1 s.d., that is between 114 and 136 as natural variation suggests, there would have been grounds for suspecting that some falsification of the figures had occurred in order to comply with standards laid down from on high. I once met such a case, involving several producers that had agreed to provide their data prior to the possible introduction of some legislation, The figures were in too good agreement. Enquiry revealed that the producers had got together and some had altered their results so that none appeared out of line.

There is more that can be said about the effect of natural variation. Consider that "bad" health authority with 150 deaths and suppose natural variation is allowed to operate. The result will be that next year the Poisson will again obtain and your probability that, with your expectation still at 125, of getting less than 150 deaths in that authority is almost 1. In other words, the authority will improve without any intervention. This gives bureaucrats a fine opportunity to castigate the apparently errant authority, to enforce changes and then sit back and think how clever they have been in reducing the rate, when nothing has been accomplished except bullying of staff. A failure to recognize natural variation may occur in many fields; education has obvious parallels to health provision. A less obvious parallel is the Stock Exchange, where some people are thought to be better at predicting the market than others. Are they; or is it natural variation? Does management recognize talent or does it just pick the best in the Poisson race?

My concern here is to emphasize that some variation is inherent in almost any system and that its presence should not be forgotten. That is not to say that all variation is natural, for one of the tasks of a statistician is to sort out the total variation into component parts, each having its proper attribution. A simple example of this was presented in §9.6. In that agricultural example there was no inherent measure of spread, as there was in the health example with the square-root rule, and the natural variation had to be separately evaluated. No doubt there are cases of death-rate variation that are causal and exceed natural variation; my plea is for uncertainty to be appreciated as a naturally arising phenomenon that can be handled by the rules of probability. It appears to be a long way from the balls in urns to the Poisson and the square-root rule but the connection is only coherence exhibiting its strength. We often say "you are lucky" but how often are we wrong and fail to recognize the skill involved?

9.12 ELLSBERG'S PARADOX

It has been emphasized in §2.5 that there is a distinction between the normative, or prescriptive, approach to uncertainty adopted in this book and the descriptive approach concerned with describing how people currently think about uncertainty. The concentration on normative ideas does not imply that descriptive analysis is without value; on the contrary, the study of people making decisions in the face of uncertainty may be very revealing in correcting any errors and persuading them of the normative view. And, recalling Cromwell (§6.8), it is possible that good decision makers may be able to demonstrate flaws in the normative theory and, like the Church of Scotland, I may be mistaken. The contrast between normative and descriptive approaches is clearly brought out in paradoxes of the Ellsberg type, discussed in this section. The results are not used in what follows and may be omitted by those who do not like paradoxes. Its presentation has been delayed until now because understanding depends on the concept of expectation developed in §9.3.

Consider our familiar urn, this one containing 9 balls. (9 is used because we want to divide by 3, which 10, or 100, do not do exactly.) 3 of the balls are red, the other 6 are either black or white, with the number b that are black uncertain for you. One ball is to be drawn from the urn in a manner that you think is random and you are asked to choose between the two options displayed in Table 9.3.

Here U is a positive number representing a prize and 0 means no prize. Thus option X gets you the prize if the withdrawn ball is red, whereas Y rewards you if it is black; there is no prize with either option if the ball is white. You are also asked to choose between two options in Table 9.4 using the same urn under the same conditions.

TABLE 9.3 First pair of Ellsberg options

Withdrawn ball	Red	Black	White
Option X	U	0	0
Option Y	0	U	0

TABLE 9.4 Second pair of Ellsberg options

Withdrawn ball	Red	Black	White
Option V	U	0	U
Option W	0	U	U

Thus option V rewards you provided the ball is red or white, whereas W rewards if black or white. Note that you are not being asked to compare an option in one table with any in the other. Everyone agrees V is better than X and W than Y because white balls pay out in the former but not in the latter. No, you are asked to choose between X and Y, and between V and W. What would you do?

Consider first the normative approach where the only thing that matters is your probability $p(U)$ that you will get the prize. Since you are uncertain about the number of black balls, you will have a distribution $p(b)$ over the 7 possible values from 0 to 6, and $p(U)$ will depend on this. For option X, $p(U)$ is $3/9$ because of your belief in the random withdrawal and the fact that the value of b is irrelevant. For option Y the calculation must be more elaborate. If you knew the number b of black balls, $p(U|b) = b/9$, so if you extend the conversation to include b, $p(U)$ will be the sum of terms $bp(b)/9$ over the 7 possible values of b. The result will be $E/9$ where E is your expectation for the number of black balls in the urn. (You may care to refresh your memory by looking at the same argument in §9.3.) Consequently you compare $p(U) = 3/9$ for X with $E/9$ for Y, and Y is preferred to X if $E > 3$; if $E < 3$ you prefer X and with $E = 3$ you are indifferent between X and Y. Exactly the same type of argument shows that W is preferred to V if $E > 3$; if $E < 3$ you prefer V and with $E = 3$ you are indifferent. Consequently your choices depend solely on the number of black balls you expect to be in the urn and you either choose both X and V, both Y and W, or express indifference in both cases.

We now pass to the descriptive approach. Several psychologists have performed experiments with subjects, most of whom have no knowledge of the probability calculus, and asked them to make the choices between the same options, with the result that most prefer X to Y and also W to V. This disagrees with the normative approach where a preference for X, because the expectation of b is less than 3, must mean a preference for V above W. The two approaches are therefore in direct conflict. When the subjects are asked why they made their choices, the usual reply is that they preferred X to Y because with X they knew the numbers of balls in the urn that would cause them to win, namely 3, whereas with Y they did not. The additional uncertainty with Y, over that with X, was thought to be enough to make X the choice. Similarly

W had 6 balls that would yield the prize, whereas with V the number, $9 - b$, is unknown, so W is preferred. We have a clear example of the dislike of uncertainty (§1.3), here sufficient to affect choices.

Which is more sensible—the normative or descriptive attitudes? Before we answer this, let us consider the nature of the disagreement and notice that it concerns coherence. Suppose you, in the technical sense, have seen a subject prefer X above Y; then you would see nothing unsound in the choice, merely noting that the subject must have an expectation for b of less than 3. Similarly if another subject has preferred W to V, then you think it sensible with their expectation being more than 3, so that both subjects are above criticism. But suppose you see the *same* subject make *both* preferences, then you think they are foolish, or incoherent, the first preference not cohering with the second. Here is a clear example of a phenomenon that I think is very common: the individual judgments, considered in isolation, are often sound, the flaw is that the judgments do not fit together, they are inconsistent, or, as we say, incoherent.

How does this incoherence arise? Both you and the subjects recognize that the key element is the unknown number, b, of black balls. The subjects worry about it and try to avoid it as much as they can by choosing X and W. You, by contrast, face up to the challenge, recognize that b is uncertain, and use a probability distribution for it. (You go further and note that the whole distribution does not matter, only the expectation is relevant, a point we return to below.) Now you have a problem; what is your distribution? I put it to you that the development in §3.4 is compelling and that you must have a distribution; the trouble is that you do not know how to assess it. Consider an analogous situation in which you do not have to make a choice between options but between two objects, the prize being awarded if you select the heavier one. There is no doubt in your mind that associated with each object there is a number called its mass but you do not know it. If you could see the objects, you could guess their masses, though the guesses would not be reliable. Were you able to handle them, better guesses could be made, and if you had an accurate pair of scales, you could do much better. Mass here is like probability in the options; you know it exists but have trouble measuring it. In other words, the normative person, you, knows how to proceed coherently with the options but has a measurement problem. In contrast, the subjects did not know how to proceed and therefore shied away from options that involved uncertainty, thereby becoming incoherent.

In the problem as presented here, it is somewhat artificial and all of us would have difficulty with the measurement of $p(b)$. A common attitude is to say that you can see no reason as to why any one of the 7 possible values is more probable than any other, so use the classical form of §7.1 with $p(b) = 1/7$ for all 7 values of b. The expectation is then 3 and you would be indifferent between X and Y, and between V and W. Most published analyses of the problem assume, sometimes implicitly, that the classical form obtains. On the contrary, you might receive information that the number of black balls had been selected by throwing a fair die and equating b with the number showing on the die. In that case the expectation is $3\frac{1}{2}$ (§9.3) leading to your choice of Y rather than X, and W rather than V. In another scenario you might have witnessed several random drawings from the urn and noticed the colors of the exposed balls. You would then have updated your knowledge by Bayes rule and have a distribution based on the data. An extreme possibility is that you are told the values of b, when X and V are selected if you are told $b = 0$, 1, or 2; Y and W if $b = 4$, 5, or 6; and you are indifferent if $b = 3$.

For anyone who is still unconvinced that the normative approach is sensible and the subjects unsound in the Ellsberg scenario, consider the following two arguments. First, suppose you were informed of the value of b, as at the end of the last paragraph, then whatever value it was you would *never* choose both X and W as the subjects did; so why choose them when uncertain about b? (We return to this point when the sure-thing principle is discussed in §10.5.) Second, in the choice between X and Y, the white balls do not matter since in neither option do they yield a prize, so that the final column of the table may be omitted. Similarly in the choice between V and W, the white balls are irrelevant, always producing the prize and again the final column may be deleted. When the final columns are removed from both tables, the remaining tables are identical. So if you choose the first row in one, you must do the same in the other, and the subjects' selection is ridiculous.

One final point before we leave Ellsberg. The paradox shows us that the only aspect of the uncertainty that matters is your expected number of black balls, and that your actions should be based solely on this number. This lesson is important because we shall see when we come to decision analysis in § 10.4 that again it is only an expectation, rather than a distribution, that is relevant. People often have difficulty with the idea of making an important decision on the basis of a single number, so let Ellsberg prepare you for this feature. Mind you, the expectation has to be carefully calculated, as we shall see.

CHAPTER 10

Decision Analysis

10.1 BELIEFS AND ACTIONS

It is early morning, you are about to set off for the day and you wonder whether to wear the light coat you took yesterday, or perhaps a heavier garment might be more suitable. Your hesitation is due to your uncertainty about the weather; will it be as warm as yesterday or maybe turn cooler? We have seen how your doubts about the weather can be measured in terms of your belief that it will be cooler, a value that has been called probability, and we have seen how uncertainties can be combined by means of the rules of the probability calculus. We have also seen how probabilities may be used, for example in changing your beliefs in the light of new information, as a scientist might do in reaction to an experimental result, or a juror on being presented with new evidence, or as you might do with the problem of the coat by listening to a weather forecast. But there is another feature of your circumstances beyond your uncertainty concerning the weather, which involves the consequences that might result from whatever action you take over the coat. If you take a heavy garment in warm weather, you will be uncomfortably hot and maybe have to carry it; whereas a light coat would be more pleasant. If you wear the light coat and the weather is

---

Understanding Uncertainty, Revised Edition. Dennis V. Lindley.
© 2014 John Wiley & Sons, Inc. Published 2014 by John Wiley & Sons, Inc.

cold, you may be uncomfortably cold. In this little problem, you have to do something, you have to act. Thinking about the act involves not only uncertainty, and therefore probability, but also the possible consequences of your action, being too hot or too cold. In agreement with our earlier analysis of uncertainty, we now need to discuss the measurement of how desirable or unpleasant the outcomes could be, to examine their calculus, and, a new feature, to explore the manner in which desirability and uncertainty may be combined to produce a solution to your problem with the coat. This is the topic of the present chapter, called *decision analysis* because we analyze the manner in which you ought sensibly to decide between taking the light or the heavy coat.

All of us are continually having to take decisions under uncertainty about how to act, often a trivial one like that of the coat but occasionally of real moment, as when we decide whether to accept a job offer, or when we act over the purchase of a house. In all such problems, apart from the uncertainties, there are problems with the outcomes that could result from the actions that might be taken. The intrusion of another aspect besides uncertainty has been touched on earlier; for example, when it was emphasized how your belief in an event E was separate from your satisfaction with E were it to happen. The need for this separation was one reason why, in §3.6, betting concepts were not used as a basis for probability, preferring the neutral concept of balls in urns, because, in the action of placing a bet, two ideas were involved, the uncertainty of the outcome and quality of that outcome. An extreme example concerned a nuclear accident (Example 13 of Chapter 1) where the very small probability needs to be balanced against the very serious consequences were a major accident to occur. In this chapter, decision analysis is developed as a method of making the uncertainties and the qualities of the outcomes combine, leading to a sensible, coherent way of deciding how to act. The method will again be normative or prescriptive, not descriptive. The distinction is here important because there is considerable evidence that people do not always act coherently, so that there is potentially considerable room for improvement in decision making by the adoption of the normative approach. Incidentally, it will not be necessary to distinguish between a decision to act and the action itself, so that we can allow ourselves the liberty of using the words interchangeably.

All of us have beliefs that have no implications for our actions, beliefs which exist purely as opinions separate from our daily activities. For example, I have beliefs about who wrote the plays ordinarily attributed to Shakespeare but they have no influence on a decision whether or not to attend a production of Hamlet, for the play is what matters, not whether Shakespeare or Marlowe or the Earl of Oxford wrote it. Sometimes beliefs can lie inactive as mere opinions and then a circumstance arises where they can be used. Recently I read an article about a person and, as a result, developed beliefs about her probity of which no use was made. Later, in an election to the governing body of a society to which I belong, her name appeared on the list of candidates for election. It was then reasonable and possible to use my opinion of her probity to decide not to vote for her. The important point about beliefs, illustrated here, is not that they be involved in action, but that they should have the potentiality to be used in action whenever the belief is relevant to the act.

This chapter shows that beliefs in the form of probability are admirably adapted for decision making. This is a most important advantage that probability and its calculus have over other ways of expressing belief that have appeared. For example, statisticians have introduced significance levels as measures, of belief in scientific hypotheses, but they can be misleading and lead to unsound decisions. Computer scientists and some manufacturers use fuzzy logic to handle uncertainty and action. This is admirable, at least in that it recognizes the existence of uncertainty and incorporates it into product design, but is mischievous in that it can mislead. The decision analysis presented here fits uncertainty with desirability perfectly like two interlocking pieces of a jigsaw puzzle. It does this by assessing desirability in terms of probability and then employing the calculus of probability to fit the two aspects of probability together.

10.2 COMPARISON OF CONSEQUENCES

The exposition of decision analysis begins by discussion of the simplest possible case, from which it will easily be possible to develop the general principles that govern more complicated circumstances. If only one action is possible, it has to be taken; there is no choice and no problem. The simplest, interesting case then is where there are two possible acts and one, and only one, of them has to be selected by you. The two acts will be denoted by A_1 and A_2; A for action, the subscripts describing the first and second acts, respectively. The simplest case of uncertainty is where there is a single event that can either be true, E, or false, E^c. Therefore the first case for analysis is one in which two acts are contemplated and the only relevant uncertainty lies in the single event. The situation is conveniently represented in the form of a contingency table (§4.1) with two rows corresponding to the two actions and two columns referring respectively, to the truth and falsity of E, giving the bare structure of Table 10.1.

It would not be right to think of A_2 as the complement of A_1 in the sense of the action of not doing A_1. If A_1 is the action of going to the cinema, A_2 cannot be the action of not going to the cinema. On the contrary, A_2 must specify what you do if attendance is not at the cinema: read a book, make love, go to the bar? Two actions are being compared. We return to this important point in §10.11. Such a table, with two rows and two columns, has four cells. Consider one of them, say that in the top, left-hand corner corresponding to action A_1 being taken when E is true. Since the only uncertainty in the problem is contained in E, the outcome, when A_1 is taken and E occurs, is known to you and no uncertainty remains. It is termed a *consequence*. This table has four possible consequences; for example, in the right-hand bottom corner, you have the consequence of taking A_2 when E does not occur. I emphasize that a consequence contains no uncertainty, it is sure, you know exactly what will happen and if you did not, you would necessarily need to include other uncertain events besides E, thereby increasing the size of the table.

It was mentioned above that consequences or, as they were called there, outcomes, vary in their desirability; some, like winning the lottery are good, others, like breaking your leg, are bad. What we seek

TABLE 10.1

	E	E^c
A_1		
A_2		

is some way of expressing these desirabilities of sure consequences in a form that will combine with the uncertainty. To accomplish this we make the assumption that any pair of consequences, c_1 and c_2, can be compared in the sense that either c_1 is more desirable than c_2, or c_2 is more desirable than c_1, or they are equally desirable. This is surely a minimal requirement, for if you cannot compare two completely described situations with uncertainty absent, it will be difficult, if not impossible, to compare two acts where uncertainty is present. Notice that the comparison is made by you and need not agree with that made by someone else, and in that respect it is like probability in being personal. We will return to this important point in §10.7. There are many cases in which the comparison demanded by the assumption is hard to determine, but recall this is a normative, not a descriptive, analysis, so that you would surely wish to do it, even if it is difficult. The point is related to the difficulty earlier encountered of comparing an event with the drawing of balls from an urn; you felt it was sensible, but hard to do. After the analysis has been developed further, a return will be made to this point and methods of making comparisons between consequences proposed in §10.3.

A further assumption is made about the comparisons, namely that they are coherent in the sense that, with three consequences, if c_1 is more desirable than c_2 and c_2 more desirable than c_3, then necessarily c_1 is more desirable than c_3. This is an innocuous assumption that finds general acceptance, when it is recalled that each consequence is without uncertainty. The following device may convince you of its necessity. It will be used again in §10.12. To avoid repeatedly saying "is more desirable than", the phrase will be replaced by the symbol $\rightarrow$. Suppose the first two comparisons hold, $c_1 \rightarrow c_2$ and $c_2 \rightarrow c_3$ but you think $c_3 \rightarrow c_1$ in violation of the third comparison. We aim to convince you that, while each of the comparisons may be sensible on their own, to hold all three of the comparisons in the last sentence at the same time is absurd. Suppose you contemplate c_3, c_2 is more desirable than it (the second comparison) and you would welcome a magician with a magic wand who could replace c_3 by c_2. Similarly by the first comparison, $c_1 \rightarrow c_2$, you might use the magician again to replace c_2 by c_1. Finally by the third comparison $c_3 \rightarrow c_1$, the magician could be employed again, to replace c_1 by c_3, which takes you back to where you started with c_3, so that the magician has been employed three times to no avail. The magician will likely have charged you for his services with the wand, so you will have paid him three times with no improvement in outcome. You are a perpetual moneymaking machine—how nice to know you. The device will be referred to as a wand. An apparent criticism of it will be discussed in connection with Efron's dice in §12.10. From now on it is supposed that consequences have been coherently compared. Notice that this is an extension of the meaning of coherence, previously used with uncertainty, to consequences.

Returning to Table 10.1 with its four cells occupied by four consequences, one of the four must be the best in the comparisons and one must be the worst. The decision problem is trivial if they are all equal. So let us attach a numerical value of 1 to the best and 0 to the worst, leaving at most two other consequences to have their numerical values found by the following ingenious method. Consider any consequence c that is intermediate between the worst and the best, better than the former, worse than the latter. We are going to replace c by a gamble that you consider just as desirable as c. Take a situation in which you withdraw a ball at random from the standard urn, full of red and white balls; if it is red, c is replaced by the best consequence, if white by the worst. Clearly your comparison of the gamble with c will be enormously influenced by the proportion of red balls in the urn, the more the red balls, the better the gamble. If they are all red, you will desert c in favor of the best; if all white, you will eagerly retain c. It is hard to escape the conclusion that there is a proportion of red balls that will make you indifferent between c and the random selection of a ball with the stated outcomes. The argument for the existence of the critical number of red balls is almost identical to the one used to justify the measurement of probability in §3.4. If accepted, you can replace c by a gamble where there is a probability, that is denoted by u, of attaining the best (and $1-u$ of the worst), where u is the probability of withdrawing a red ball equal, by randomness, to the proportion of red balls in the urn. The number so attached to a consequence is called its *utility*; the best consequence having utility 1, the worst 0, and any intermediate consequence a value between these two extremes. What this device does is to regard the sure consequence c as equivalent to a value between the best and the worst, this value being a probability u, thereby providing a numerical measure of the desirability of c. The nomenclature and the importance of utility is discussed in §10.7, for the moment let us see how it works in the simple table above and, to make it more intelligible, consider some special acts and events.

Before we do so, the reader should be warned that some people use the term utility merely as a description of worth, without specifying how the measurement is to be made. It is common, particularly in the humanities, to dismiss such a view as utilitarian. The measurement of utility by means of a gamble, as proposed here, is essential in justifying the use of expectation (§10.9) and the reduction of a complex system to a single number.

10.3 MEDICAL EXAMPLE

Suppose that you have a past history of cancer, you are currently sick and it is possible that your cancer has returned and spread. This is the uncertain event E, for which you will have a probability $p(E \mid K)$ based on the knowledge K that you currently have, a probability that will be abbreviated to p because both E and K will remain unaltered throughout the analysis and the results will thereby become easier to appreciate. Notice that p has nothing to do with the probabilities conceptually involved in the determination of your utilities; it purely describes your uncertainty about the spread of cancer in the light of what the doctors and others have told you. The complementary event E^c is that you have no cancer. Suppose further that there are two medical procedures, or actions, that might be taken. The first A_1 is a comparatively mild method, whereas A_2 involves serious surgery. Your problem is whether to opt for A_1 or A_2. In practice there will be other uncertainties present, such as the surgeon's skill but, for the moment, let us confine ourselves to E and the two procedures, leaving until later the elaboration needed to come closer to reality.

With two acts and a single uncertain event, there are four consequences that we list:

A_1 and E: The mild treatment with the cancer present will leave you seriously ill with low life expectancy.

A_1 and E^c: There is no cancer and recovery is rapid and sure.

A_2 and E: The surgery will remove the cancer but there will be some permanent damage and months of recovery from the operation.

A_2 and E^c: No cancer but there will be convalescence.

The next stage is to assign utilities to each of these consequences.

First, you need to decide which is the best of these four consequences. Since this is an opinion by "you" and people sensibly differ in their attitudes toward illness, we can only take one possibility, but here A_1 and E^c is reasonably the best with a happy outcome from a minor medical procedure. Similarly A_1 and E is reasonably the worst. Notice that all these judgments are by "you" and not by the doctors. You may well like to listen to their advice when they may recommend one action above the other, but you are under no obligation to adopt their recommendation. This emphasizes the point we have repeatedly made that our development admits many views; it merely tells you how to organize your views, and now your actions, into a coherent whole.

Having determined the best and worst of the four consequences and assigned values 0 and 1, you need, using the procedure described above, to assess utilities for the remaining two consequences arising from action A_2. The result will be a table (Table 10.2), as the earlier one, but with probabilities and utilities included.

Here u and v are the utilities for the two consequences that might arise from A_2, p is the probability that you have cancer. Consider the value u assigned to the consequence of serious surgery A_2, which removes the cancer E but leads to months of recovery. The method of §10.2 invites you to consider an imaginary procedure that could immediately take you to the best consequence (A_1 and E^c) of rapid,

TABLE 10.2

	E	E^c
A_1	0	1
A_2	u	v
	p	$1-p$

sure recovery, but could alternatively put you in the terrible position of having low life expectancy with the cancer (A_1 and E). Your choice of the value u means that you have equated your present state (A_2 and E) to this imaginary procedure in which u is your probability of the best, and $1-u$ of the worst, consequence. A similar choice with the consequence A_2 and E^c leads to the value v. Of course, the procedure is fanciful in being able to restore the cancer but we often wish we had a magic wand to give us something we greatly desire, while literature contains many examples where the magic goes wrong. We return to "wand" procedures in §10.12. Again you might find it hard to settle on u and v but it is logically compelling that they must exist. Furthermore, once they are determined, the solution to your decision problem proceeds easily, the utilities and probabilities can be combined, unlike chalk and cheese, and the better act found, in a way now to be described.

Consider the serious option A_2, which, in its original form, can lead to two consequences of utilities u and v but, by the wand device, can each conceptually lead to either the worst or the best consequence with utilities 0 and 1. Surely you would prefer the act that has the higher probability of achieving the best, and thereby lower for the worst, so let us calculate $p(\text{best} \,|\, A_2)$, the probability of the best consequence were A_2 selected. We do this by extending the conversation (§5.6) to include the uncertain event E, giving

$$p(\text{best} \,|\, A_2) = p(\text{best} \,|\, E \text{ and } A_2) p(E \,|\, A_2) + p(\text{best} \,|\, E^c \text{ and } A_2) p(E^c \,|\, A_2), \tag{10.1}$$

where all the probabilities on the right-hand side are known, either from the utility considerations or from the uncertainty of E. Thus $p(\text{best} \,|\, E \text{ and } A_2) = u$ by the wand and $p(E \,|\, A_2) = p$ by your original uncertainty for E. Inserting their values, we have

$$p(\text{best} \,|\, A_2) = up + v(1-p). \tag{10.2}$$

It is possible to do the same calculation with A_1 but it is obvious there that A_1 only leads to the best consequence if E^c holds, so

$$p(\text{best} \,|\, A_1) = (1-p). \tag{10.3}$$

Since you want to maximize your probability of getting the best consequence, where the only other possibility is to obtain the worst, you prefer A_2 to A_1 and undergo serious surgery if (10.2) exceeds (10.3). That is if

$$up + v(1-p) > (1-p).$$

Recall that the symbol $>$ means "greater than" (§2.9). Bravely doing a little mathematics by first subtracting $v(1-p)$ from both sides and simplifying, yields

$$up > (1-v)(1-p),$$

and then dividing both sides by $(1-v)p$, we obtain

$$\frac{u}{1-v} > \frac{1-p}{p} \qquad (10.4)$$

as the condition for preferring the serious surgery. This inequality relates an expression on the left involving only utilities to one on the right with probabilities, namely the odds against (§3.8) the cancer having spread, and says that the serious surgery A_2 should only be undertaken if the odds against the cancer having spread are sufficiently small, the critical value $u/(1-v)$ involving the utilities. The odds against are small only if the probability of cancer is large, so you would undertake the serious operation only then. (Equation (10.4) can be expressed in terms of probability, rather than odds, as $p > (1-v)/(1-v+u)$.) This result, in terms of either odds or probability, is intuitively obvious, the new element the analysis provides is a statement of exactly what is meant by large. There are several aspects of this result that deserve attention.

10.4 MAXIMIZATION OF EXPECTED UTILITY

The method just developed has the important ability to combine two different concepts, uncertainty and desirability. It demonstrates how we might simultaneously discuss the small probability of a nuclear accident and the serious consequences were it to happen. In our little medical example, it combines the diagnosis with the prognosis. These combinations have been effected by using the language of probability to measure the desirabilities, or utilities, and then employing the calculus of probabilities, in the form of the extension of the conversation, to put the two probabilities together. It is because utility has been described in terms of probability that the combination is possible. Some writers have advocated utilitarian concepts in which utility is merely regarded as a numerical measure of worth, the bigger the number, the better the outcome is. Our concept is more than this, it measures utility on the scale of probability. To help appreciate this point, consider a utilitarian who attaches utilities 0, $\frac{1}{2}$, and 1 to three consequences. This clearly places the outcomes in order with 0 the worst, 1 the best, and $\frac{1}{2}$, the intermediate, but what does it mean to say that the best is as much an improvement over the intermediate, as that is over the worst, $1 - \frac{1}{2} = \frac{1}{2} - 0$? It is clear what is meant here, namely that the intermediate is halfway between the best and the worst in the sense that it is equated to a gamble that has equal probabilities of receiving the best or the worst.

Having emphasized the importance of combining uncertainty with desirability, let us look at how the combination proceeds, returning to (10.2) above, which itself is an abbreviated form of (10.1), and concentrating on the right-hand side, here repeated for convenience,

$$up + v(1 - p).$$

Expressions like this have been encountered before. In discussing an uncertain quantity, which could assume various values, each with its own probability, we found it useful to form the products of value and probability and sum the results over all values, calling the result the *expectation* of the uncertain quantity as in §9.3. The expression here is the expectation of the utility acquired by taking action A_2, or briefly the *expected utility* of A_2, since it takes the two values of utility, u and v, multiplies each by its associated probability, p and $1-p$, and adds the results. Similarly (10.3) above is the expected utility of A_1, as is easily seen by replacing the utilities, u and v, in the second row of Table 10.2 corresponding to A_2 with those, 0 and 1, in the first row for A_1. Consequently the choice between the two acts rests on a comparison of their two expected utilities, the recommendation being to take the larger. This is an example of the general method referred to as *maximum expected utility*, abbreviated to MEU, in which you select that action, which, for you, has the highest expected utility.

10.5 MORE ON UTILITY

In obtaining the utilities in the medical example, attention was confined to the four consequences in the table. It is often useful to fit a decision problem into a wider picture and use other comparisons, partly because it thereby provides more opportunities for coherence to be exploited. Here we might introduce perfect health as the best consequence and death as the worst. (Let it be emphasized again, this may not be your opinion, you may think there is a fate worse than death.) The four consequences in the table could then be compared with these extremes of 1 and 0, with the result as shown in Table 10.3.

Here s and t replace 0 and 1, respectively; u and v will change but the same letters have been used. Then A_2 has the same expression for its expected utility but that of A_1 becomes $sp + t(1-p)$. Consequently

TABLE 10.3

	E	E^c
A_1	s	t
A_2	u	v
	p	$1-p$

10.5 MORE ON UTILITY

A_2 is preferred over A_1 by MEU if

$$up + v(1-p) > sp + t(1-p)$$

or $$(u-s)p > (t-v)(1-p)$$

on subtracting $v(1-p) + sp$ from each side of the inequality. Dividing both sides of the latest inequality by $(t-v)p$, A_2 is preferred if, and only if,

$$\frac{u-s}{t-v} > \frac{1-p}{p}, \tag{10.5}$$

that is if the odds against cancer having spread are less than a function of the utilities. This is the same as (10.4) when $s=0$ and $t=1$. Let us look at this function carefully. Suppose each of the four utilities, s, t, u, and v had been increased by a fixed amount, then the function would not have changed since it involves differences of utilities. Suppose they had each been multiplied by the same, positive number, then again the function would be unaltered since ratios are involved. In other words, it does not matter where the origin 0 is, or what the scale is to give 1 the best, the relevant criterion for the choice of act, here $(u-s)/(t-v)$ is unaffected. We say that utility is invariant under changes of origin or scale. In this it is like longitude on the earth; we use Greenwich as the origin, but any other place could be used; we use degrees east or west as the scale but we could use radians or kilometers at the equator. Probability is firmly pinned to 0, false, and 1, true, but utility can go anywhere and is fixed only when 0 and 1 have been fixed.

In the medical example we took a situation in which the best and worst of the four consequences both pertained to the same action. This need not necessarily be true, so let us take an example in which they are relevant to different actions. The resulting table (Table 10.4) might look like this:

TABLE 10.4

	E	E^c
A_1	s	1
A_2	0	v
	p	$1-p$

where A_1 with E^c is the best and A_2 with E the worst, the other two consequences having utilities s and v where the same letters are retained, and they will be between 0 and 1, intermediate between the worst and the best. Now something interesting happens. Suppose E were true, then A_1 is better than A_2 since s exceeds 0; suppose E^c were true, then A_1 is still better than A_2 since 1 exceeds v; as a result, whatever happens A_1 is better than A_2 and, adopting a charming Americanism, you are on to a *sure thing*. (Notice that in the original Table 10.2, A_2 was better when E was true since $u > 0$, but A_1 was better when E was false since $1 > v$. There was a real problem in choosing between the acts.) A sure thing avoids MEU although MEU would give the same result as the reader can easily verify. I was once in the position of deciding whether to buy a new house or stay where we were, and judged that a relevant factor was whether I was likely to stay in my present job for the next 5 years or change jobs. If staying, it was clearly better to buy, but after some thought we decided that purchase was more sensible even if I did change jobs. We were on to a sure thing.

10.6 SOME COMPLICATIONS

To appreciate another point about MEU let us return to (10.1) and notice that it contains $p(E|A_2)$, the probability of E were A_2 to be selected; similarly in considering A_1, $p(E|A_1)$ would arise. In the medical example it was tacitly assumed that these two probabilities were equal; the choice of action, rather than the action itself, not influencing your cancer. There are situations in which they can be different. Consider the action of buying a new washing machine, where there is a choice between two models, A_1 being cheap and A_2 more expensive. The prime uncertain event E for you is a serious failure within a decade. Ordinarily $p(E|A_1) > p(E|A_2)$ on the principle that the more expensive machine is less likely to fail. (Notice that this is a likelihood comparison, so "likely to fail" is correct.) If this were not so, A_1 is a sure thing under reasonable conditions. Even when the choice of act influences the uncertainty, MEU still obtains, as can be seen from (10.1). If p_1 and p_2 are your probabilities of E given A_1 and given A_2, respectively then, generalizing Table 10.3, A_1 is preferred to A_2 if, and only if,

$$sp_1 + t(1-p_1) > up_2 + v(1-p_2), \qquad (10.6)$$

which does not simplify in any helpful way.

In this general case, the sure-thing principle does not necessarily obtain. To see this take the utilities in Table 10.4, where we observed the principle, but replace p by $p_1 = p(E|A_1)$ and $p_2 = p(E|A_2)$ as in (10.6). Consider the numerical values

$$s = 1/4, v = 3/4, p_1 = 4/5, p_2 = 1/5$$

with $u = 0$ and $t = 1$ as in Table 10.4. Inserting all six values into (10.6) we have

$$1/4 \times 4/5 + 1/5 > 3/4 \times 4/5$$

that simplifies to $8 > 12$ that is not true, so that A_2 is preferred over A_1. This despite A_1 being preferred both when E is true ($s = 1/4, u = 0$) and when false ($t = 1, v = 3/4$), so the principle is violated. Notice that this is despite the losses $s - u$ and $t - v$, being the same at $1/4$, whether E is true or false. The explanation lies in the fact that E^c, which is where the larger utilities, t and v arise, has greater probability when A_2 is taken, 4/5, than when A_1 is selected, 1/5. Short cuts, like the sure-thing principle, can be dangerous, only MEU can be relied upon.

A serious limitation of the decision analysis so far presented is that it only involves one uncertain event. However, the extension to any number is straightforward. Suppose there are two events, E and F. These yield four exclusive and exhaustive (§9.1) possibilities EF, EF^c, E^cF, and E^cF^c, and the decision table has four columns and hence eight consequences. Assign 4 probabilities to the events, in the case where choice of action does not affect the uncertainty, or 8 when it does. Also assign 8 utilities, when the expected utility for an action, corresponding to a row, can be calculated by multiplying the utility in each column by its corresponding probability and adding the 4 results, as in the general extension of the conversation in §9.1. This calculation is done for each row and that action (row) is selected of higher expected utility. Clearly this method extends to any number of actions and we may omit the mathematics. Generally MEU covers all situations where a single person "you" is involved. We have seen the difficulties with two persons, exemplified by the prisoners' dilemma in §5.11.

Finally a warning that is addressed to pessimists. There are many treatments of decision analysis that do not speak in terms of utility but rather use losses. To see how this works, suppose you knew which event was true, equivalently which column of the decision table obtained. Then, all uncertainty being removed, you can choose among the decisions, the rows of the table, naturally selecting that of highest utility, any other act resulting in a loss in comparison. Thus the general form of Table 10.3 supposes, in accord with Table 10.2, that if E is true, A_2 is the better act. Then u exceeds s and A_1 would incur a loss $(u - s)$ in comparison. Similarly, if E^c is true and A_1 the better act, again as in Table 10.2, t exceeds v and A_2 will incur a loss $(t - v)$. A loss is what you suffer, in comparison with the best, by not doing the best. The attractive feature of losses is that the general solution we found, expressed in (10.5), only involves the losses, not the four separate utilities, resulting in a reduction from 4 utilities to only 2 losses and even then only their ratio is relevant. As a result of the simplicity of losses over utilities, the former have become popular; unfortunately they have a serious disadvantage. To see this, notice that the general solution (10.5) only applies when the events have the same uncertainty, expressed through the probability p there, for all acts. When this is not true and the uncertainties are different at p_1 and p_2, the general solution is provided by (10.6), which is not expressible solely in terms of losses. Readers might like to convince themselves of this, either by doing a little mathematics or by choosing two different sets of utilities (s, t, u, v) with the same losses $(u - s, t - v)$ and observing that (10.6) will not yield the same advice in the two sets despite the identity of the losses. It is usually better to assign utilities directly to consequences, rather than relate consequences by considering differences.

10.7 REASON AND EMOTION

Let us leave the more technical considerations of utility and how it is used in decision analysis; instead let us contemplate the concept itself. The first thing to note is that utility applies to a consequence, which itself is the outcome of a specific act when a specific event is true. A consequence, alternatively called an outcome, can have many aspects.

For example, in the cancer problem discussed above, there was a consequence, there described as A_1 and E^c, where the mild treatment had been applied and no cancer found, so that recovery is sure. But you may wish to take into account other aspects of this outcome besides the simple recovery, like the occurrence of your silver wedding anniversary next month that you would now be able to enjoy. If a decision was to go to the opera, the quality of the performance would enter into your utilities, as well as the cost of the ticket. Generally, you can include anything you think relevant when contemplating a consequence. For example, people often bet when, on a monetary basis, the odds are unfavorable. This may be coherent if account is taken in the utility of the thrill of gambling, where a win of 10 dollars is not just an increase in assets but is exciting in the way that a 10 dollar payment of an outstanding debt would not be. There are connections with the confusion between uncertainty and desirability (§3.6). In summary, a consequence can include anything; in particular it can include emotions and matters of faith.

Throughout this book we have applied reasoning, first to the uncertainties and now to decision making in the face of that uncertainty. We have avoided concepts like faith and emotions, concentrating entirely on coherence, which is essentially reason. Coherence generalizes the logic of truth and falsity to embrace uncertainty and action. But in utility, a concept derived entirely by reasoning, we see that it is possible, even desirable, to include ideas beyond reason. We can take account of the silver wedding, the thrill of a gamble, or my preference for Verdi over Elton John (§2.4). Indeed, we not only can take, but must take, if our decision making is truly to reflect our preferences. It has repeatedly been emphasized that probability is personal; we now see the same individuality applies to utility. The distinction between the two is that probability includes beliefs, whereas utility incorporates preferences. The distinction between the two is not sharp and I may say that I believe Verdi is a better artist than John, though the contrast is more honestly expressed by saying that I prefer Verdi to John. A key feature is that an approach using pure reason has led to the conclusion that something more than pure reason must be included. This may be expressed in an epigram:

Pure reason shows that reason is not enough.

My personal judgment is that this result is very important. The reasoning process is essentially the same throughout the world, whereas emotions and faiths vary widely. What is being claimed here is that persons of all faiths can use the reasoning process, expressed through MEU, to communicate. This is done by each faith incorporating its own utilities and probabilities into MEU. On its own, MEU does not eliminate differences between emotions, as has been seen in the prisoners' dilemma (§5.11), but it may lessen the impact of the differences by providing a common language of communication, so important if several faiths are to coexist in peace.

We have seen that your uncertainties can, and indeed should, be altered by evidence, and that the formal way to do this is by Bayes rule. Utilities can also be affected by evidence, though the change here is less formal. For example, your utility for classical music will typically be influenced by attendances at performances of it. Or your love of gambling will respond to experiences at the casino. Evidence therefore plays an important role in MEU. This will be discussed in more detail when the scientific method is studied in Chapter 11. Evidence is especially important when it can be shared, either by direct experience, or through reliable reporting. It was seen in §6.9 that the shared experience of drawing balls from an urn led to disparate views of the constitution of the urn approaching agreement. It is generally true that shared evidence, coherently treated, brings beliefs and preferences closer together. In contrast, there are beliefs and preferences that are not based on shared evidence. Orthodox medicine is evidence-based, but alternative medicine relies less on evidence and so does not fit so comfortably within MEU. This is not to dismiss alternative medicine, only to comment that individual uncertainties and utilities will necessarily differ among themselves more than when shared evidence is available.

10.8 NUMERACY

There is a serious objection to our approach that deserves to be addressed. We have seen that a consequence may be a complicated concept involving many different features, some, like money, being tangible, but others, like pleasure derived from a piece of music, intangible. These features may be important but imprecise. The objection questions whether it is sensible to reduce such a collection of disparate ideas to a single number in the form of utility; is not this carrying simplicity too far? We have encountered in §3.1 a similar objection to belief being reduced to a number, probability. Here the idea is extended even further to embrace utility and the combination of utility and probability in expected utility. A complicated set of ideas is reduced to a number; is it not absurd? If we set aside those people who hate arithmetic and cannot do even simple mathematics, rejoicing in their innumeracy, there are three important rejoinders to these protests.

The first is the one advanced in §3.1 when countering the similar objection in respect of probability; namely that, in any situation save the very simplest, one has to combine and contrast several aspects. Numbers combine more easily, and according to strict rules, than any other features. In decision analysis, it is necessary to deal with several consequences that have to be contrasted and combined. Thus in the medical example of §10.3 there were four, rather different, consequences that had first to be compared, and then some combinations calculated so that you could choose between the two actions contemplated. Numbers do the combining more effectively than any other device. A sensible strategy would therefore try reducing the complicated consequences to numbers and see what happens. The result of doing this, MEU, has much to recommend it and works very well provided some limitations, explored in §10.11, are appreciated.

This is certainly the most powerful argument in support of numeracy but there is a second argument that depends on the recognition that the utility is not, and does not pretend to be, a complete description of the consequence. It is only a summary that is adequate for its purpose, namely to act in a particular context. Similarly, the price of a book is a numerical description that takes into account tangibles, like the number of pages, but also intangibles like its popularity. Nevertheless it is adequate for the purpose of distribution among the public, without describing all aspects of the book. Neither utility nor price, which may well be different, capture the total concept of a consequence or a book; they provide a summary that is adequate for their intended purpose.

The third reason for reducing all aspects of decision analysis to numbers is that, properly done, it overcomes the supreme difficulty, not just of combining beliefs, or contrasting preferences in the form of consequences, but of combining beliefs with preferences. This has historically proved a hard task. The solution proposed here is to measure your preferences in terms of gambles on the best and worst, so introducing probabilities, the measure that has already been used for beliefs. By doing this, the two numerical scales, for beliefs and for preferences, are the same and can, therefore, be combined in the form of expected utility, where the expectation incorporates your belief probabilities and the utility includes your preferences. Notice that the amalgam of beliefs and preferences comes about through a rule of probability; namely the extension of the conversation, as displayed in Equation (10.1) of §10.3. It is the ingenious idea of measuring preferences on a scale of probability that enables the combination to be made, and the manner of its making is dictated by the calculus of probability. It is not necessary to introduce a new concept in order to achieve the combination, for the tool is already there. Alternatively expressed, the use of expectation arises naturally and its use does not involve an additional assumption.

The proceedings in a court of law show how these numerate ideas might be used. The legal profession wisely separates the two aspects of belief and decision (§10.14). In the trial, it is the responsibility of the jury to deal with the uncertainty surrounding whether or not the defendant is guilty. It is usually the judge who decides what to do when the verdict is "guilty". Our solution, which has considerable difficulties in implementation but is sound in principle, is to have the jury express a probability for guilt, instead of the apparently firm assertion. The judge would then incorporate society's utilities with the probability provided and decide on the sentence by maximizing expected utility. The key issue here is the combination of two different concepts.

Underlying these ideas is the assumption that the jury acts as a single person, a single "you". The agreement is normally effected in the jury room. We have little to say formally about the process of reaching agreement, beyond remarking that the members will have shared evidence that, as with the urns (§6.9), encourages beliefs to converge. A similar problem on a larger scale arises when society presents a view from among the diverse opinions of its members. Democracy currently seems the best way of achieving this, leading to the majority attitude often being accepted. We might note that some legal systems have moved toward the acceptance of majority, rather than unanimous, verdicts by a jury.

10.9 EXPECTED UTILITY

The analysis in this chapter has introduced two ideas: utility and expected utility. Returning to Equation (10.1), here repeated for convenience,

$$p(\text{best}\,|\,A_2) = p(\text{best}\,|\,E \text{ and } A_2)p(E\,|\,A_2) + p(\text{best}\,|\,E^c \text{ and } A_2)p(E^c\,|\,A^2),$$

the first probability on the right-hand side is a utility, namely that of the consequence E and A_2, whereas the lone probability on the left we called an expected utility. (Equation (10.2) may provide further clarification.) We now demonstrate that the utility is itself an expectation. Because, in the demonstration, the conditions, E and A_2, remain fixed throughout, let them effectively be forgotten by incorporating them into the knowledge base so that the utility $p(\text{best}\,|\,E \text{ and } A_2)$ above is written $p(\text{best})$. Now suppose that, in your formulation of the decision problem, you felt that E was not the only relevant, uncertain event but that you ought to think about other uncertainties. Thus, in the medical example of §10.3, you might, in addition to the uncertainty about your cancer, feel the surgeon's expertise is also relevant. In other words, you feel the need to extend the conversation to include F, the event that the surgeon was skilled. This gives

$$p(\text{best}) = p(\text{best}\,|\,F)p(F) + p(\text{best}\,|\,F^c)p(F^c). \qquad (10.7)$$

Now let us look at the first probability on the right-hand side which, in full, is

$$p(\text{best}\,|\,EF \text{ and } A_2)$$

on restoring E and A_2. This is your utility of the consequence of taking decision A_2 with E and F both true. Similar remarks apply to $p(\text{best} | F^c)$, with the result that the utility $p(\text{best})$, on the left-hand side of (10.7) is revealed as an expected utility found by taking the product of a utility $p(\text{best} | F)$ with its associated probability $p(F)$ and adding the similar product with F^c.

The argument is general and the conclusion is that any utility, taking into account only E, is equal to an expected utility when additional notice is taken of another event F. In fact, any utility is really your expectation over all the uncertainties you have omitted from your decision analysis. Thus the two terms, utility and expected utility are synonymous. It is usual to use the former term when uncertainty is not emphasized and use the adjective only when it is desired to emphasize the expectation aspect. Whether you include F, or generally how many uncertainties you take into account, is up to you and is essentially a question of how small or large (§11.7) is the world you need.

10.10 DECISION TREES

This is a convenient place to introduce a pictorial device that is often very useful in thinking about a decision problem, using Table 10.3 as an example. The fundamental problem is a choice between A_1 and A_2. This choice is represented by a decision node, drawn as a square, followed by two branches, one for A_1 and one for A_2, as in Figure 10.1. If A_1 is selected, either E or E^c arises, where the outcome, unlike a decision node, is not in your hands but rests on uncertainty and is therefore represented by a random node, drawn as a circle, followed by two branches, one for E, one for E^c (Figure 10.1). Their respective probabilities may be used as labels for the branches. The case (§10.6) where these may depend on the act has been drawn. Similar nodes and branches follow from A_2. Finally, at the ends of the last 4 branches we may write the utilities of the 4 consequences, like fruit on the tree, and Figure 10.1 is complete. It is called a decision tree but, unlike nature's trees, it grows from left to right, rather than upright, the growth reflecting time, the earlier stages on the left, the final ones, the

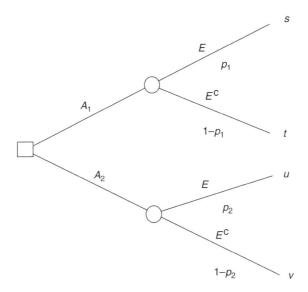

FIGURE 10.1 Decision tree for the situation in §10.6.

consequences, on the right. Clearly any number of branches, corresponding to acts, may proceed from a decision node, not just two as here, and any number, corresponding to events, from a random node. Although time flows from left to right, the analysis proceeds in reverse time order, from right to left, from the imagined, uncertain future back to now, the choice of act. To see how this works, consider the upper, random node, that flowing from A_1, where the branches following, to the right, can be condensed to provide the expected utility $sp_1 + t(1 - p_1)$ (cf. Equation (10.6)) on multiplying each utility on a branch by its corresponding probability and adding the results. Similarly at the random node from A_2 there is an expected utility $up_2 + v(1 - p_2)$ and, going back to the decision node, the choice between A_1 and A_2 is made by selecting that with the larger expected utility. The general procedure is to move from right to left, taking expectations at a random node and maxima at a decision node.

It is easy to see how to include another event in the tree as in §10.9. Consider the upper branches in Figure 10.1 proceeding through A_1 and E. Another random node followed by two branches, corresponding to the extra events F and F^c, may be included as in Figure 10.2 with utilities s_1 and s_2 at their ends, replacing the original utility s. Again we

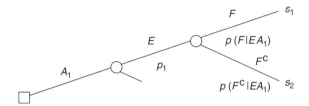

FIGURE 10.2 Decision tree with event F included.

proceed from the far right, obtaining at the random node for F and F^c, expected utility

$$s_1 p(F \mid E) + s_2 p(F^c \mid E)$$

that, we saw in §10.9, equals s, and we are back to Figure 10.1. Similar extensions may be made at the three other terminations of Figure 10.1. Notice again, the equivalence between the expected utility and the original utility.

The real power of a decision tree is seen when there is a series of decisions that have to be made in sequence, one after another, with uncertain events occurring between. Without going into detail and exhibiting the complete, large tree, consider a medical example where, as above, there is initially a choice between two treatments A_1 and A_2. Let us follow A_1 and suppose event E occurs, that the patient develops complications, when a further decision about treatment may need to be made. Suppose treatment B is selected and event F then occurs. The corresponding part of the tree is given in Figure 10.3. Probabilities may be placed on the branches proceeding from the random node. In principle the tree could continue forever with a contemplated series of acts and events but, in practice, it will be expedient to stop after a few branches. When it does, utility evaluations may be inserted at the right-hand ends. In the example, it is natural to stop after F. Analysis of the tree is simple in principle: proceed from right to left, at each random node calculate an expected utility, at each decision node select

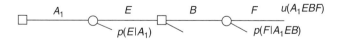

FIGURE 10.3 Decision tree with event E, B and F included.

the branch with maximum expected utility. In the example of Figure 10.3, at the final, random node the probability on each branch, of which only one is shown, is multiplied by the terminal utility and the results added, giving the expected utility of B. With a similar procedure for the actions alternative to B, that act among them of maximum expected utility may be selected. This maximum effectively replaces the branches labeled B and F in Figure 10.3, and we are back to the simple form of Figure 10.1, except that we have only drawn the uppermost sequences of branches, and a choice made, as there, between A_1 and A_2.

Notice how the analysis of the tree proceeds in reverse time order, from the acts and events in the future back to the present decisions. This is reflected in an issue that applies generally in life and is captured succinctly in the epigram:

you cannot decide what to do today until you have decided what to do with the tomorrows that today's decisions might bring.

A beautiful example of this is to be found in §12.3 where a decision is taken that, in the short term is disadvantageous but, in the long term, yields an optimum result. The medical example of Figure 10.3 illustrated this, for a choice between A_1 and A_2 now depends on events like E and what act, like B, will be necessary to take tomorrow. The construction of a decision tree demands that you think not solely in terms of immediate effects but with serious consideration of longer term consequences. Of course the tree will have to stop somewhere but the timescale depends very much on the nature of the problem. The little problem of which coat to wear need scarcely go beyond that day, but decision problems about nuclear waste may need to consider millennia.

10.11 THE ART AND SCIENCE OF DECISION ANALYSIS

The construction of a decision tree is an art form of real value, even when separated from the numeracy of the science of probabilities and utilities, and the analysis through maximization of expected utility.

Thinking within the framework of the tree encourages, indeed almost forces, you to think seriously about what might happen and what the consequences could be if it did. Then again, like all good art, it is a fine communicator in that it clearly presents to another person the problem laid out in a form that is easily appreciated. Even if the reader is uncomfortable with numeracy, despite the persuasive arguments that have been used here, such a person can value the clarity of the tree. They might also be impressed by the power and convenience that trees offer. Decisions today affect decisions tomorrow. Events today affect events tomorrow. The numerical approach offers a principled way to combine these factors to make a sound recommendation for what to do now. I look forward to an enlightened age when it will be thought mandatory for any proposal for action to be accompanied by its decision tree.

Unfortunately, the happy situation of the last sentence will not easily come about because people in power perceive a gross disadvantage in trees and their associated probabilities and utilities; namely that trees expose their thinking to informed criticism. Partly this arises for the reason just given, that a decision tree is good art and therefore a good communicator exposing the decision maker's thinking to public gaze. But there is more to it than that, for the study of the tree reveals what possible actions have been considered and how one has been balanced against another. Also the tree tells us what uncertain events have been considered: did a firm take into account accidents to the work force, or only shareholders' profits? This is before numeracy enters and the uncertainties and consequences measured. Were they to be included, then the exposure of the decision maker's views would be complete. Although probabilities have made some progress toward acceptance so that, for example, one does see statements about the chance of dying from lung cancer, utilities are hardly ever mentioned, in my view because they expose the real motivations behind a recommendation for action. An example will be met in §14.6, where the current financial crisis is discussed. The bankers' decision-making may not have used MEU, but even if it did the bankers have been silent about their utilities.

The introduction of decision trees, while it would go some way to make society more open, would expose a more fundamental difficulty, the difference between personal and social utility, between the desires of the individual and those of society. This is a conflict that has always been with us and is clearly exposed by the apparatus of a decision tree. Here is an example. An individual automobile driver, unencumbered by speed limits, may feel that his utility is maximized by driving fast. Partly to protect others on the road and partly because such a driver may underestimate the danger to himself of driving fast, we have agreed democratically to speed limits, and to fines for speeders. We pay police to enforce these laws, and we pay fines if we speed and are caught. We do so in order to change the individual utilities of drivers, to make it individually optimal for them to drive more slowly.

The two utilities, of the individual and of society, are in conflict and my own view is that a major unresolved problem is how to balance the wishes of the individual against those of society. This is another aspect of the point mentioned in §5.11 that the contribution that the methods here described make toward our understanding of uncertainty and its use in decision analysis, do not apply to conflict situations. Our view is personalistic. This is not to say that the ideas cannot be applied to social problems, they can; but they do not demonstrate how radically different views may be accommodated. The way we proceed in a democratic society is for each party to publish its manifesto, or platform, and for the electorate to choose between them. An extension, in the spirit of this book, would be for the platforms to include probabilities but especially utilities. While this system may be the best we have, it has defects and there is a real need for a normative system that embraces dissent and is not as personalistic as that presented here, though recall that the "you" of our method could be a government, at least when it is dealing with issues within the country. It is principally in dealing with another government that serious inadequacies arise.

Our study of decision analysis does reveal one matter that is often ignored, especially in elections. However you structure decision making, either in the form of a table or through a tree, the choice is always between the members of a list of possible decisions from which you select what you think is the best. To put it differently, it makes no sense to include another row in your table, or another branch in your tree, corresponding to "do something else". Nor, when the uncertain events are listed, does it make sense to include "something else happens." In both cases it is essential to be more specific, for otherwise the subsequent development along the tree cannot be foreseen, nor the numeracy included. It is not sensible, as a politician suggested, to distinguish between known unknowns and unknown unknowns. Everything is a choice between what is available. We have mentioned that the construction of a decision tree is an art form and one of the main contributions to good art is the ability to think of new possibilities. Scientific method is almost silent on this matter, except to make one aware of the need for innovation, yet it is surely true that some good decision making has come about through the introduction of a possibility that had not previously been contemplated. However, once ingenuity has been exhausted, only choice remains:

One does not do something because it is good, but because it is better than anything else you can think of.

In particular, you should vote in an election and choose the party that you judge to be the best, for to deny yourself the choice allows others to select.

A related merit of decision trees is that they encourage you to think of further branches, either relating to an uncertain event or to another possible action. For example, it has been suggested that good decision makers are characterized by their ability to think of an act that others have not contemplated. It is even possible that the art of making the tree is more important than the science of solving it by MEU. Of course, one has to balance the complication that arises from including extra branches against the desire for simplicity.

10.12 FURTHER COMPLICATIONS

Before we leave decision analysis, there is one matter, more technical in character, that must be mentioned. To appreciate this, return to Figure 10.3, which is part of a decision tree in which action A_1 resulted in an event E, to which the response was a further act B, followed by an event F, so that the time order proceeds from left to right. Here A_1 is the first, and F the last, feature. At the end of the tree, on the right, it is necessary to insert a utility in the form of a number. The point to make here is that this utility, describing the consequence of acts A_1 and B with events E and F, could depend on all four of these branches, though not, of course, on other branches like A_2, which do not end at the same place. Mathematically the utility of a consequence is a function of all branches that lead to that consequence; here $u(A_1, E, B, F)$. Thus it would typically happen that A_1, a medical treatment, would be costly in time, money, and equipment, resulting in a loss of utility in comparison with a simpler treatment like A_2. Exactly how this is incorporated into the final utility is a matter for further discussion; all that is being said here is that the cost should be incorporated. Similarly E may have costs, both in terms of hospital care and through long-term effects.

Similar remarks apply to the probabilities on the branches emanating from random nodes. They can depend on all branches that precede it to the left, before it in time. We have repeatedly emphasized that probability depends on two things: the uncertain event and the conditions under which the uncertainty is being considered. The latter includes both what you know to be true and what you are supposing to be true. This applies here and, for example, at the branch labeled with the uncertain event F, the relevant probability is $p(F \mid A_1, E, B)$ since A_1, E, and B precede F and you are supposing them to be true. Similarly you have $p(E \mid A_1)$. It often happens that some form of independence obtains, for example that given E and B, F is independent of A_1. This can be expressed in words by saying that the outcome of the second act does not depend on the original act but only on its outcome. We may then write $p(F \mid E, B)$, omitting A_1. Such independence conditions play a key role in decision analysis, in particular, making the calculations much simpler than they otherwise would be.

Many people are unhappy with the wand device that was used to construct our form of utility, so let us look at it more carefully and use as an example a situation where you are trying to assess the utility of your present state of health. Here you are asked to contemplate a magic wand, which would restore you to perfect health but might go wrong and kill you. You are being asked to compare your present state with something better and with something worse, the comparison involving a probability u that the wand will do its magic, and $1 - u$ that it will cause disaster. With perfect health having utility 1, disaster utility 0, u is the least value you will accept before using the wand and is the utility of your present state. Many people object to the use of an imaginary device, or what is often called a thought experiment, namely an experiment that does not use materials but only thinking. Since you have of necessity to think about a consequence, the procedure may not be unreasonable. Recall too, the point made above, that we have to make choices between actions, so that anyone who objects to wands must produce an alternative procedure. Indeed, there are two questions to be addressed:

1. How would you assess the quality of a consequence?
2. How would you combine this with the uncertainty?

As has been said before, utility as probability answers the second question extremely well. As to the first, notice that the wand at least provides a sensible measure. If your current state of health is fairly good, the passage to perfect health would not be a great improvement, so only worth a small probability of death. The last phrase means $1 - u$ is small and therefore u is near one. On the contrary, if you are in severe pain, perfect health would be a great advance, worth risking death for, and $1 - u$ could be large, u near 0. So things go in the right direction, but there is more to it than that, for the probability connection enables us to exploit the powerful, basic device of coherence.

To see how this works consider four consequences labeled A, B, C, and D, the more advanced the letter in the alphabet, the better it is, so A is the worst, utility 0; D is the best, utility 1. B and C are the intermediates with utilities u and v, respectively, with u less than v. (See Figure 10.4). These values will have been obtained by the wand device using A and D as before. Now another possibility suggests itself; since B is an intermediate between A and C, why not consider replacing B by a wand that would yield C with probability p and A with

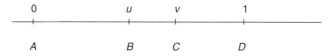

FIGURE 10.4 Comparison of four utilities.

probability $1-p$. How should p relate to u and v? This is easily answered, for you have just agreed to replace B by a probability p of C, and previously you have agreed to replace C by a probability v of D. Putting these two statements together by the product rule, you must agree that B can be replaced by a probability pv of D. (In all these replacements, the alternative is A.) But earlier B had been equivalent to a probability u of D, so therefore

$$u = pv \quad \text{or} \quad p = u/v.$$

You may prefer to use a tree as in Figure 10.5 with random nodes only and the probabilities, necessarily adding to 1, at the tips of the tree. These considerations lead to the following practical device: Use three wands to evaluate u, v, and p; then check that indeed $p = u/v$. If it does, you are coherent; if not, then you must need to adjust at least one of u, v, and p so that it is true for the adjusted values and coherence is obtained. Without coherence, you would be a perpetual, money-making machine (see §10.2).

There is more, for consider replacing C by a wand with probability q of D and $1-q$ of B. How is q related to u and v? Since B can be replaced by a probability $1-u$ at A, C must be equivalent to a probability $(1-q)(1-u)$ of A by the product rule again. But you previously agreed that C could be replaced by $1-v$ at A, so

$$(1-q)(1-u) = 1-v \quad \text{or} \quad 1-q = (1-v)/(1-u),$$

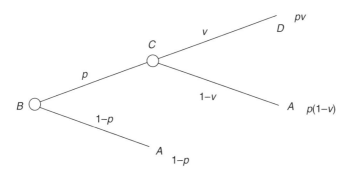

FIGURE 10.5 Tree representation for utility comparison.

and a further coherence check is possible. The important, general lesson that emerges from these considerations is that if you need to contemplate several consequences (at least four) then there are several wands that you can use, not just to produce a utility for each consequence but also to check on coherence. Indeed, as with probability, there is a very real advantage in increasing the numbers of events and consequences, because you thereby increase the opportunities for checks on coherence. The argument is essentially one for coherence in utility, as well as in probability, resulting in coherence in decision analysis.

10.13 COMBINATION OF FEATURES

The ability to combine utility assessment with coherence becomes even more important when the consequences involved concern two, or more, disparate features. To illustrate, consider circumstances that have two features, your state of health and your monetary assets. To keep things simple, suppose there are two states of health, good and bad; and two levels of assets, high and low. These yield four consequences conveniently represented in Table 10.5.

The consequence of both good health and high assets is clearly the best; that of bad health with low assets the worst, so you can ascribe to them utilities 1 and 0, respectively and derive values u and v for the other two, as shown in the table. Notice this table differs from earlier ones in that no acts are involved. Suppose you are in bad health with high assets, utility v, and that $u = v$, then you would be equally content, because of the same utility, with good health and low assets. Expressed differently, you would be willing to pay the difference between high

TABLE 10.5 Consequences with two features

Health	Assets	
	Low	High
Good	u	1
Bad	0	v

and low assets to be restored to good health. If v exceeds u, $v > u$, you would not be prepared to pay the difference; but if u exceeds v, $u > v$, you would be willing to pay even more.

In reality assets are on a continuous scale and not just confined to two values; similarly health has many gradations. It is then more convenient to describe the situation as in Figure 10.6 with two scales, that for assets horizontally, increasing from left to right; that for health vertically, the quality increasing as one ascends. In this representation you would want to aim top-right, toward the northeast, whereas unpleasant consequences occur in the southwest. Without any consideration of uncertainty or any wands, you could construct curves, three of which are shown in the figure, upon any one of which your utility is constant, just as u might equal v in the tabulation. Moving along any one of these curves, as from A to B in the figure, your perception of utility remains constant and the loss in assets results in an improvement in health. Movement in the contrary direction might correspond to deteriorations in health caused by working hard in order to gain increased assets. The same type of figure will be found useful when financial matters are studied in §14.6.

The further northeast the curves are, the higher your utility on them, and to compare the values on different curves you could use a thought experiment of the type already considered. For example, suppose you are at P with high assets but intermediate health, you

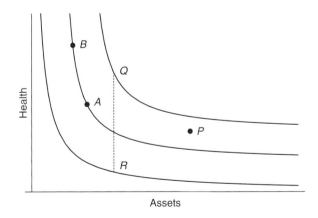

FIGURE 10.6 Curves of constant utility with increasing health and increasing assets.

might think of an imaginary medical treatment that could either improve your health to Q or make it worse at R, but costing you money, so, losing you assets. What probability of improvement would persuade you to undertake the treatment? There are many thought experiments of this type that would both provide a utility along a curve and also check on coherence. What this analysis finally achieves is a balance between health and money, effectively trading one for the other. Modern society uses money as the medium for the measurement of many things, whereas decision analysis uses the less materialistic and more personal concept of utility. People who have a lot of money often say "money isn't everything", which is true, but utility is everything because, in principle, it can embrace the enjoyment of a Beethoven symphony or the ugliness of a rock concert, thereby revealing an aspect of my utility. One aspect that is too technical to discuss in any detail here is the utility of money or, more strictly, of assets. Typically you will have utility for assets like that shown in Figure 10.7, where you attach higher utility to increased assets, but the increase of utility with increase in assets flattens out to become almost constant at really high assets. For example, the pairs (A, B) and (P, Q) in the figure correspond to the same change of assets but the loss in utility in passing from A to B is greater than that from P to Q. For most of us, the loss of 100 dollars (from P to Q, or A to B) is more serious when we are poor at A, than when we are rich at P. A utility function like that of Figure 10.7 can help us understand lots of monetary behavior and is the basis, often in a disguised form, of portfolio

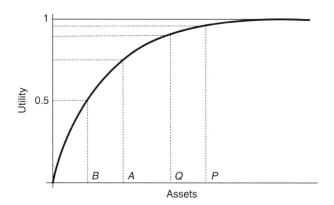

FIGURE 10.7 Curve of increasing utility against increasing assets.

analysis when one spreads one's assets about in many different ventures. Notice that we have used assets, not as often happens, gains or losses, in line with the remarks in §10.6. It is where you are that matters, not the changes. Another feature of Figure 10.7 is that utility is always bounded, by 1 in the figure. There are technical reasons why this should be so and the issue rarely arises in practice.

The discussion around Figure 10.6 revolved around two features, there health and money, but the method extends to any number of features, though a diagrammatic representation is not possible. The idea is to think of situations that, for you, have the same utility, forming curves in Figure 10.6, but imagined surfaces with more than two features, finally using thought experiments to attach numerical values to each surface. Here is an example which arose recently in Britain, where "you" is the National Health Service (NHS). Thus we are talking about social utility, rather than the utilities of individuals (see §10.11). The three features are money, namely the assets of the NHS, the degree of multiple sclerosis (MS), and the degree of damage to a hip. Notice that we do not need to measure these last two features, any more than we did health in the earlier example, utility will do that for us in the context of a decision problem. The decision problem that arose was inspired by the introduction of a new drug that was claimed to be beneficial to those with a modest degree of MS but was very expensive. There were doubts concerning how effective it was, but let us ignore this uncertainty while we concentrate on the utility aspects. The NHS decided the drug was too expensive to warrant NHS money being spent on it. This decision naturally angered sufferers from MS who pointed out that the expected improvement in their condition would enable them to work and thereby save the NHS on invalidity benefit. Where, you may ask, does the hip damage come in? It enters because the money spent on one patient with MS could be used to pay for 10 operations to replace a hip. So the NHS effectively had to balance 10 good hips against one person relieved of MS. Other features in lieu of hips might have been used, for the point is that in any organization like the NHS, there are only limited resources and, as a result, hips and MS have to be compared. Our proposal is that the comparison should be effected by utility, and the suggestion made that this utility be published openly for all to see and comment upon. One can understand the distress to sufferers from MS by the denial of the drug but equally the discomfort of 10 people with painful hips has to be thought about. People are very reluctant to admit that there is a need for a balancing act between MS and hips, but it is so. Utility concepts are a possible way out of the dilemma, though, as mentioned before, they do not resolve the conflict between personal and social utility.

10.14 LEGAL APPLICATIONS

Consider the situation in a court room, where a defendant is charged with some infringement of the law, and suppose it is a trial by jury. There is one uncertain event of importance to the court—Is the defendant guilty of the offence as charged?—which event is denoted by G. Then it is a basic tenet of this book that you, as a member of the jury, have a probability of guilt, $p(G \mid K)$, in the light of your background knowledge K. (There are many trials held without a jury, in which case "you" will be someone else, like a magistrate, but we will continue to speak of "juror" for linguistic convenience.) We saw in §6.6 how evidence E before the court would change your probability to $p(G \mid EK)$ using Bayes rule. The calculation required by the rule needs your likelihood ratio $p(E \mid GK)/p(E \mid G^c K)$, involving your probabilities of the evidence, both under the supposition of guilt and of innocence, G^c. It was emphasized how important it was to consider and to compare evidence in the light both of guilt and of innocence.

Before evidence is presented, it is necessary to consider carefully what your background knowledge is. As a member of the jury, you are supposed to be a representative of society and to come to court with the knowledge that a typical member of society might possess. As soon as the trial begins, you will learn things, like the formal charge; and you will also see the defendant, so enlarging K. At this point you may be able to contemplate a numerical value for your probability. For example, if the charge is murder, where all admit that a person was killed, you may feel it reasonable to let $p(G \mid K) = 1/N$, where N is the population of the country to which the law pertains, on the principle that someone did the killing and until specific evidence is produced, no person is more probable than any other. (The law says all are innocent until proved guilty but that is not satisfactory since it says $p(G^c | K) = 1$ in default of Cromwell's rule, for someone did the killing.) If evidence comes that the killing was particularly violent and must have been committed by a man, you may wish to replace N by the number of adult males.

There are cases where the assignment of the initial probabilities, $p(G | K)$, is really difficult. Suppose the charge is one of dangerous driving in which all accept that a road accident occurred with the defendant driving. Also suppose the only point at issue is whether the defendant's behavior was dangerous or whether some circumstance arose which he could not reasonably have foreseen. One suggestion is to say that initially you have no knowledge and both possibilities are equally likely, so $p(G | K) = \frac{1}{2}$. But this is hardly convincing since several different circumstances might equate to the defendant's innocence, so why not put $p(G | K) = 1/(n+1)$ if you can think of n different circumstances?

A possible way out of difficulties like these is to recognize that your task as a juror is to assess the defendant's guilt in the light of *all* the evidence, so that fundamentally all that you need is $p(G | EK)$, where E is the totality of the evidence. The point of doing calculations on the way as pieces of evidence arrive is to exploit coherence and thereby achieve a more reasonable final probability than otherwise. As a result of these ideas, one possibility is to leave $p(G | K)$ until K has been inflated by some of the evidence, sufficient to give you some confidence in your probability, and only then exploit coherence by updating with new evidence. There is no obligation to assess every probability; we have a framework which can be as big or as small as you please, increased size having the advantage of more coherence. The situation is analogous to geometry. You might judge that a carpet will fit into a room, or you may measure both the carpet and the room and settle the issue. The measurement uses geometrical coherence, direct judgment does not but may be adequate. The ideas here are related to the concepts of small and large worlds (§11.7).

In using the coherence argument in court, a difficulty can arise when two pieces of evidence are presented. Omitting explicit reference to the background knowledge in the notation, because it stays fixed throughout this discussion, the first piece of evidence, E_1, will change your probability to $p(G|E_1)$. When the second piece of evidence E_2 is presented, a further use of Bayes rule will update it to $p(G|E_1E_2)$ and the relevant likelihood ratio will involve $p(E_2|GE_1)$ and $p(E_2|G^cE_1)$. (To see this apply Bayes rule with all probabilities conditional on at least E_1.) To appreciate what is happening, take the case where the two pieces of evidence are of different types. For example, E_1 may refer to an alibi and E_2 to forensic evidence provided by a blood stain. If you judge them to be independent (§4.3) given guilt and also given innocence, so that $p(E_2|GE_1) = p(E_2|G)$, the updating by E_2 is much simpler since E_1 is irrelevant. In contrast, take the position where they are both alibi evidence, then you may feel that the two witnesses have collaborated and the independence condition fails. In which case you might find it easier to consider E_1E_2 as a single piece of evidence and update $p(G)$ to $p(G|E_1E_2)$ directly without going through the intermediate stage with only one piece of evidence. Independence is a potentially powerful tool in the court room but it has to be introduced with care.

At the end of the trial the jury is asked to pronounce the defendant "guilty" or "not guilty"; in other words, to decide whether the charge is true or false. According to the ideas presented in this book, the pronouncement is wrong, for the guilt is uncertain and therefore what should be required of the jury is a final probability of guilt. Hopefully this might be near 0 or 1, so removing most of the uncertainty, but society would be better served by an honest reflection of indecision, such as probability 0.8. Actually the current requirement for "guilt" is "beyond reasonable doubt" in some cases and "on the balance of probabilities" in others. For us, the latter is clear, probability in excess of one half, but the former, like most literary expressions, is imprecise and senior judges have been asked to say what sort of probability is needed to be beyond reasonable doubt; essentially what is "reasonable"? The value offered may seem low, a probability of 0.8 frequently being proposed. A statistician might say at least 0.95. I think the question of guilt is wrongly put and that the jury should state their probability of guilt.

There is an interesting separation of tasks in an English court, where the jury pronounces on guilt but the judge acts in passing sentence, which is automatic in the case of a "not guilty" verdict. This is somewhat in line with the treatment in this book, the jury dealing with probability, the judge dealing in decision making. If judges are to act coherently they will need utilities to combine with the probability provided by the jury. The broad outlines of these utilities could be provided by statute, though the judge would surely need some freedom in interpretation since no drafting can cover all eventualities. As an example, I suggest that instead of saying that a maximum fine for an offence should be 100 dollars, perhaps 1% of assets might be a more reasonable maximum, so that a rich person's illegal parking could have a significant effect on reducing taxes. The point here is that a fine is not a way of raising money but a deterrent, so that 100 dollars deters the poor more than the rich. Utility considerations could also reflect findings in penology.

The thesis of this book impinges on court practice in other ways. The law at the moment rules that some types of evidence are inadmissible, so that they are denied to the jury, though the judge may be aware of them in passing sentence. However, it was seen in §6.12 that data, or evidence, is always expected to be of value, in the sense that your expected value of the information provided by evidence is always positive; so that, as a member of the jury, you would expect the inadmissible evidence to help you in your task. Evidence has a cost that needs to be balanced against the information gain, using utility considerations as in §10.13. Hence the recommendation that flows from our thesis is that the only grounds for excluding evidence are on grounds of cost. It has been argued by lawyers that evidence should be excluded because jurors could not handle it sensibly. This is a valid argument in the descriptive mode but ceases to be true in the normative position. When the jurors are coherent, all evidence might be admitted.

Another way in which probability could affect legal practice is in respect of the double jeopardy rule whereby someone may not be tried twice for the same offence. If new evidence arises after the completion of the original trial and is expected to provide a lot of information, then the court's probability will be expected to be changed. The present rule may partly arise through the jury's being forced to make a definite choice between "guilty" and "not guilty" and the law's natural reluctance to admit a mistake. With an open recognition of the uncertainty of guilt by the jury stating a probability, what had been perceived as a mistake becomes merely an adjustment of uncertainty. The case for every juror, and therefore every citizen, having an understanding of uncertainty and coherence becomes compelling.

There is one aspect of the trial that our ideas do not encompass and that is how the individual jurors reach agreement; how do the twelve "yous" become a single "you"? This was mentioned in §10.8. A rash conjecture is that if coherence were exploited then disagreement might be lessened. Another feature of a trial that needs examination in the light of our reasoned analysis is the adversarial system with prosecution and defense lawyers; a system that has spread from the law to politics in its widest sense where we have pressure groups whose statements cannot be believed because they are presenting only one side of the case. After all, there is another method of reaching truth that has arguably been more successful than the dramatic style that the adversarial system encourages. It is called science and is the topic of Chapter 11.

The above discussion only provides an outline of how our study of uncertainty could be used in legal contexts, namely as a tool that should improve the way we think about the uncertain reality that is about us. While it is no panacea, it is a framework for thinking that has the great merit of using that wonderful ability we have to reason, which yet enables our emotional and other preferences to be incorporated. The calculus of probability has claim to be one of the greatest of human kind's discoveries.

CHAPTER 11

Science

11.1 SCIENTIFIC METHOD

The description of uncertainty, in the numerical form of probability, has an important role to play in science, so it is to this usage that we now turn. Before doing so, a few words must be said about the nature of science, because until these are understood, the role of uncertainty in science may not be properly appreciated.

The central idea in our understanding of science, and one that affects our whole attitude to the subject, is that

The unity of all science consists alone in its method, not in its material.

Science is a way of observing, thinking about, and acting in the world you live in. It is a tool for you to use in order that you may enjoy a better life. It is a way of systematizing your knowledge. Most people, including some scientists, think that science is a subject that embraces topics like chemistry, physics, biology but perhaps not sociology or education; some would have doubts concerning fringe sciences like psychology, and all would exclude what is ordinarily subsumed under the term "arts". This view is wrong, for science is a method, admittedly

Understanding Uncertainty, Revised Edition. Dennis V. Lindley.
© 2014 John Wiley & Sons, Inc. Published 2014 by John Wiley & Sons, Inc.

a method that has been most successful in those fields that we normally think of as scientific and less so in the arts, but it has the potentiality to be employed everywhere. Like any tool, it varies in its usefulness, just as a spade is good in the garden but less effective in replacing a light bulb. The topic of this chapter is properly called the scientific method, rather than science, but the proper term is cumbersome and the shorter one so commonly used that we shall frequently speak of "science" when precision would really require "scientific method".

In the inanimate world of physics and chemistry, the scientific method has been enormously successful. In the structure of animate matter, it has been of great importance, producing results in biology and allied subjects. Scientific method has had less impact in fields where human beings play an important role, such as sociology and psychology, yet even there it has produced results of value. Until recently it has had little effect on management, but the emergence of the topic of management science testifies to the role it is beginning to play there, a role that may be hampered by the inadequate education some managers have in mathematical concepts. We have already noted in §6.5 how probability, which as we shall see below is an essential ingredient in scientific method, could be used in legal questions where its development is influenced by the conservatism of the legal profession. Politics and warfare have made limited use of the method. The method has made little contribution to the humanities, though indirectly, through the introduction of new materials developed by science, it has had an impact on them. For example, the media would be quite different without the use of the technological by-products of science, like television and printing. Even the novel may have changed through being produced on a word processor rather than with a pen.

11.2 SCIENCE AND EDUCATION

The recognition that science is a method, rather than a topic, has important consequences, not least in our view of what constitutes a reasonable education. It is not considered a serious defect if a chemist knows nothing about management, or a manager is ignorant of chemistry, for some specialization is necessary in our complicated world. But for anyone, chemist or manager, to know nothing of the scientific method is to deprive them of a tool that could prove of value in their work. It used to be said that education consisted of the three R's, reading, writing, and arithmetic, but the time may have come to change this, and yet preserve the near-alliteration, to reading, writing, and randomness, where randomness is a substitute for probability and scientific method; for, as will be seen, uncertainty, and therefore probability, is central to the scientific approach. To lack a knowledge of scientific method is comparable in seriousness to a lack of an ability to read, or to write. Just as the ability to write does not make you a writer, neither does the understanding of scientific method make you a scientist; what it does is enable you to understand scientific arguments when presented to you, just as you can understand writing. All of us need to understand the tool, many of us need to use it.

It is not only the ability to use the scientific method that is lacking but also the simpler ability to understand it when it is used. It is easy to appreciate and sympathize with a mother whose young child is given a vaccine and then, a few months later, develops a serious illness, attributing the former as the cause of the latter. Yet if that mother has a sound grasp of the scientific method (not of the science of vaccination) she would be able to understand that the evidence for causation is fragile, and she would see the social implications of not giving infants the vaccine. We must at least understand the tool, even if we never use it. Here is a tool that has been of enormous benefit to all of us, has improved the standard of living for most of us, and has the potentiality to enhance it for all, yet many people do not have an inkling of what it is about. We do not need to replace our leaders by scientists, that might be even worse than the lot we do have, but we do require leaders who can at least appreciate, but better still use, the methodology of science.

In this connection it is worth noting that many scientists do not understand scientific method. This curious state of affairs comes about because of the technological sophistication of some branches of science, so that some scientists are essentially highly skilled technicians who understandably do not deal with the broad aspect of their field but rather, are extremely adept at producing data that, as we shall see, play a central role in science. There is a wider aspect to this. Some philosophers of science see the method as a collection of paradigms; when a paradigm exhibits defects, it is replaced by another. Scientific method is seen at its height during a paradigm shift but many scientists spend all their lives working within a paradigm and make little use of the method. My own field of statistics is currently undergoing a paradigm shift that is proving difficult to make. An eminent scientist once said that it is impossible to complete the introduction of a new paradigm until the practitioners of the old one have died. Another, when asked advice about introducing new ideas, said succinctly "Attend funerals". But this is descriptive; we shall see later that change is integral to normative science.

The appreciation that science is a tool can help to lessen the conflict that often exists between science and other attempts to understand and manage the world, such as religion. There is no conflict between a saw and an axe in dealing with a tree; they are merely different tools for the same task. Similarly there is no conflict at a basic level between a scientist and a poet. Conflict can arise when the two tools produce different answers, as when the Catholic religion gave way and admitted, on the basis of scientific evidence, that the sun, and not the earth, was the center of the solar system. Poetry can conflict with science because of its disregard of facts, as when Babbage protested at Tennyson's line "Every moment dies a man, every moment one is born," arguing that one and one-sixteenth is born; or when a modern poet did not realize there is a material difference between tears of emotion and those caused by peeling onions.

Let us pass from general considerations and ask the question: If science is a method, of what does it consist and, more specifically, what has it to do with uncertainty?

11.3 DATA UNCERTAINTY

One view is that scientific pronouncements are true, that they have the authority of science and there is no room for uncertainty. Bodies falling freely in a vacuum do so with constant acceleration irrespective of their mass. The addition of an acid to a base produces salt and water. These are statements that are regarded as true or, expressed differently, have for you probability 1. Many of the conclusions of science are thought by all, except by people we would regard as cranks, to be true, to be certain. Yet we should recall Cromwell's rule (§6.8) and remember that many scientific results, once thought to be true, ultimately turned out to need modification. The classic example is Newton's laws that have been replaced by Einstein's, though the former are adequate for use on our planet. Then there are statements about evolution like "Apes and men are descended from a common ancestor", which for almost everyone who has contributed to the field are true, but where others have serious doubts, thereby emphasizing that probability is personal. There may be serious differences here between the descriptive and normative views. However, most readers will have probability nearly 1 for many scientific pronouncements. The departure from 1, that you recognize as necessary because of Cromwell, is so slight as to be ignored almost all the time. It is not here that uncertainty enters as an important ingredient of the scientific method but elsewhere.

Anything like a thorough treatment of scientific methodology would take too long and be out of place in a book on uncertainty. Our concern is with the role of uncertainty, and therefore of probability, in science. A simplified version of the scientific method notes three phases. First, the accumulation of data either in the field or in the laboratory; second, the development of theories based on the data; and third, use of the theories to predict and control features of our world. In essence, observe, think, act. None of these phases is productive on its own, their strengths come in combination, in thinking about data and testing the thoughts against reality. The classic example of the triplet is the observation of the motions of the heavenly bodies, the introduction of Newton's laws, and their use to predict the tides. Some aspects of the final phase are often placed in the realm of technology, rather than science, but here we take the view that engineering is part of the scientific method that embraces action as well as contemplation. The production of theories alone has little merit; their value mainly lies in the actions that depend on them.

Immediately the three aspects are recognized, it becomes clear where one source of uncertainty is present—in the data—for we have seen in §9.1 that the variation inherent in nature leads to uncertainty about measurements. Scientists have always understood the variability present in their observations and have taken serious steps to reduce it. Among the tools used to do this are careful control in the laboratory, repetition in similar circumstances, and recording associated values in recognition of Simpson's paradox. The early study of how to handle this basic variation was known as the theory of errors because the discrepancies were wrongly thought of as errors; as mistakes on the scientist's part. This is incorrect and we now appreciate that variation, and hence uncertainty, is inherent in nature. Reference was made above to the mother whose child became seriously ill after vaccination. To attribute cause and effect here may be to ignore unavoidable variability. The scientific method therefore uses data and recognizes and handles the uncertainty present therein.

To a scientist, it is appalling that there are many people who do not like data, who eschew the facts that result from careful observation and experiment. Perhaps I should not use the word "facts" but to me the observation that the temperature difference is 2.34 °C is a fact, despite the next time the observation is 2.71 °C. The measurement is the fact, not the true temperature. We are all familiar with the phrase, "Do not bother me with facts, my mind is made up". It is a natural temptation to select the "facts" that suit your position and ignore embarrassing evidence that does not. Some scientists can be seen doing this, my position is that this is a description of bad science; normative scientists are scrupulous with regard to facts, as will be demonstrated in §11.6. The best facts are numerical, because this permits easier combination of facts (§3.1), but there are those who argue that some things cannot be measured. Scientific method would dispute this, only admitting that some features of life are hard to measure. Older civilizations would have used terms like "warm", "cold" to describe the day, whereas nowadays we measure using temperature expressed in degrees, and when this proves inadequate we include wind speed and use wind chill. Yes, some things are hard to measure, and until the difficulty is overcome it may be necessary to use other methodologies besides science. We do not know how to measure the quality of a piece of music, though the amount of money people are prepared to pay to hear Elton John, rather than Verdi (§2.4), tells us something, though surely not all, about their respective merits.

From a scientific viewpoint, the best facts, the best data, are often obtained under carefully controlled conditions such as those that are encountered in a laboratory. This is perhaps why laypersons often associate science with white-coated men and women surrounded by apparatus. There are many disciplines where laboratory conditions would not tell all the story and work in the field is essential; then, as we have seen in §8.2, other factors enter. Herein lies an explanation of why physics and chemistry have proceeded faster in scientific development than botany or zoology, which have themselves made more progress than psychology or sociology, and why the scientific method finds little application in the humanities. Recall that science is a method and does not have a monopoly on tools of discovery, so its disappointing performance in the humanities is no reflection on its merit elsewhere, anymore than a spade is unsound because it is of little use in replacing a light bulb.

11.4 THEORIES

When scientists have obtained data, what do they do with them? They formulate theories. This is the second stage in the method, where the hard evidence of the senses is combined with thought; where data and reason come together to produce results. To appreciate this combination it is necessary to understand the meaning of a theory and the role it plays in the analysis. One way to gain this appreciation is to look at earlier work before the advent of modern science.

Early man, the hunter, must have been assiduous in the gathering of data to help him in the hunt and in his survival. He will have noted the behavior of the animals, of how they responded to the weather and the seasons, of how they could be approached more effectively to be killed. All these data will have been subject to variation, for an animal does not always respond in the same way, so that the hunter will have had to develop general rules from the variety of observations in the field. From this synthesis, he must have predicted what animals were likely to do when engaged in the future. Indeed, we can say that one of the central tasks of man must always have been to predict the future from observations on the past. Let us put this remark into the modern context of the language and mode of thinking that has been developed in this book. Thinking of future data, F say, which is surely uncertain now, and therefore described by probability, is dependent on past data, D, in the form $p(F|D)$, your probability of the future, given your experience of the past. Expressed in modern terms, a key feature of man's endeavor must always have been, and remains so today, to assess $p(F|D)$.

The same procedure can be seen later in the apprenticeship system where a beginner would sit at the foot of the master for a number of years and steadily acquire the specific knowledge that the latter had. In the wheelwright's shop, he would have understood what woods to use for the spokes, that different woods were necessary for the rim and how they could be bent to the required shape, so that eventually he could build as good a wheel as his mentor. Again we have $p(F|D)$ where F is the apprentice's wheel and D those of the master that he has watched being built, using past data on the behavior of the woods to predict future performance of the new wheel.

The situation is no different today when a financial advisor tries to predict the future behavior of the stock market on the basis of its past performance; or when you go to catch the train, using your experience of how late it has been in the past; or when a farmer plants his seed using his experience of past seasons. Prediction on the basis of experience is encapsulated in a probability, though it is not a probability you can easily assess or calculate with. Conceptually it is an expression of your opinion of the future based on the past. How does this differ from science? As a start in answering this question, let us take a simple example that has been used before but we look at it somewhat differently. The example, as a piece of science, is ridiculously simple, yet it does contain the basic essentials upon which real science, with its complexity, can be built. Remember that simplicity can be a virtue, as will later be seen when we consider real theories, rather than the toy one of our example.

We return to the urn of §6.9, containing a large number of balls, all indistinguishable except that some are colored red, some white, and from which you are to withdraw balls in a way that you think is random. This forms part of your knowledge base and remains fixed throughout what follows. Suppose you have two rival theories, R that the urn contains two red balls for every white one, and W that the proportions are reversed with two white to every red (§6.9), conveniently calling the first the red urn, the second the white one. Suppose that you do not know whether the urn before you is the red or the white one. It will further be supposed that your uncertainty about which urn it is, is captured by your thinking that they are equally probable, though this is not important for what follows. In other words, for you $p(R) = p(W) = 1/2$. Now suppose that you have taken a ball from the urn and found it to be white, this being the past data D in the exposition above, and enquire about the event that the next ball will be red, future data F. Recall that earlier we used lowercase letters for experiences with the balls, reserving capital letters for the constitutions of the urns, so that past data here is w and you are interested in the future data being r. In probability terms, you need $p(r|w)$. In §7.6 we saw how to calculate this by extending the conversation to include R and W, the rival possibilities, so that

$$p(r|w) = p(r|R,w)p(R|w) + p(r|W,w)p(W|w). \qquad (11.1)$$

When this situation was considered in §6.9, our interest lay in $p(R|w)$, your probability, after a white ball had been withdrawn, that the urn was red, and its value was found, by Bayes rule, to be $1/3$, down from its original value of $1/2$. Similarly $p(W|w) = 2/3$, is up from $1/2$. These deal with two of the probabilities on the right-hand side of (11.1) but here our concern is primarily with the other two. Let us begin with $p(r|R,w)$; in words, the probability that the future drawing will yield a red ball, given that the urn is the red one from which a white ball has been drawn. Now it was supposed that the urn contains a large number of balls, so that the withdrawal of one ball, of whatever color, has no significant effect on the withdrawal of another ball and the probability of getting a future red ball remains the same as before, depending only on whether it was the red R or the white W urn. In our notation, $p(r|R,w) = p(r|R)$ or, better still, using the language of §4.3, r and w are independent, given R. Once you know the constitution of the urn, successive withdrawals are independent, a result which follows from your belief in the randomness of the selection of the balls. The same phenomenon occurs with the white urn and the remaining probability on the right, $p(r|W,w)$ will simplify to $p(r|W)$. It is this conditional independence that we wish to emphasize, so let us display the result:

$$p(r|R,w) = p(r|R). \qquad (11.2)$$

Now translate this result back into the language of the scientific method, where we have already met past data D, which in the urn example is w, and the future data F, here r, so that all that is lacking is the new idea of a theory. It does not stretch the English language too far to say that you are entertaining the theory that the urn is red, and comparing it with an alternative theory that it is white. If we denote a theory by the Greek letter theta, θ, we may equate R with θ and W with θ^c. Here θ^c is the complement of θ, meaning θ is false, or $\theta^c = W$ is true. Accepting this translation, (11.2) says

$$p(F|\theta,D) = p(F|\theta),$$

or that past and future data are independent, given θ. The same result obtains with θ^c in place of θ.

This is the important role that a theory plays in the manipulation of uncertainty within the scientific method, in that it enables the mass of past data D to be forgotten, in the evaluation of future uncertainty, and replaced by the theory. Instead of $p(F|D)$ all you need is $p(F|\theta)$. This is usually an enormous simplification, as in the classic example mentioned above where past data are the vast number of observations on the heavenly bodies, the theory is that of Newton, and parts of the future data concern prediction of the tides. We have emphasized in §2.6 the great virtue of simplicity. Here is exposed a brilliant example where all the observations on the planets and stars over millennia can be forgotten and replaced by a few simple laws that, with the aid of mathematics and computing, can evaluate the tides.

There is a point about the urn discussion that was omitted in order not to interrupt the flow but is now explored. Although r and w are independent, given R, according to (11.2), they are not independent, given just K. The reader who cares to do the calculations in (11.1) will find that $p(r\,|\,w) = 4/9$, down from the original value of $p(r) = \frac{1}{2}$, the withdrawal of a white ball making the occurrence of a red one slightly less probable. This provides a simple, yet vivid, example of how independence is always conditional on other information. Here r and w are dependent given only the general knowledge base, which is here that the urns are either 2/3 red or 2/3 white and that the withdrawals are, for you, random. Yet when, to that knowledge base, is added the theory, that R obtains, they become independent. Much writing about probability fails to mention this dependence and talks glibly about independence without reference to the conditions, so failing to describe an essential ingredient of the scientific method. The urn phenomenon extends generally when F and D are dependent on the apprentice's knowledge base, but are independent for the scientist in possession of the theory.

Returning to the scientific method, it proceeds by collecting data in the form of experimental results in the laboratory, as in physics or chemistry, or in the field, as in biology, or by observation in nature, as in sociology or archaeology. It then studies what is ordinarily a vast collection of information. Next, by a process that need not concern us here because it hardly has anything to do with uncertainty, a theory is developed, not by experimentation or observation, but by thought. In this process, the scientist considers the data, tries to find patterns in it, sorts it into groups, discarding some, accepting others; generally manipulating data so that some order becomes apparent from the chaos. This is the "Eureka!" phase where bright ideas are born out of a flash of inspiration, Most flashes turn out, next day, to be wrong and the idea has to be abandoned but a few persist, often undergoing substantial modification and ultimately emerge as a theory that works for the original set of data. This theory goes out into the world and is tested against further data. Neither observation nor theory on their own are of tremendous value. The great scientific gain comes in their combination; in the alliance between contact with reality and reasoning about concepts that relate to that reality. The brilliance of science comes about through this passage from data to theory and then, back to more data and, more fruitfully, action in the face of uncertainty. Let us now look at the role of uncertainty not only in the data, where its presence has been noted, but in the theory.

11.5 UNCERTAINTY OF A THEORY

As mentioned in the first paragraph of §11.3, many people think that a scientific theory is something that is true, or even worse, think that any scientific statement has behind it the authority of science and is therefore correct. Thus when a scientist recently said that a disease, usually known by its initials as BSE, does not cross species from cattle to humans, this was taken as fact. What the scientist should have said was "my probability that it does not cross is . . . ", quoting a figure that was near one, like 0.95. There is a variety of reasons why this was not done. That most sympathetic to the scientist, is that we live in a society that, outside gambling, abhors uncertainty and prefers an appearance of truth, as when the forecaster says it will rain tomorrow when he really means there is a high probability of rain. A second reason why the statement about BSE was so firm is that scientists, at the time, had differing views about the transfer across species, so that one might have said 0.95, another 0.20. We should not be surprised at this because, from the beginning, it has been argued that uncertainty, and therefore probability, is personal. The reason for the possible difference is that the data on BSE were not extensive and, to some extent, contradictory and, as we will see below, it is only on the basis of substantial evidence that scientists reach agreement. Scientists do not like to be seen disagreeing in public, for much of what respect they have derives from an apparent ability to make authoritative statements, which might be lost if they were to adopt an attitude less assertive. A third, related reason for scientists often making a firm statement, when they should incorporate uncertainty, is that they need coherently to change their views yet they like to appear authoritative, or feel the public wants them to be. This change comes about with the acquisition of new data and the consequent updating by Bayes rule. If the original statement was well supported, then the change will usually be small, but if, as with BSE, the data are slight, then a substantial shift in scientific opinion is reasonable. It is people with rigid views who are dangerous, not those who can change coherently with extra data. I was once at a small dinner party when a senior academic made a statement, only to have a young lady respectfully point out that was not what he had said a decade ago. He asked what he had said, she told him, he thought for a while and then said "Good, that shows I have learnt something in the last ten years". Scientists, and indeed all of us, do not react to new information as much as we ought, instead adhering to outmoded paradigms. A fourth, and less important, reason for making firm statements is that scientists commonly adopt the frequency view of probability (§7.7), which does not apply to a specific statement about a disease because there is no obvious series in which to embed it. This reason will be discussed further when significance tests are considered in §§11.9 and 14.4.

The truth of the matter is that when it is first formulated, and in the early stages of its investigation, any theory is uncertain with the originator of the theory having high probability that it is true, whereas colleagues, even setting aside personal animosities, are naturally sceptical. It is only with the accumulation of more data that agreement between investigators can be attained and the theory given a probability near to 0 or 1, so that, in the latter case, it can be reasonably asserted to be true, whereas in the former, it can be dismissed. To see how this works let us return to the urns with two "theories", R and W. In §6.8 we saw that in repeated drawings of balls from the urn, every red ball doubled the odds in favor of it being the red urn and every white ball halved the odds. If it really was the red urn, R, with twice as many red as white balls, in the long run there would be twice as many doublings as halvings and the odds on R would increase without limit. Equivalently, your probability of R would tend to 1. Similarly were it the white urn W, its probability would tend to 1. In other words, the accumulation of data, in this case repeated withdrawals of balls from the urn, results in the true theory, R or W, being established beyond reasonable doubt, to use the imprecise, legal terminology, or with probability almost 1.

Another illustration of how agreement is reached by the accumulation of data, even though there was dispute at the start, is provided by the evaluation of a chance in §7.7. Recall that, in an exchangeable series of length n, an event had occurred r times and it was argued that $(nf + mg)/(n + m)$ might be your probability that it would occur next time; here m and g referring to your original view and $f = r/n$. As the length n of the series increases, nf and n become large in comparison with mg and m so that the expression reduces almost to $nf/n = f$, irrespective of your original views.

11.6 THE BAYESIAN DEVELOPMENT

To see how this works in general, take the probability of future data, given the theory and past experience, $p(F \mid \theta, D)$, which, as was seen in § 11.4, does not depend on the past data, so may be written $p(F \mid \theta)$. We now examine how this coheres with your uncertainty about the theory, $p(\theta \mid D)$. Since D, explicitly or implicitly, occurs everywhere as a conditional, becoming part of your knowledge base, it can be omitted from the notation and can (nearly) be forgotten, so that you are left with $p(F \mid \theta)$ and $p(\theta)$. Applying Bayes rule in its odds form (§6.5, with a change of notation)

$$o(\theta \mid F) = \frac{p(F \mid \theta)}{p(F \mid \theta^c)} o(\theta).$$

Here your initial odds $o(\theta)$—which depends on past data omitted from the notation—is updated by future, further data F, to revise your odds to $o(\theta \mid F)$. "Future" here means after the theory θ has been formulated on the basis of past data. Suppose now that F is highly probable on θ, but not on θ^c; then Bayes rule shows that $o(\theta \mid F)$ will be larger than $o(\theta)$, because the likelihood ratio will exceed one, so that the theory will be more strongly supported. (Recall, again from §6.5, that the rule can be written in terms of likelihoods.) The result in the last sentence may alternatively be expressed by saying that if θ is more likely than θ^c on the basis of F, its odds, and therefore its probability, will increase. This is how science proceeds; as data supporting a theory grows, so your probability of it grows. If the data do not support the theory, your probability decreases. In this way Bayes and data test a theory. Science proceeds by checking the theory against increasing amounts of data until it can be accepted and BSE asserted to cross species, or rejected, showing that it cannot. It is not until this stage that scientific authority is really authoritative. Before then, the statements are uncertain.

There are some details to be added to the general exposition just given about the establishment of a theory. Notice that a theory never attains a probability of 1. Your probability can only get very close to 1, as a consequence of which scientific theories do not have the force of logic. It is true that $2 \times 2 = 4$ but it is only highly probable, on the evidence we have, that Einstein's analysis is correct. This is in accordance with Cromwell's rule and scientists should always remember they might be mistaken as they were with Newton's laws. These worked splendidly and with enormous success until some behavior of the planet Mercury did not agree with Newtonian predictions, leading ultimately to Einstein replacing Newton. In practice, the distinction between logical truth and scientific truth does not matter, only occasionally, as with Mercury, does it become significant.

To appreciate a second feature of the acquisition of knowledge through Bayes, return to the urns and suppose the red theory, R, is correct so that more red balls are withdrawn than white, with every red ball doubling the odds, every white ball halving it. In the numerical work it was supposed that you initially thought the two theories were equally probable, $p(R) = p(W) = \frac{1}{2}$. Consider what would happen had you thought the red theory highly improbable, say $p(R) = .01$, odds of about 1 in 100; then the doubling would still occur twice as often as the halving and the odds would grow just the same and truth attained. The only distinction would be that with $p(R) = 0.01$ rather than 0.50, you would take longer to get there. Suppose the scientist whose initial opinion had been equally divided between the two possibilities had reached odds of 10,000 to 1, then his sceptical colleague would be at 100 to 1 since the former has the odds of 1 to 1 multiplied by 10,000, whereas the latter with only one hundredth but the same multiplication will be at 100 to 1. The two odds may seem very different, but the probabilities 0.9999 and 0.9900 are not so very different on a scale from 0 to 1. The same happens with a general theory θ, where people vary in their initial assessments $p(\theta)$, and it takes more evidence to convince some people than others, but all get there eventually, except for the opinionated one with $p(\theta) = 0$ who never learns since all multiplications of zero, remain zero.

An important question to ask is what constitutes a good theory? We have seen that it is necessary for you to assess $p(F \mid \theta)$ in order to use Bayes rule and update your opinion about the theory in a coherent manner, so you would prefer a theory in which this can easily be done. In other words, you want a theory that enables you to easily predict future outcomes. One that fails to do so, or makes prediction very difficult, is useless. This is another reason for preferring simple theories. But there is more to it than that for you need the likelihood ratio to update your odds. Recall, from Bayes rule as displayed above, that this ratio is

$$p(F\,|\,\theta)/p(F\,|\,\theta^c),$$

where θ^c is the complement of θ. (There are some tricky concepts involved with θ^c but these are postponed until §11.8.) What happens is that as each data set F is investigated, your odds are multiplied by the ratio, so what you would like would be ratios that are very large or very small, for these will substantially alter your odds, whereas a ratio around 1 will scarcely affect it. Concentrating on very large values of the ratio, what you would like would be a theory θ that predicts data F of high probability, but with smaller probability were the theory false, θ^c. A famous example is provided by the general theory of relativity, which predicted that the trajectory of a beam of light would be perturbed by a gravitational field as it passed by a massive object. Indeed, it predicted the actual extent of the bending in an eclipse, so that when an expedition was sent to observe the eclipse and found the bending to be what had been predicted, $p(F\,|\,\theta)$ was near 1, whereas other theories predicted no bending, $p(F\,|\,\theta^c)$ near 0. The likelihood ratio was enormous and relativity became not proved, but most seriously considered. A good theory is one that makes lots of predictions that can be tested, preferably predictions that are less probable were the theory not true. Bearing in mind the distinction between probability and likelihood, what is wanted are data that are highly likely when the theory is true, and unlikely when false. A good theory cries out with good testing possibilities.

There are theories that lack possible tests. For example, reincarnation, which asserts that the soul, on the death of one animal, passes into the birth of another. I cannot think of any way of testing this, even if we remove the notion of soul and think of one animal becoming another. The question that we have met before, "How do you know that?", becomes relevant again. There are other theories that are destructible and therefore of little interest, such as "there are fairies at the bottom of my garden". This is eminently testable using apparatus sensitive to different wavelengths, to sound, to smell, to any phenomenon we are familiar with. The result is always negative. The only possibility left is that fairies do not exhibit movement, emit light or sound, do not do anything that is in our ken. If so, the fairy theory is untestable and is as unsatisfactory as reincarnation. Of course, one day we may discover a new sense and then fairies may become of interest because they can be tested using the new sense but not for the moment.

11.7 MODIFICATION OF THEORIES

The above development of scientific method is too simple to cover every case but our contention is that the principle, demonstrated with the urn of two possible constitutions, underlies every scientific procedure. There are many technical difficulties to be surmounted, which are unsuitable for a book at this level, but uncertainty is ever present in the early stages of the development of any theory. Uncertainty must be described by probability if the scientist is to be coherent. Probability updates by Bayes, so the ideas already expounded are central to any investigation bearing the name of science. Here we discuss two extensions of the Bayesian logic.

It often happens that, in testing a theory against future data, one realizes that the theory as stated is false but that a modification might be acceptable, so that the old theory is replaced by a new one. For example, suppose that when taking the balls from the urn that might be red R or white W, we find a blue ball. One immediate possibility is that the blue has slipped into the urn by accident, so that this piece of data can be ignored. Scientists, often with good reason, reject outliers, the name given to values that lie outside what the theory predicts. Here the blue ball might be regarded as an outlier. But if further blue balls appear then both theories seem doubtful and it would be better to have theories that admit at least three colors, or even four. There is a fascinating problem that concerns how many different colors of balls there are in the urns. This is a simplified version of how many species are there, not in the urn, but on our planet.

Let us pursue another variant of the urn scenario in which 100, say, balls have been taken, of which 50 are found to be red, and 50 white. This is most improbable on both theories but immediately suggests the possibility that the numbers of red and white balls in the urn are equal, a "theory" that will be referred to as the intermediate possibility I.

There are now three theories, R, W, and I, and it is necessary to revise your probabilities. There are no difficulties with the data uncertainties, thus $p(w|R) = 1/3$ and $p(w|W) = 2/3$, as before, and the new one, $p(w|I) = 1/2$, but the uncertainties for the three theories need care. You need to assess your probabilities for them and, in doing so, to ensure that they cohere with your original assessments for R and for W before I was introduced. As an aid in reducing confusion, let your new values be written with asterisks, $p^*(R), p^*(W)$, and $p^*(I)$, necessarily adding to 1, and consider how these must cohere with the values $p(R)$ and $p(W)$, also adding to 1, before the desirability of including I arose. In the former scenario, you were supposing that the event, R or W, was true, probability 1, so if, in the extended case you were to condition on this event, the new values should be the same as the old. That is, in the new setup with I included, the probability of it being the red urn, conditional on it being either the red or white urn—not the intermediate—must equal your original probability of it being red. In symbols

$$p^*(R|R \text{ or } W) = p(R).$$

Similarly $p^*(W|R \text{ or } W) = p(W)$, though the first equality will automatically make the second hold. These are the only coherence conditions that must obtain when the scenario is extended to include I. The condition may more simply be expressed by saying that $p^*(R)$ and $p^*(W)$ must be in the same ratio as the original $p(R)$ and $p(W)$. (A proof of this result is provided at the end of this section.) Here is a numerical example. Suppose you originally thought $p(R) = 1/3$. $p(W) = 2/3$, the white urn being twice as probable as the red. Suppose the introduction of the intermediate urn suggests $p^*(I) = 1/4$. Then $p^*(W)$ must still be twice $p^*(R)$ as originally. This is achieved with $p^*(R) = 1/4, p^*(W) = 1/2$ and the three new probabilities add to one.

The need to include additional elements in a discussion often arises, and the technique applied to the simple, urn example is frequently used to ensure coherence. The original situation is described as a *small world*. In the example it includes R, W, and the results, like r, of withdrawing balls. The inclusion of additional elements, here just I, gives a *large world* of which the smaller is a part. Coherence is then reached by making probabilities in the large world, which are conditional on the small world, agree with the original values in the small world, as expressed in the displayed equation above.

Even the most enthusiastic supporter of the thesis of this book cannot simultaneously contemplate every feature of the universe. A user of the scientific method cannot consider the butterfly flapping its wings in the Amazon rain forest when studying human DNA in the laboratory. It is necessary to embrace a small world that is adequate for the purpose. We have seen from Simpson in §8.2, the dangers of making the world too small. If it is made too big then it may become so complex that the scientist has real difficulty in making sense of it. Somewhere there is a happy, medium-sized world that includes only the essentials and excludes the redundant. Here our normative approach has little to say. There is art in finding an appropriate world. Probability, utility, and MEU are powerful tools that need to be used with discretion. Even a practitioner of the scientific method needs art. There is further reference to this matter at the end of §13.3.

Here is a proof of the result stated above. The multiplication rule (§5.3), says that, for any two events E and F, $p(F \mid E) = p(EF)/p(E)$ provided $p(E)$ is not zero. If F implies E, in the sense that F being true necessarily means E is also true, then the event EF is the same as F and the equation becomes $p(F \mid E) = p(F)/p(E)$. Now R implies the event "R or W", so the displayed equation above can be written

$$p^*(R)/p^*(R \text{ or } W) = p(R).$$

Interchanging the roles of R and W, we similarly have

$$p^*(W)/p^*(R \text{ or } W) = p(W).$$

Dividing the term on the left-hand side of the first equation by the similar term in the second equation, the probability $p^*(R \text{ or } W)$ cancels, and the result may be equated to the result of a similar operation on the right-hand sides, with the result

$$p^*(R)/p^*(W) = p(R)/p(W),$$

as was asserted above.

11.8 MODELS

One difficulty that often arises in applying the methods just expounded is that, although a theory θ may be precise and well defined, its complement θ^c may not be. More correctly, although the theory predicts future data in the form $p(F \mid \theta)$, it is not always clear what data to anticipate if the theory is false and $p(F \mid \theta^c)$, needed to form the likelihood ratio, may be elusive. An example is provided by the theory of relativity—what does it mean for future data at the eclipse if it is false? One possibility is to see the eclipse experiment of §11.6 as a contest between Einstein and Newton so that the two predictions are compared, just as the red and white urns were, and Newton is thought of as the complement of Einstein. This is hardly satisfactory because it was already realized at the time of the experiment that something could be wrong with Newton as a result of observations on the movement of Mercury. A better procedure, and the one that is commonly adopted, goes like this. In the eclipse experiment, relativity predicted the amount of the bending of light to be 6 degrees, so that other possibilities are that the bending is any value other than 6; 5 or 7, or even 0 that was Newton's value. Consequently it is possible to think of the theory saying the bending is 6, and the theory being false meaning the bending is not 6. This means that the value 6 is to be compared, not with just one value, but with several. We saw in §7.5 how this could be done with a few alternatives and technical sophistications make it possible to handle all values besides 6. Generally it happens in the context of a particular experiment that the theory θ implies a quantity, let us denote it by another Greek letter ϕ (phi), which has a definite value. You can take all values of ϕ other than this to constitute θ^c. In most experiments, the data will contain an element of uncertainty so that you will need to think about $p(F \mid \phi)$, rather than $p(F \mid \theta)$, and recognize that θ implies a special value for ϕ. It is usual to denote this special value by ϕ_0. The theory says $\phi = \phi_0$, the complement, or alternative, to the theory says $\phi \neq \phi_0$. We refer to the use of ϕ as a *model* and ϕ is called the *parameter* of the model. In general, for an experiment the theory suggests a model and your uncertainty is expressed in terms of the parameter, rather than the theory. It will be seen how this is done in the next section, but for the moment let us look at the concept of a model.

The relationship between a theory and its models is akin to that between strategy and tactics: strategy describing the overall method or theory, tactics dealing with particular situations or models. In our example, relativity is supposed to apply everywhere, producing models for individual scenarios. One way of appreciating the distinction between a theory and a model is to recognize that a theory incorporates an explanation for many phenomena, whereas a model does not and is specific. The theory of general relativity applied to the whole universe and, when applied to the eclipse, predicted a bending of 6 degrees. There was no theory that predicted 3 degrees, for example. The model, by contrast, only applies to the eclipse and, in that specific context, embraced both 6 degrees and 3 degrees. Scientists have found models so useful as a way of thinking about data that they have been extensively used even without a theory, just as a military battle can use tactics without an overall strategy. Here is an example. Consider a scientist who is interested in the dependence of one quantity on several others, as when a manufacturer, using the scientific method, enquires how the quality of the product depends on temperature, quality of the raw material, and the operator of the manufacturing process (§4.7). We refer to the *dependent* quantity, and the *explanatory* quantities and seek to determine how the latter influences the former, or equivalently, how the former depends on the latter. The dependent quantity will be denoted by y and we will, for ease of exposition, deal with two explanatory quantities, w and x. Many writers use the term *variable* where we have used quantity.

Within the framework developed in this book, the dependence of y on w and x is expressed through your probability of y, conditional on w and x (and your knowledge base), $p(y|w,x)$ and it is usual to refer to this probability structure as the model. Notice that there are a lot of distributions here, one for each value of w and x, so that the model is quite complicated and some simplification is desirable. It was seen in §9.3, and again in §10.4, that an important feature of a quantity is its expectation, which is a number, rather than a possibly large collection of numbers that is a distribution, so interest has centered on $E(y|w,x)$, what you expect the quantity to be for given values of the explanatory quantities. Even the expectation is hard to handle in general and a simplification that is often employed is to suppose the expectation of the dependent quantity is *linear* in the explanatory quantities; that is

$$E(y \mid w, x) = \alpha w + \beta x,$$

where α (alpha) and β (beta), the first two letters of the Greek alphabet, are parameters, like ϕ above. A useful convention has arisen that the Roman alphabet is used for quantities that can be observed and the Greek for quantities that are not directly observable but are integral to the model, like the parameters. What the displayed equation says is that if w is changed by a unit amount, and x remains constant, then you expect y to change by an amount α, whatever value x takes or whatever value w had before the change. A similar conclusion holds with the roles of x and w reversed, but here the change in y is β.

We have continually emphasized the merits of simplicity, provided it is not carried to excess, and here we have an instance of possible excess, because the effect on y of a change in x may well depend on the value of w at the time of the change, a possibility denied by the above model. For example, suppose you are interested in the dependence of the amount, y, the product of a chemical process, on two explanatory quantities, w the temperature of the reaction and x the amount of catalyst used. It could happen that the efficacy of the catalyst might depend on the temperature, a feature not present in the above model, so that the simplicity of the model would then be an inadequate description of the true state of affairs. A valuable recipe is to keep things simple, but not too simple. Another feature of the above model that requires watching relates to the distinction made in §4.7 between "do x" and "see x". Does the model reflect what you expect to happen to y when you see x change, or your expectation were you to make the change, and instead "do x"?

When discussing Simpson's paradox in §8.2, it was seen how the relationship between two quantities could change when a third quantity is introduced. A similar phenomenon can arise here, where the relationship between y and w can alter when x is included. It does not follow that even if

$$E(y \mid w, x) = \alpha w + \beta x,$$

and therefore, if $x = 0$

$$E(y\,|\,w, x = 0) = \alpha w,$$

that, when x is unstated,

$$E(y\,|\,w) = \alpha w.$$

Even if $E(y\,|\,w)$ is linear, as in the original model with w and y, so that

$$E(y\,|\,w) = \alpha^* w,$$

it does not follow that $\alpha = \alpha^*$. Recall that α is the change you expect in y were w to change by a unit amount to $w + 1$, when x is held fixed. In contrast, α^* is the change you expect in y were w to change by a unit amount, when nothing is said about x. It is easy to construct examples in which the quantities, α and α^*, are different, but it suffices to remark that our original example with Simpson's paradox will do, for the change in recovery (y) went one way when only treatment (w) was considered, but the opposite way when sex (x) was included as well. Thus in that case α and α^* had opposite signs, an apparently effective treatment becoming harmful.

Models have been used with great success in many applications of the scientific method and I have no desire to denigrate them, only to issue a word of caution that they need careful thought and can carry the virtue of simplicity too far. Models are no substitute for a theory, any more than tactics are for a strategy; it is best to have an overall strategy or theory that, in particular cases, provides tactics or model. This is why a scientist, properly "a user of the scientific method", likes to have a general explanation of a class of phenomena, of how things work, rather than just an observation of it working. The model may show y increases with w, but it is preferable to understand why this happens.

11.9 HYPOTHESIS TESTING

It was mentioned in the last section that a theory θ often leads, in a particular experiment, to a model incorporating a parameter ϕ, the truth of the theory implying that the parameter has a particular value, ϕ_0, so that investigating the theory becomes a question of seeing whether ϕ_0 is a reasonable value for ϕ. The same situation arises with models that are not supported by theory, when a particular parametric value assumes especial importance. For example, in the linear model

$$E(y \mid w, x) = \alpha w + \beta x,$$

$\alpha = 0$ might be such a special value, saying that w has no effect on your expectation of y, assuming x held fixed. Such situations have assumed considerable importance in some branches of science, so much so that some people have seen in them the central core of the scientific method. From the viewpoint developed here, this centrality is wrong. Nevertheless the topic is of considerable importance and therefore merits serious attention. We begin with an example.

In Example 4 of Chapter 1, the effect of selenium on cancer was discussed. In order to investigate this a clinical trial is set up with some patients being given selenium and others a placebo. It is not an easy matter to set up a trial in which one can be reasonably certain that any effects observed are truly due to selenium and not due to other spurious causes. A considerable literature has grown up on the design of such trials, and it can be taken that a modern clinical trial takes account of this work and is capable of proper interpretation. The design need not concern us here; all we need is confidence that any observed difference between the two sets of patients is due to the selenium and not due to anything else. This difference is reflected in the parameter, referred to previously as ϕ, of interest. In this formulation the value $\phi = 0$ is of special interest because, if correct, it would mean that selenium had no effect on cancer, whereas a positive value would indicate a beneficial effect and a negative one a harmful effect. Incidentally, the trial would hardly have been set up if the negative value was thought reasonably probable. In our notation $p(\phi < 0 \mid K)$ is small. All procedures that are widely used develop their own necessary nomenclature, which is now introduced.

The value of special interest ϕ_0 is called the *null* value and the assertion that $\phi = \phi_0$, the *null hypothesis*. In what follows it will be supposed that $\phi_0 = 0$, as in the selenium example. The nonnull values of the parameter are called the *alternatives*, or *alternative hypotheses*, and the procedure to be developed is termed a *test* of the null hypothesis. A convenient way of thinking about the whole business is to regard the null hypothesis as an Aunt Sally, or straw man, that the trial attempts to knock down. In the selenium trial, the hope is that the straw man will be overthrown and the metal shown to be of value. If a theory has provided the null value, then every attempt at an overthrow that fails, thereby enhances the theory, and some philosophers have considered this the main feature of the scientific method. Notice that the use of the straw man does not make explicit mention of other men, of alternative hypotheses. With these preliminaries, we are ready to test the null, that the null hypothesis is true, the parameter assumes the null value, $\phi = 0$; against the alternative that it is false, the parameter is not zero, written $\phi \neq 0$.

We know how to do this, for if we think of $\phi = 0$ as corresponding to the red urn, R, and $\phi \neq 0$ to the alternative possibility of a white urn, W, then

$$o(\phi = 0 \,|\, D) = \frac{p(D \,|\, \phi = 0)}{p(D \,|\, \phi \neq 0)} o(\phi = 0),$$

where o denotes the odds. The ratio of probabilities is the likelihood ratio and D denotes the data from the trial. In words, the equation expresses your opinion, in the form of odds of the null hypothesis on the basis of the results of the trial, appearing on the left, in terms of the same odds before the trial, on the right. The latter, multiplied by the likelihood ratio, expresses how the probability of the data on the null hypothesis differs from that on the alternative. In other words, the analysis for the two possible urns is analogous to null against alternative, showing how your opinion is altered by the withdrawal of balls, here replaced by looking at the patients in the trial. Remember that all odds and probabilities are also conditional on an unstated knowledge base, here dependent on the careful design of the clinical trial.

It was seen in the case with the urns in §11.4 that every withdrawal of a red ball, more probable on R than on W, enhanced your probability that it was the red urn; while every white ball reduced that probability. Similarly here, data more likely on the null than on the alternative give a likelihood ratio in excess of one and the odds, or equally, the probability, of the null is increased; whereas if the alternative is more likely, the probability is decreased. (Notice the distinction between likely and probable.) And just as you eventually reach assurance about which urn it is by taking out a lot of balls, you eventually learn whether the null is reasonably true by performing a large trial. You eventually learn if the selenium is useless, or effective; being either beneficial $\phi > 0$, or harmful $\phi < 0$. Many clinical trials do not reach such assurance and many tests of a theory are not conclusive. However repeated trials and tests, like repeated experiences with balls, can settle the issue. It is important to bear in mind, as was seen in §11.4, that some people will be more easily convinced than others. Bayes's result displayed above describes the manner in which a null value, or a theory, can be tested. Before the subject is left, three matters deserve attention.

The first has already been touched upon, in that people will start from differing views about the null, some thinking it highly probable, others having severe doubts, yet others being intermediate. This reflects reality but their differences will, as we have seen, be ironed out by the accumulation of data and the multiplying effect of the likelihood ratio.

The second matter is a variant of the first in that people will differ in their initial probabilities for the data when the alternative is true, the denominator in Bayes rule, $p(D \mid \phi \neq 0)$, and may also have trouble thinking about it. To see how this may be handled, consider the selenium trial where there are two possibilities, that the selenium is beneficial, $\phi > 0$, or harmful, $\phi < 0$. Replace these by $\phi = +1$ and $\phi = -1$, respectively, a simplification that in practice is silly and is introduced here only for ease of exposition, the realistic case involving mathematical technicalities. The procedure in the silly case carries over to realism. Now extend the conversation to include the possible alternative values of ϕ,

$$p(D \mid \phi \neq 0) = p(D \mid \phi = +1)p(\phi = +1 \mid \phi \neq 0) \\ + p(D \mid \phi = -1)p(\phi = -1 \mid \phi \neq 0).$$

It ordinarily happens that the two probabilities of the data on the right are easily obtained for they are, in spirit, similar to the numerator of the likelihood ratio $p(D \mid \phi = 0)$. It is the other two probabilities on the right that can cause trouble. In the selenium case, $p(\phi = +1 \mid \phi \neq 0)$ is presumably large because the trial was set up in the expectation that selenium is beneficial. Necessarily $p(\phi = -1 \,|\, \phi \neq 0)$ is small, the two adding to one. With them in place, the calculation can proceed. People may disagree, but again Bayes rule can eventually lead to reasonably firm conclusions.

11.10 SIGNIFICANCE TESTS

The third matter is quite different in character, for although the setting up of a null hypothesis and its attempted destruction, occupies a central role in handling uncertainty, most writers on the topic do not use the methods based on Bayes rule just described, instead preferring a technique that commits a variant of the prosecutor's fallacy (§6.6), terming it not just a test (of the null hypothesis) but a *significance* test. (Recall that mathematicians often use a common word in a special setting, so it is with "significance" here, so do not attach much of its popular interpretation to this technical usage.) To see how a significance test works, stay with the selenium trial but suppose that the relevant data, that were written D, consist of a single number, written d. For example, d might be the difference in recovery rates between patients receiving selenium and those on the placebo. The discussion of sufficiency in §6.9 is relevant. There it was seen that not all the data from the urns were needed for coherent inference, rather a single number sufficed. Again the argument to be presented extends to cases where restriction to a single number is unrealistic. If the null hypothesis, $\phi = 0$, is true, then your probability distribution for d is $p(d \,|\, \phi = 0)$ and is usually available. Indeed it is the numerator of the likelihood ratio just used. This distribution expresses your opinion that some values of d have high probability, whereas others are improbable. For example, it will usually happen, bearing in mind you are supposing $\phi = 0$ and the selenium is ineffectual, that you think values of the difference d near zero will be the most probable, while large values of d of either sign will be improbable. The procedure used in a significance test is to select, before seeing the data, values of d that, in total, you think have small probability and to declare the result "significant" if the actual value of d obtained in the trial is one of them. Figure 11.1

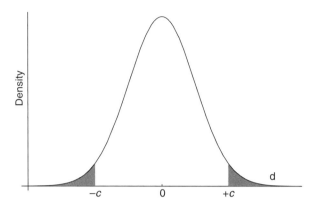

FIGURE 11.1 Probability distribution of d on the null hypothesis, with the tails for a significance test shaded.

shows a possible distribution for you, centered around $d = 0$, with a set of values in the tails that you deem improbable (see also Fig. 9.6). The actual probability you assign to this set is called the *significance level* and the result of the trial is said to be *significant* if the difference d actually observed falls in this set. For historical reasons, the significance level is denoted by the Greek letter alpha, α. To recapitulate, if the trial result is one of these improbable values then, on the idea that improbable events don't happen, or at least happen rarely, doubt is cast on the assumption that $\phi = 0$ or that the null hypothesis is true. Referring to the figure, if d lies in the tails by exceeding $+c$, or being less than $-c$, an improbable event for you has happened and doubt may be cast on the null hypothesis or, as is often said, either an improbable event has occurred or the null hypothesis is false.

Let us look at some features of this popular method. The most attractive is that the approach uses your probabilities for d only when $\phi = 0$, the alternatives $\phi \neq 0$ never occur and the difficulties mentioned above of assessing probabilities for nonzero values do not arise. This makes the significance test rather simple to use. A second feature is that the only probability used is α, the level. Some users fix this before the trial results and, for historical reasons again, use three values 0.05, 0.01, and 0.001. Others let α be the least value that produces significance for the observed value of d, corresponding in the example to c being selected to be $+d$ or $-d$. Evidence against the null is held to be strong only if the value of α produced this way is 0.05 or less. A third feature is that the test does not only use your probability of the difference observed in the trial, as does Bayes, but instead your probability of the set of improbable values, in the example those exceeding c without regard to sign. This has been elegantly expressed by saying that a significance test does not use only the value of d observed, but also those values that might have occurred but did not.

The first two features make a significance test simple to use and perhaps account for its popularity. It is this popularity that has virtually forced me to include the test in a book about uncertainty. Yet, from our perspective, the third feature exposes its folly because it uses the probability of an aspect of the data, lying in the tails of your distribution, when the null hypothesis is true, rather than what our development demands, your probability that $\phi = 0$ given the data. This is almost the prosecutor's fallacy, confusing $p(d \mid \phi = 0)$ with $p(\phi = 0 \mid d)$, replacing d in the first probability with values of d in the set. The contradiction goes even deeper than this because the significance test tries to make an absolute statement about $\phi = 0$, whereas Bayes makes statements comparing $\phi = 0$ with alternatives $\phi \neq 0$. There are no absolutes in this world, everything is comparative; a property that a significance test fails adequately to recognize. This section has dealt only with one type of significance test. There are other significance tests, that also employ the probability distribution of the data on the null hypothesis, where the null hypothesis has a more complicated structure than that treated here. Their advantages and disadvantages are similar to those expounded here. There is more on significance tests in §14.4.

11.11 REPETITION

An essential ingredient of the scientific method is the interaction between observation and reason. The process begins with the collection of data, for example, in the form of experiments performed in a laboratory, which are thought about, resulting in the production of a theory, which is then tested by further experimentation. The strength of science lies in this seesaw between outward contact with reality and inward thought. It is not practical experience on its own, or deep contemplation in the silence of one's room, that produces results, but rather the combination of the two, where the practitioner and the theorist meet. A typical scenario is one in which a scientist performs an experiment and develops a theory, which is then investigated by other scientists who attempt to reproduce the original results in the laboratory. It is this ability to repeat, to verify for yourself, that lies at the heart of the scientific method. The original experiments may have been done in Europe, but the repetitions can be performed in America, India, China, or Africa, or anywhere else, for science is international in methodology and ultimately everywhere the same after sufficient experimentation. Of course, since the results are developed by human beings, there will be differences in character between the sciences of Pakistan and Brazil, but Newton's laws are the same in the dry deserts of Asia as in the humidity of the Amazon.

The simplest form of repetition, exemplified by the tossing of a coin, is captured by the concept of exchangeability (§7.3), where one scientist repeats the work of another, tossing the coin a further time. It has been shown in §11.5 how each successful repetition enhances the theory by increasing its probability, or odds, by the use of Bayes rule. Pure repetition, pure exchangeability, rarely happens and more commonly the second scientist modifies the experiment, testing the theory, trying in a friendly way to destroy it and being delighted when there is a failure to do so. Experience shows that exchangeability continues to be basic, only being modified in ways that need not concern us here, to produce concepts like partial exchangeability. Often the repetition will not go as expected, and in extreme cases the theory will be abandoned. More often the theory will be modified to account for the observations and this new theory itself tested by further experimentation. It has been seen how this happens in the example of §11.7. Repeatability is a cornerstone of the scientific method and the ability of one scientist to reproduce the results of another is essential to the procedure.

It is this ability to repeat earlier work, often in a modified form, that distinguishes beliefs based on science from those that do not use the rigor of the scientific method. An illustration of these ideas is provided by the differences between Chinese and Western medicines, with acupuncture, for example, being accepted in the former but regarded with suspicion in the latter. If A is the theory of acupuncture, then roughly $p(A)$ is near 1 in China and small in the West, though the actual values will depend on who "you" are that is doing the assessing. The scientific procedure is clear; experiences with the procedure can be examined and trials with acupuncture carried out. The results of some trials have recently been reported and suggest little curative effects save in relief from dental pain and in the alleviation of unpleasant experiences resulting from intrusive cancer therapies. These have the effect of lowering $p(A)$ by Bayes, or modifying the theory, limiting its effect to pain relief. The jury is still out on acupuncture, but there is no need for China and the West to be hostile. The tools are there for their reconciliation. Incidentally, this discussion brings out a difference between a theory and a model. The evidence about the benefits to dental health of acupuncture is described by a model saying how a change in one quantity, the insertion of a needle, produces a change in another, pain; but there is only the vaguest theory to explain how the pain relief happens or how acupuncture works.

The preceding argument works well where laboratory or field experiments are possible but there are cases where these are non-existent or of limited value. Let us take an example that is currently giving rise to much debate, the theory of evolution of life on this planet, mainly developed by Darwin. The first point to notice is that Darwin followed the procedure already described in that he studied some data, part of which was that from the journey on the "Beagle", developed his theory, and then spent several years testing it, for example, by using data on pigeons, before putting it all together in his great book, "The Origin of Species"; a book which is both magnificent science and great literature. The greater part of the book is taken up with the testing, the theory occupying only a small portion of the text. This commonly happens because a good theory is simple, as when three rules describe uncertainty or $E = mc^2$ encapsulates relativity. However, Darwin's examples were mostly on domestic species. More complete testing involved extensive investigations of fossils that cannot be produced on demand, as can results in a laboratory. One would have liked to have a complete sequence from ape to man, whereas one was dependent on what chance would yield from digs based on limited knowledge. The result is that although it was known what data would best test the theory of evolution, in the sense of giving a dramatic likelihood ratio, these data were not available. Nevertheless data have been accumulating since the theory become public, likelihood ratios evaluated, and probabilities updated. The result is the general acceptance of the theory, at least in modified forms which are still the subjects of debate. Incidentally, support for evolution was provided by ideas of Mendelian genetics that supplied a mechanism to explain how the modification of species could happen.

Creationists, and others opposed to Darwin, often say that evolution is only a theory. In this they are correct but then so is relativity or any of the other ideas that make science so successful, producing results that the creationists enjoy. A distinction between many theories and that of evolution is that the data available for testing the latter cannot be completely planned. Evolution is not a faith because it can be, and has been, tested, whereas faith is largely immune to testing. It is public exposure to trial, this attempted destruction of hypotheses, that helps make science the great method that it is.

11.12 SUMMARY

This chapter is concluded by a recapitulation of the role of uncertainty in scientific method, followed by a few miscellaneous comments. The methodology begins with data D, followed by the development using reason of a theory θ, or at least a model, and the testing of theory or model on further data F. There is then an extra stage, discussed in Chapter 10, of action based on the theory or model. The initial uncertainty about θ is described by $p(\theta | D)$, your probability of the theory based on the original data. Ordinarily this probability will vary substantially from scientist to scientist but will be updated by further data F using Bayes rule

$$p(\theta | D, F) = p(F | \theta) p(\theta | D) / p(F | D).$$

(Recall that typically F and D will be independent given θ.) As data F accumulate with successive updatings, either θ comes to be accepted, or is modified, or is destroyed. In this way general agreement among scientists is reached. At bottom, the sequence is as follows: Experience of the real world, thought, further experience, followed by action. The strength of the method lies in its combination of all four stages and does not reside solely in any subset of them.

The simple form of Bayes rule just given hides the fact that, in addition to $p(F \mid \theta)$, you also need $p(F \mid \theta^c)$, your probability of the data assuming the theory is false. The odds form shows this more clearly:

$$o(\theta \mid F) = \frac{p(F \mid \theta)}{p(F \mid \theta^c)} o(\theta),$$

absorbing D into the knowledge base. The scientific method is always comparative and there are no absolutes in the world of science. It follows from this comparative attitude that a good theory is one that enables you to think of an experiment that will lead to data that are highly probable on θ, highly improbable on θ^c, or vice versa, so that the likelihood ratio is extreme and your odds substantially changed. One way to get a large likelihood ratio is to have $p(F \mid \theta) = 1$, because since $p(F \mid \theta^c)$ is less than 1, and often substantially less, the ratio must then exceed 1. To get $p(F \mid \theta) = 1$ requires logic. The simplest way to handle logic is by mathematics. This explains why mathematics is the language of science. It is why we have felt it necessary to include a modicum of mathematics in developing the theory that probability is the appropriate mechanism for the study of uncertainty. It helps to explain why physics has advanced more rapidly than biology. Physical theories are mathematical and traditional, biological ones less so, although modern work on Mendelian genetics and the structure of DNA use more mathematics, often of a different character from that used in the applications to physics. The scientific method has, by contrast, made less progress in economics because the intrusion of erratic human behavior into economic systems has previously prevented the use of mathematics. Economic theories tend to be normative, based on rational expectation, or MEU; whereas they could try to be descriptive, to reflect the activities of people who are incoherent and have not been trained in maximizing expected utility.

There are some areas of enquiry that seem ripe for study by the scientific method, yet it is rarely used. Britons are, because of suitable soils and moderate climate, keen gardeners; yet the bulk of gardening literature has little scientific content. An article will extol the beauty of some variety of tree and make modest reference to suitable soils and climate but the issue of how the topmost leaves get nutrition from the roots many feet below receives no mention. The suggestion here is not that the many handsome articles in newspapers abandon their artistic attitude and discuss osmosis but rather that the balance between science and the arts needs some correction. Another topic that needs even more corrective balance is cookery. There are many books and television programs with numerous recipes, yet when a lady recently discussed how to boil an egg, taking into account the chemistry of albumen and yolk, several chefs howled in anger. To hear about the science of cookery go to the food chemists, who mostly work for the food industry, and they will explain the science of boiling, frying, and braising. Recently one chef has entered the scene using scientific ideas and, as a result, received great praise from Michelin and others, so all is not despair. We have seen in §10.14 how the scientific method is connecting with legal affairs. This is happening in two ways. First by the increasing use of science-based evidence like DNA. Second by examining the very structure of the legal argument, using Bayes rule to incorporate evidence, and MEU to reach a decision.

Scientific method is one, successful way of understanding and controlling the world about us. It is not the only method but it deserves more attention and understanding than it has hitherto received. One reason for its success is that it can handle uncertainty through a proper use of probability.

CHAPTER **12**

Examples

12.1 INTRODUCTION

My purpose of writing this book is to introduce you to modern methods of handling uncertainty, so that you can live comfortably with the concept and perhaps treat simpler cases using the basic rules of probability, rather than resort to spurious claims of certainty or inappropriate, illogical procedures. The aim is not to turn you into a probabilist; for that would need mathematical skills that go beyond the view of mathematics as a language used in this book. It would also require extensive practice in handling probability, practice that is ordinarily provided in text books by the inclusion of exercises. Nevertheless, it does help appreciate the power of probability to see it being exercised, to see problems being solved using the ideas that have been developed here. So a few uncertain situations are now examined with the tools we already have. When I told a colleague what I proposed to do, she expressed disquiet remarking that the examples I was using gave surprising results that left the recipient with the feeling that probability was too subtle for them; people having a fondness for common sense and an understandable distaste for conclusions that disagree with it. My colleague's view is sound, so if you feel you can

---

Understanding Uncertainty, Revised Edition. Dennis V. Lindley.
© 2014 John Wiley & Sons, Inc. Published 2014 by John Wiley & Sons, Inc.

dispense with the illustrations, feel free to do so, because none are used in the remaining material. But if you like puzzles, or feel you would like more experience in using probability, then read on for here are some that have entertained and instructed many people. §12.2 is an aperitif, while §12.3 discusses the optimum strategy in a game show on TV. §§12.4 and 12.5 deal with problems that have been much discussed in the literature and on radio. §§12.7 and 12.8 concern social problems. §§12.9 and 12.10 tidy up a couple of technical difficulties, and in doing so reveal delightful surprises.

12.2 CARDS

Our usual urn contains three cards, rather than balls. One card is red on both sides, a second is white on both sides, while the third has one side red and the other white. One of the cards is drawn at random from the urn and you are shown, again at random, one of the sides. All these facts constitute part of your knowledge base. You see that the exposed side is red, this is the datum, and you need to evaluate your probability that the other side is also red. When people are presented with this problem, it is not uncommon for them to argue, by what seems to them to be common sense, that the datum has eliminated the possibility that the withdrawn card is entirely white, so only two possibilities remain. They were equally probable originally and will remain so; hence your probability that the card has both sides red is 1/2. This is an example of Cardano's method that will be studied in §12.5.

To provide the coherent answer using the rules of probability, some notation is needed. Denote the three cards by RR, WW, and RW, in an obvious fashion, and the datum, the red side seen, by r. Then your knowledge base provides you with the following probabilities:

$$p(RR) = p(WW) = p(RW) = 1/3,$$
$$p(r \mid RR) = 1, \quad p(r \mid WW) = 0, \quad p(r \mid RW) = 1/2.$$

You require $p(RR \mid r)$, since the other side of the card is red only if the card is RR. Here is a transposed conditional (§6.1) and Bayes rule is immediately indicated. Since the WW possibility has been eliminated by the sight of the red side, only two possibilities remain and the odds form of §6.5 may be used. This gives

$$\frac{p(RW\,|\,r)}{p(RR\,|\,r)} = \frac{p(r\,|\,RW)}{p(r\,|\,RR)} \times \frac{p(RW)}{p(RR)}.$$

All the probabilities on the right of the equality are known from your knowledge base. Inserting their values into the equation, we easily have

$$\frac{p(RW\,|\,r)}{p(RR\,|\,r)} = \frac{1/2 \times 1/3}{1 \times 1/3} = 1/2.$$

Using the connection between odds and probabilities in §3.8, $p(RR\,|\,r) = 2/3$, so that your probability that the other side of the card is also red is 2/3, not 1/2 as common sense might suggest. What the commonsense argument forgets is that if the card withdrawn is RR, you are twice as likely to see a red side than if it were RW.

This is an artificial problem but has been presented here to demonstrate the value of the notation in organizing your thinking and your employment of the rules of probability. The next example is real and arose out of a popular game show on TV in the United States, where it is known as the Monty Hall problem.

12.3 THE THREE DOORS

The scene is a TV show, the participants a contestant and a host. Before them are three outwardly identical doors. The host tells the contestant truthfully that behind one of the doors is a valuable prize and behind the other two there is nothing; he, the host, knowing where the prize is. One door, at the contestant's choosing, will be opened and she will receive the prize only if it is thereby revealed. He then invites the contestant to choose, but not open, one of the doors. This the contestant does, whereupon the host opens one of the other two doors, revealing that there is nothing there, and invites the contestant to alter her choice to the one remaining door that has neither been opened nor selected. Should she change? It might be added that this is a long-running show that the contestant has often seen, but never participated in, and she has noticed that the door first opened by the host never has the prize.

One answer argues that presumably the contestant initially had no reason to think the prize lay behind one door rather than any other, so that her probabilities for a door hiding the prize is the same for all doors and, there being three doors, each has probability 1/3. (This is the classical interpretation of probability, §7.1.) After the opening, one door is eliminated, two remain and their probabilities for containing the prize are still equal but now 1/2. (Again there is a connection with §11.7.) Consequently, her probabilities for the two doors being equal, it does not matter whether or not she changes her choice of door.

This was the popular view until a journalist put forward a different solution in her column, arguing that the contestant should change. The outcome from the column's publication was a burst of correspondence from mathematicians saying she was wrong, going on to remark that she just did not understand probability, piling on the condemnation by deploring the lack of knowledge of mathematics among the public. Unfortunately the journalist was right and academe had egg on its face. Let us analyze the situation carefully, for it needs only a minimal use of probability; just the addition rule. Incidentally, among the mathematicians who had the wrong answer was one of the most original and prolific of his time, who went on to remark later that "probability is the only branch of math in which a brilliant mathematician can make an elementary error". Quite why this should be so is not clear to me because the subject is only the working out of the logical consequences of the three rules. I suspect that the difficulties arise, not in the math, but in the application of the rules to the real world. §§12.4 and 12.5 illustrate the difficulty and suggest how it might be overcome.

We begin by supposing, as does the naïve analysis already given, that the contestant's probabilities for the prize being behind any door are 1/3. Let us number the doors, 1, 2, and 3 for identification purposes, attaching the number 1 to the door selected by the contestant. Then there are three possibilities:

(a) The prize is behind door 2 and the host opens door 3,
(b) The prize is behind door 3 and the host opens door 2,
(c) The prize is behind the chosen door 1 and the host opens either door 2 or 3.

12.3 THE THREE DOORS   303

Notice that in cases (a) and (b), the host has no choice as to which door to open since there is only one door available, besides that chosen, which has nothing behind it. In case (c), either unselected door may be opened. Next observe that, by the assumption made in the first sentence of this paragraph, the three cases, (a), (b), and (c), each has probability 1/3 for the contestant, yet in (a) and (b) she will get the prize by changing. For example, in (a), door 3 having been opened, a change means choosing door 2, which is where the prize is; similarly in (b). In case (c) the prize will be lost by the change since her first choice was correct. Since (a) and (b) both result in the prize if she changes and both have probability 1/3, by the addition rule the probability that a change results in the prize is 2/3 and of not getting it, case (c), 1/3. There is therefore a substantial expectation of gain by changing, doubling the original probability of 1/3 to 2/3, and the journalist was correct.

It is of some interest to look back and see why the first answer was incorrect. The error lies in thinking of the host's action as random when selecting the door to be opened. Were it truly random and the opened door found to reveal nothing, then the answer would be correct but, as we have seen in the correct analysis, it is far from random in cases (a) and (b), the host having no choice. Only in case (c) could it be random. It is not uncommon for people to tacitly assume that something is random when, in fact, it is not.

The use of probability in solving this problem is minimal, the real difficulty lies in connecting the reality of the TV show with the calculus. The naïve argument does this carelessly, whereas the journalist made the connection correctly and simply. It is often this way with problems in real life, where the mathematics is often straightforward (though straightforward is a relative term). What perplexes people is turning reality into a convenient model, (a) to (c) above, to which the calculus may be applied. There is a real art in constructing the model within which the science can be employed. One way of lessening, but not entirely removing, the difficulty is to take the calculus as primary and force the problem into it. With the three doors, the uncertainty concerns the prize door, which can be 1, 2, or 3. The evidence is the empty, opened door, I, II, or III. Supposing the numbering is chosen so that 1 is the door selected initially by the contestant and II the door opened by the host, we have likelihoods $p(\text{II} \mid 1) = 1/2, p(\text{II} \mid 2) = 0, p(\text{II} \mid 3) = 1$, with priors

$p(1) = p(2) = p(3) = 1/3$. Since the evidence rules out 2, there are only two possibilities and the odds form of Bayes rule gives

$$o(3 \mid \text{II}) = \frac{p(\text{II} \mid 3)}{p(\text{II} \mid 1)} o(3) = 2$$

on inserting the numerical values. Hence $p(3 \mid \text{II}) = 2/3$ as before and change is optimum. Although the first solution is the one usually given, I prefer this second one because it reduces the need to think, replacing it by the automatic calculus. Thinking is hard, so only use it where essential.

It is instructive to consider what would be the correct choice, change door or not, when the contestant did not think the prize lay at random. That is, the contestant does not believe that all doors have probability 1/3 of hiding the prize. Suppose she has probability p_i that it lies behind door i, with $p_1 + p_2 + p_3 = 1$. Further suppose that, as before, she selects door 1. The three cases, (a), (b), and (c) above, will still arise but, instead of being equally probable, will now have probabilities p_2, p_3, and p_1, respectively. If she changes her choice of door, she will win in cases (a) and (b) with total probability $p_2 + p_3$; staying with the selected door 1 will have probability p_1 of gaining the prize. Now $p_2 + p_3 = 1 - p_1$ because the three probabilities add to 1, so the best she can obtain is the larger of $(1 - p_1)$ and p_1. Similarly, were she initially to select door 2, the better is the larger of $(1 - p_2)$ and p_2; for door 3, $(1 - p_3)$ and p_3. Her overall best strategy for selection and possible change corresponds to the largest of these 6 values. To see which is the largest, let the doors be renumbered in such a way that $p_1 < p_2 < p_3$, with the consequence that $1 - p_1 > 1 - p_2 > 1 - p_3$. (Other inequalities are possible but they lead to similar conclusions) The choice lies between p_3, the largest of the first three, and $1 - p_1$ for the last three. Now it cannot happen that p_3 exceeds $1 - p_1$, for then $p_1 + p_3$ would exceed 1, which is impossible; so $1 - p_1$ must be the largest, corresponding to an initial selection of door 1, followed by a subsequent change. But door 1 was initially the least likely to hide the prize, so the contestant's optimum strategy is

> Select the door least likely to hide the prize and then change when the host opens the empty door.

12.4 THE PROBLEM OF TWO DAUGHTERS   305

Her probability of obtaining the prize is 1 minus the least probability. The extreme case when the least value is 0 drives home the idea for if, on entering the studio, she accidentally saw that door 1 was empty, the above strategy would make it certain that she would obtain the prize from the sole unselected and unopened door.

We have here a beautiful example of a point made before in § 10.10 that when you make a decision today (the initial selection), it is essential to take into account the tomorrows (the opened door) that the decision might influence. Here it pays to do an apparently ridiculous thing today (choose the least likely door) in order that the opened door may be very revealing tomorrow. Chess players are aware of this, for sometimes it pays to sacrifice a piece in order to obtain an enhanced position for future moves.

(The alert reader may have noticed that there is a further strategy that might be considered, which we illustrate by the initial selection of door 1 and the cases (a), (b), and (c) above. That is to change if door 2 is opened but not with door 3. This will get the prize with probability p_3; case (b), and $\frac{1}{2} p_1$, case (c), assuming the host opens at random in the latter case. The total probability is $p_3 + \frac{1}{2} p_1$, which is less than $p_3 + p_1 = 1 - p_2$, which could be obtained by one of the strategies already considered, namely select and then change. So the additional strategies are not optimum.)

12.4 THE PROBLEM OF TWO DAUGHTERS

(A version of this problem was present in the first edition under the title "The Newcomers to Your Street". The math is essentially the same in both versions but the scenarios differ. There are two reasons for a change between editions. The first is that the problem has appeared often, both in print and on the radio, under the two-daughter title. The second is that the analyses given there are often defective, if not wrong, so that there is an opportunity here to provide what is hopefully a sound discussion and to clear up some ambiguities.)

There is a mass of evidence to support the theory that, excluding multiple births, the chance of a human baby at birth being male is $\frac{1}{2}$, independent of all other human beings. In the language of §7.4, human beings at birth form a Bernoulli series of chance $1/2$. If you accept this theory, then your probability that any particular child will be male is $1/2$, and that a second child will be female is also $1/2$, independent of the first. So, for you, a family of two children will be BB, BG, GB, or GG, each having probability $1/4$ by the multiplication rule for independent events. Here B means boy, G means girl, and the order of the letters is the order of birth. In the light of these remarks, consider the following scenario.

You meet a lady whom you are reliably informed has two children who are not identical twins. This, together with the probabilities described in the last paragraph, constitute your background knowledge that will be fixed throughout the discussion that follows and therefore not included in the notation. In conversation with her she says that she has a daughter; this is the new evidence. What is your probability now that her other child is also a girl? Common sense suggests that is still $1/2$ since the births are independent in respect of sex. Is this so? A book, described as a best seller, says it is 1/3; so which is correct? It may seem extravagant to do so but we here use the full force of the probability calculus that has been developed in this book. The results are interesting and surprising, so that the extravagance pays off.

Let g denote the new information that one of her children is a girl. The question asked is what is your value of $p(GG\,|\,g)$, your probability that she has a family of two girls, given that she has at least one. This is clearly a case for Bayes rule, which says

$$p(GG\,|\,g) = p(g\,|\,GG)p(GG)/p(g). \qquad (12.1)$$

(See §6.3 with g for E, GG for F, and K omitted.) Here $p(g)$ is found by extending the conversation to include the constitution of the family, as in §9.1, giving

$$p(g) = p(g\,|\,GG)p(GG) + p(g\,|\,GB)p(GB) + p(g\,|\,BG)p(BG) + p(g\,|\,BB)p(B). \qquad (12.2)$$

Assuming your informant is truthful, $p(g\,|\,BB) = 0$. Almost as clearly $p(g\,|\,GG) = 1$ though you may consider a declaration of 1 daughter, when she has 2, impossible. By the Bernoulli theory, all the unconditional

12.4 THE PROBLEM OF TWO DAUGHTERS

probabilities on the right-hand side are $\frac{1}{4}$. Combining all these ideas, we have

$$p(g) = \frac{1}{4} + \frac{1}{4}p(g|GB) + \frac{1}{4}p(g|BG),$$

and (12.1) becomes

$$p(GG|g) = 1/[1 + p(g|GB) + p(g|BG)], \qquad (12.3)$$

the values $\frac{1}{4}$ cancelling from the numerator and denominator in (12.1). In words, your probability that the other child is also a girl, so that the family is GG, is 1 divided by the expression in square brackets [. . .]. There is nothing so far in the analysis to say what values you associate with the probabilities therein, $p(g|GB)$ and $p(g|BG)$. What is your probability that the lady will declare the existence of a daughter, rather than a son, when the family is of mixed sex, either GB or BG? Until these have been settled there is no unique, numerical answer to the sex of the other child. We investigate several possibilities.

1. A natural one is that if the family is of mixed sex, the lady is as likely to admit a son as a daughter, when $p(g|GB) = p(g|BG) = \frac{1}{2}$. In this case (12.3) yields $p(GG|g) = \frac{1}{2}$, in agreement with common sense.
2. Suppose that, at the time you receive the additional information about her daughter, you had your daughter with you and your informant knew this. Then you might feel that, in the case of a mixed sex family, the lady would be more likely to announce the existence of a daughter than a son. For example she might have said, "I have a daughter too". A possibility here is that you assess $p(g|GB) = p(g|BG) = 1$. In this case (12.3) yields $p(GG|g) = \frac{1}{3}$, in agreement with the book mentioned above.
3. Another possibility is that you think the lady refers to her elder child, when $p(g|GB) = 1$ but $p(g|BG) = 0$, taking us back to the $\frac{1}{2}$ of common sense.
4. There are societies in which only male children are respected. If the new information arose in such a society, you might feel that if the lady had a boy, she would proudly announce that fact. Then $p(g|GB) = p(g|BG) = 0$, so that $p(GG|g) = 1$. This last result

does not need probability, for it follows by logic and hence does not violate Cromwell's rule (§6.8).

There are several lessons to be learned from this apparently simple problem, which apply more widely to real-life problems of importance. The first is that here is a problem that is both well defined and in which there appears to be enough evidence to reach a unique answer. In fact this is not so. One cannot say that $1/2$ is right and $1/3$ wrong. Additional material is needed for a coherent response. Important problems often have this element of ambiguity that is not recognized, perhaps because at school we were taught that things are either "right" or "wrong". This is not so and pupils who are so taught are being misled.

Another feature of the two-daughter problem is that it is often not enough to appreciate what you know—one child is a girl—but it is also necessary to ask yourself how you know it. In the second possibility above, you came to know of one daughter because your daughter was with you; this affected your assessments of $p(g \mid GB)$ and $p(g \mid BG)$. As this is being written, a politician is reported as providing a fact, the equivalent of g here, only to be challenged by a rival "how do you know that?" The reply was that a lobbyist had told him, to which the rival responded that the fact might have been different had it come from a neutral source.

Another comment is more technical but is relevant to all statistical analyses of data, here the data is g. Once the genetic facts, such as $p(GG) = 1/2$, have been admitted, the resolution of the problem rests on four probabilities of g, conditional on the four possible genetic configurations, GG, GB, BG, and BB. This is most clearly seen in (12.2). Also (12.3) reveals that it is only the ratios of these four that matter. For example, if each value was halved, the result $p(GG \mid g)$ would not be affected. These four values constitute the likelihoods for the data. Recall that generally $p(E \mid F)$ as E varies, provides a probability distribution given F, whereas as F varies it provides the likelihood function for data E. We see from this little problem that the only contribution from the data g is the likelihood function for g. This is the likelihood principle: that the only contribution from the data to the analysis is the likelihood function for the data. As we shall see in §14.4, many statistical procedures violate the principle. Such violation would occur here if the solution used your probability that the lady would have announced a boy.

12.5 TWO MORE DAUGHTERS AND CARDANO

The problem to be considered here is the same as that in §12.4 except that your informant tells you her daughter's name, Helen, in addition to her existence. The question remains: what is your probability that the other child is also a girl? The common sense response is that name does not affect the sex of the other child, so that the answer is the same as that given previously. Some writers have argued that this conclusion is wrong and have provided different answers from the ones they gave for the original problem of §12.4. One writer, who gave $1/3$ as the answer there, claimed that $1/2$ was now the correct answer. We investigate these claims again using the full force of the probability calculus.

The notation remains the same except that g, for the information about the girl, is replaced by gh, a girl called Helen. We then have two pieces of information, g for a girl and h for Helen, so that the event gh is the conjunction (§5.1) of g and h. The argument of §12.4 up to equation (12.3) can be used in our new problem, with g there everywhere replaced by gh. In particular the four probabilities $p(g \mid GG)$ and so on in (12.2) are replaced by $p(gh \mid GG)$ and so on; that is, a different likelihood is required. We saw that the previous analysis required careful thought and several possibilities for the likelihood were explored, following (12.3). Consider one of the new probabilities $p(gh \mid GB)$ and use the multiplication rule of probability to obtain

$$p(gh \mid GB) = p(g \mid GB)p(h \mid gGB). \qquad (12.4)$$

To help appreciate this, replace E, in the statement of the rule (5.5), by g, and F there by h; K there is replaced by GB in addition to the original genetics background knowledge. In (12.4) $p(g \mid GB)$ is familiar from the original form of the problem. The new consideration is $p(h \mid gGB)$, your probability that the daughter is called Helen, given that the family is GB and "the lady has told you about the daughter". It is the part in quotes that distinguishes *gGB* from *GB* considered in §12.4. This probability requires careful thought, thought that is not a question of mathematics but of your uncertainty about a girl's name. Contrast $p(h\,|\,gGB)$ with $p(h\,|\,g)$, where your uncertainty is still about the girl's name but you are unsure about the genetic constitution of the family. It seems reasonable to me that these two probabilities are the same; that the genetics does not influence the name of the girl whose existence you have been told about. You, dear reader, may disagree and feel that Helen is a more common name for girls when they are firstborn than a later addition to the family. You are entitled to that belief but it seems unjustified to me, so I will assume that Helen has the same chance of occurring for a girl wherever the family has a girl. That is

$$p(h\,|\,gGB) = p(h\,|\,g), \tag{12.5}$$

or, in the terminology of §4.3, h and *GB* are independent, given g. If so it seems sensible to say $p(h\,|\,g)$ is the frequency of "Helen" among all girls, or rather your assessment of that chance (§7.8). Notice that (12.5) is a new assumption; it does not logically follow from the assumptions already made.

Exactly the same reasoning applies when the family is *BG*, the order of the births being irrelevant, so that (12.5) is acceptable when *GB* is replaced by *BG*. The case of two boys *BB* does not arise since the supposition of truth makes the event *gBB* impossible. The case of two girls does require further consideration because there are two possibilities for the name "Helen" to be present in the family. I judge this irrelevant because you are saying something about the name of the girl you have been told about, not about whether the family of two girls has one called Helen. If this irrelevance is accepted as a further assumption, (12.5) continues to hold when *GB* is replace by *GG*. Putting these three results into (12.4) we have

$$p(gh\,|\,GB) = p(g\,|\,GB)p(h\,|\,g),$$

and similarly for *BG* and *GG*. Consequently the relevant form of the right-hand side of (12.2), with g replaced by gh, is the same as (12.2) except that every term is multiplied by $p(h\,|\,g)$. It follows that $p(gh) = p(g)p(h\,|\,g)$ on the left-hand side.

12.5 TWO MORE DAUGHTERS AND CARDANO

Where does this take us? Recall that the mathematics here is the same as that in §12.4 except that g is everywhere replaced by gh. Hence Bayes rule here is, following (12.1) with the replacement,

$$p(GG\,|\,gh) = p(gh\,|\,GG)p(G)/p(gh).$$

But $p(gh\,|\,GG)$ has been shown to be $p(g\,|\,GG)p(h\,|\,g)$ and now $p(gh) = p(g)p(h\,|\,g)$, so that $p(h\,|\,g)$ cancels from the numerator and denominator on the right-hand side, so that finally $p(GG\,|\,gh) = p(GG\,|\,g)$. In words, knowledge of the girl's name does not affect your probability that the other child is a girl. Remember some assumptions have been made in reaching this conclusion. For example, if you thought firstborn girls were never called Helen, then you might not agree with the conclusion. Recall also that the original calculation of $p(GG\,|\,g)$ in §12.4 presented several scenarios. The results here are not "right" and alternatives "wrong", rather they may be the most acceptable to most readers. While the mathematics is common to all of us, some would say objective, the beliefs are your concern and they do not necessarily agree with mine, they are subjective.

Before leaving the problem of the two daughters, it is worthwhile asking how it happened that the author of a best seller thought that the knowledge of one daughter's existence changed the probability that the other was also a girl from $1/2$, the genetic value, to $1/3$, and then learning the name, returned it to $1/2$. This is not because the beliefs were different from mine but because he replaced Bayes rule by an older method, usually credited to Cardano two centuries before Bayes. Let us explore Cardano's method and see why it can often go wrong.

Suppose you are faced with an uncertain situation with a number of exclusive and exhaustive events (§9.1) that are all equally uncertain. The two-daughter problem, before the lady gives you any information, provides an example with GG, GB, BG, and BB each with probability $1/4$. Next suppose you are given some information that could not possibly arise were a particular one of the events true, so ruling that event out. This happens in our problem with the information g that one child was a girl, when $p(g\,|\,BB) = 0$, so ruling out BB as a possibility. Then Cardano's method says that you are left with only three possibilities, still exclusive and exhaustive, so each has probability $1/3$. Hence $p(GG \mid g) = 1/3$, whereas $p(GG) = 1/4$. Generally, with one possibility ruled out, the other possibilities have their equal probabilities increased equally to make their total one. This is Cardano's rule.

To see that the method can fail, apply Bayes rule for both the event GG (12.1) and the event GB, giving

$$p(GG \mid g) = p(g \mid GG)[p(GG)/p(g)] \text{ and}$$
$$p(GB \mid g) = p(g \mid GB)[p(GB)/p(g)].$$

Next notice that the two ratios that have been put in square brackets [...] are equal, since $p(GG) = p(GB) = 1/4$ and, as a result, if we take the ratio of the two left-hand sides, these terms will cancel and we are left with

$$\frac{p(GG \mid g)}{p(GB \mid g)} = \frac{p(g \mid GG)}{p(g \mid GB)}. \tag{12.6}$$

According to Cardano's rule the ratio on the left-hand side of (12.6) is 1, the original value $p(GG)/p(GB)$ before evidence g was received, and as a consequence, so is the right. But we saw that the right-hand side could take several values. In the first form considered in §12.4, we had $p(g \mid GG) = 1$, because the lady could not truthfully say she had a boy, and $p(g \mid GB) = 1/2$, because she could equally have said she had a boy. There the ratio is 2 and $p(GG \mid g)$ is twice $p(GB \mid g)$ in violation of Cardano. The rule would be correct in the second form in which you had your daughter with you, where $p(g \mid GG) = p(g \mid GB)$ is reasonable. Remember it is not only *what* you know but also *how* you know it that matters. In general Cardano's rule only works when, in addition to ruling out one case, the data has equal likelihood in all the others.

We won't go into details, because they can be somewhat complicated, but similar arguments apply when you are given the girl's name; or that she was born on a Tuesday, as arose in a radio program. In natural situations, like the first in §12.4, the name or the day make no difference but if you had introduced your daughter and given her name, Helen, to the lady, then Cardano may be sound but only Bayes is 100%

reliable. This is because Bayes follows from the three basic rules of probability (§5.4), Cardano's does not.

12.6 THE TWO ENVELOPES

You are presented with two indistinguishable envelopes and told truthfully that one of them contains twice as much money, in the form of checks payable to you, as the other. You open one of them at random and find it contains an amount of money that will be denoted by C. Like the problem of the three doors, you are invited to change and receive the contents of the unopened envelope instead of the contents C of the opened one. To decide what to do, consider the following argument: the unopened envelope either contains $½C$ if you were lucky in your choice, or $2C$ if unlucky. Since you chose at random, each of these possibilities has probability $½$, so that your expected return (§9.3), were you to change, is

$$½ × ½C + ½ × 2C = 5C/4,$$

which exceeds the current amount you have of C and therefore you should change. But this seems ridiculous because it implies that whatever envelope you select, you expect the other to contain more money. The world is always better the other side of the fence. What has gone wrong?

Again we need some notation, in addition to C, the amount in the opened envelope, and write L for the event that you have opened the envelope with the larger amount therein; L^c is then the event that you have selected the smaller amount. L is uncertain for you and your probability after you have opened an envelope is $p(L|C)$. Now $p(L)$ is $½$ as your choice was at random but is it true that $p(L|C) = ½$ as was assumed in the last paragraph? Again we calculate it by Bayes rule.

$$p(L|C) = p(C|L)p(L)/p(C), \qquad (12.7)$$

with $p(C)$ given by the extension of the conversation as

$$p(C) = p(C|L)p(L) + p(C|L^c)p(L^c). \qquad (12.8)$$

Now if the envelope with the smaller amount has been chosen and it contains amount C, then it logically follows, with no uncertainty, that the other envelope contains $2C$, hence $p(C|L^c) = p(2C|L)$. Since $p(L) = p(L^c) = \frac{1}{2}$, (12.8) becomes

$$p(C) = \frac{1}{2}[p(C|L) + p(2C|L)],$$

and (12.7) yields

$$p(L|C) = \frac{p(C|L)}{p(C|L) + p(2C|L)}. \qquad (12.9)$$

This is $\frac{1}{2}$, the value adopted in the previous paragraph, only if

$$p(C|L) = p(2C|L); \qquad (12.10)$$

that is, if you think that the envelope containing the larger amount is equally likely to contain an amount C, as twice that amount, $2C$. If you really feel this, then you should always change.

But do your probabilities satisfy equation (12.10)? To answer the question it is necessary to leave the narrow world of little problems and escape into the reality where you have been presented with this delightful offer. In that context, suppose, when you open the chosen envelope, you find C is rather larger than you had anticipated. Then $p(C|L)$ is rather small because you had not anticipated such a big check and $p(2C|L)$, for twice that amount, is even smaller. As a result, $p(L|C)$, from (12.9), exceeds $\frac{1}{2}$ and you feel it more probable that the opened envelope is the one with the larger amount rather than the smaller, so there is a case for retention. Let us put in some numbers; suppose $p(C|L) = 0.1$, only 1 chance in 10, or odds against of $9 - 1$, and $p(2C/L) = 0.01$. Then $p(L|C)$, from (12.9) is 10/11 and the content of the unopened envelope is $\frac{1}{2}C$, with probability $p(L|C)$, and $2C$ with the complementary probability $p(L^c|C) = 1 - p(L|C)$, giving your expected value for the contents of the unopened envelope to be

$$\frac{10}{11} \times \frac{1}{2}C + \frac{1}{11} \times 2C = \frac{14}{22}C.$$

Since this is seriously less than C, the amount you have already, you should not change envelopes.

In contrast, suppose C is very small in your view, so that $p(C|L)$ is small, then $p(2C|L)$ will typically be larger and, from (12.9), $p(L|C)$ will be less than $\frac{1}{2}$ and change seems sensible. The values of $p(C|L)$ for different values of C form a distribution (§9.2), and a more complete analysis than has been given here shows that you will typically have a distribution such that there is a unique value of the contents C of the opened envelope, such that if C exceeds this value, you should not change but if less than it, then change is advisable and you expect to do better by changing. Imagine you had initially anticipated getting about 10 dollars, then seeing 20 dollars would encourage you to stay whereas 2 dollars would suggest a change; surely an appealing resolution.

Though this problem is artificial, the analysis has consequences for other situations that do occur in practice, which are technically too complicated to present here, so that it is worthwhile to explore our scenario a little further. The naïve argument used in the first paragraph of the discussion, expressed in the notation of the second paragraph, claims that $p(L|C) = \frac{1}{2}$ for all C, and therefore from (12.9), $p(C|L) = p(2C|L)$ for all C. The amount C is an uncertain quantity (§9.2) and it often happens that when people are asked about an uncertain quantity for which they have little information, they will respond with a phrase like "I haven't a clue". (Here we are in descriptive, rather than normative, mode.) When pressed to be more precise, they might formulate their near ignorance by saying that every value of the uncertain quantity has the same probability for them. Even more interestingly, experienced statisticians routinely make the assumption that all values of some uncertain quantities are equally probable, sometimes openly but often tacitly. To them, ignorance of an uncertain quantity means all values have the same uncertainty. They do this despite the fact that, as in our envelope example, taking the values to be equally probable can lead to unsatisfactory conclusions and generally to incoherent analyses. To be fair, it is often a good approximation to a coherent analysis and a little incoherence can be forgiven. The true situation is that you are never ignorant about an uncertain quantity that is meaningful to you and some values of it are more probable than others. You may well have difficulty in saying exactly how much more probable but equality of all values is not a realistic option. The discussion in §12.9 is relevant.

12.7 Y2K

During 1999, and even earlier, there were concerns that computer systems would fail on January 1, 2000 because they identified years by the last two digits only and might therefore confuse 2000 with 1900. The feature was termed the millennium bug and denoted Y2K. As a result of the fears, computer programs were investigated, any Y2K defects hopefully found and removed. January 2000 duly arrived and nothing happened; computers worked satisfactorily. Some people congratulated computer experts on removing the bug, while others said it had all been a con and that it was now clear the bug had not existed. This is clearly a problem of uncertainty concerning the bug's existence, so let us see what probability has to contribute.

Denote by B the event that the bug existed in 1998, say, before the remedial action was contemplated, and by $p(B)$ your probability then, dependent on some unstated knowledge base. The decision was taken to act, A, and the result was F, a world essentially free of computer troubles in 2000. The immediate question is how is your uncertainty about the bug's existence affected by initiating A and obtaining the reaction F; what is your $p(B|AF)$? This is easily found using Bayes rule with data F and conditioning everything on A, with the result

$$\frac{p(B|AF)}{p(B^c|AF)} = \frac{p(F|BA)}{p(F|B^cA)} \times \frac{p(B|A)}{p(B^c|A)}, \qquad (12.11)$$

using the odds form of §6.5, with the required uncertainty on the left. Consider the probabilities on the right. Since action A, without its consequences, will not affect your uncertainty about the bug, $p(B|A) = p(B)$. If the bug does not exist, B^c, a new century free of trouble will surely result, so $p(F|B^cA) = 1$. Consequently, (12.11) says that the effect of A and F is to multiply the original odds of B by $p(F|BA)$. Let us look at this uncertainty, your probability of no trouble when the bug exists and remedial action has been applied. There are two extreme possibilities.

In the first, you think that the remedial action is thorough and that any bad effects of the bug will likely be removed. Then $p(F|BA)$ is near 1 and your original odds for B are multiplied by a number near 1, so that they are scarcely altered by the action and the outcome. The effect is only slightly to diminish your uncertainty about the bug, so that you are none the wiser as a result of a trouble-free 2000.

The second extreme possibility is that you have a low opinion of software engineers and anticipate trouble in 2000 whatever they do. Then $p(F \mid BA)$ is small and the original odds for the bug's existence are substantially diminished. Despite the action, which you consider inefficient, there has been no trouble, so the explanation is that the bug did not exist.

People who wrote to the press in January 2000, saying that because life was free of trouble, the bug did not exist and the whole thing was a con, could hold that view only if they have a low opinion of software engineers. This is an example of coherence. People with the contrary view about software, hardly alter their views about the bug and the happy outcome provides them with little, or no, information about whether the bug existed.

There is an aspect of the above analysis that deserves more attention. We said the initiation of action A would not affect your opinion of B and put $p(B \mid A) = p(B)$. This is reasonable provided it is the same "you" making all the uncertainty judgments. To illustrate, consider a large business deciding whether or not to act against the bug; they are the "you" in the language here but, in order to avoid subsequent confusion, refer to them as "they". Thus $p(B)$ is their probability that the bug exists. If $p(B \mid A)$ is similarly "theirs", then it will be $p(B)$. In contrast, consider your probability for B, and later you learn that the large business has initiated action A, then you may well change your probability for B, arguing that if the business has acted, perhaps the bug is more probable than you had thought. The difficulty arises because two probabilists are involved, "them" and "you". The whole question of your using their uncertainties is a tricky one and consequently will not be discussed.

12.8 UFOs

There are people who are thought of as cranks but often they are merely people who have probabilities somewhat different from the majority.

One might anticipate that they are incoherent, so let us take a look at a group whom many consider cranks and investigate their incoherence. There are some who think our Earth has, during the past 50 years, been visited by aliens from space, leading to the presence of unidentified flying objects (UFOs). About 1000 claimed cases of UFOs have been witnessed by them. As a result of the publicity these claims have received, and perhaps also because of the real importance such a visit might have on our civilization, a scientific investigation was carried out with the result that only about 20 cases were found to be lacking a simple, natural explanation. Of the 20, none led to a confirmed alien visit being classed as "doubtful". As a result of this, UFO watchers were excited, saying that the existence of these 20 cases supported their contention. Is this coherent? The analysis that follows is carried out for n cases investigated with r found to be doubtful and $n - r$ confirmed as natural. In this way the effect of changing numbers can be assessed.

Before the investigation begins, it seems sensible to regard the n cases as exchangeable (§7.3), so that they are, to the scientists, a Bernoulli series with chance θ, say, of any one of them being explicable as a natural phenomenon. Furthermore, θ will have a distribution $p(\theta)$ on some knowledge base. Naturally the scientist's distribution may differ from that of the UFO watchers, so will be left unspecified for the moment. If a particular case is natural, chance θ, the result of the investigation is either to classify it as "natural" or to leave it as "doubtful". Again it is reasonable to assume exchangeability and suppose there is a chance α, say, of the mistake of a natural phenomenon being classed as doubtful. On the other hand, if a case is truly that of a UFO, it can either be correctly classified or thought doubtful. For a third time, exchangeability will be invoked with a chance β, say, of the mistake of a UFO being classed as doubtful. That a natural phenomenon be classed as a UFO, or vice-versa, will be supposed impossible. With these assumptions in place, there are four possibilities with their associated chances:

Truly natural, classed as natural	$\theta(1 - \alpha)$
Truly natural, but doubtful	$\theta\alpha$
UFO correctly classified	$(1 - \theta)(1 - \beta)$
UFO cast as doubtful	$(1 - \theta)\beta$.

12.8 UFOs

In explanation of the chances, consider the second situation of two events both occurring, being natural and being classed as doubtful. By the multiplication rule of §5.3, this is the chance of being natural, multiplied by the chance of classification as doubtful, given that it is truly natural; chances that are respectively θ and α, hence $\theta\alpha$ as stated.

Although there are four possibilities, the second and fourth, in both of which the judgment is doubtful, cannot be observed separately, so the data reduce to three possibilities with their chances now listed, together with the observed numbers of each:

Natural	$\theta(1-\alpha)$	$n-r$
Doubtful	$\theta\alpha + (1-\theta)\beta$	r
UFO	$(1-\theta)(1-\beta)$	0.

The chance of doubtful is obtained by the addition rule, taking into account the two ways that a doubtful result can arise. There were no confirmed sightings of UFOs. As with a Bernoulli series, the n separate studies are independent, given the parameters θ, α, and β, so that the chance of the set of results, which is identifiable as your probability given the parameters, is

$$[\theta(1-\alpha)]^{n-r} \times [\theta\alpha + (1-\theta)\beta]^r. \qquad (12.12)$$

(Anything raised to power 0 is 1.) This complicated expression is your likelihood function for θ, α, and β given the data $(n-r, r, 0)$ and its multiplication by your original probabilities for the three chances and division by your probability of the data gives, by Bayes rule, your final probabilities for them.

The situation is complicated, so let us make a simplifying assumption that the two chances of misclassification are the same, $\alpha = \beta$, and see what happens then, returning to the general case later. The chance of a doubtful conclusion in the second table above is now $\theta\alpha + (1-\theta)\alpha = \alpha$, irrespective of θ. The likelihood (12.12) then becomes

$$\theta^{n-r}(1-\alpha)^{n-r}\alpha^r. \qquad (12.13)$$

This likelihood has to be multiplied by your original probabilities for θ and α. It is reasonable to assume that these are independent, the former referring to the true state of affairs, the latter to the scientific procedure. Independence means that the joint probability factorizes (§5.3) into that for θ, times that for α; but the likelihood (12.13) similarly factorizes and when the two are multiplied, as required by Bayes, the product form persists, so that θ and α remain independent after the data are taken into consideration and θ, the parameter of interest, may be studied separately from α, which is not of interest. Ignoring α, your probability for θ, given the data, is therefore

$$\kappa \theta^{n-r} p(\theta),$$

where $p(\theta)$ is your original uncertainty, κ a number, and the effect of the data is to change your opinion by multiplying by θ, $n - r$ times. The number r of doubtful observations, to which the watchers attached importance, is irrelevant. The value of κ can be found by noting that the sum of the probabilities over all values of θ must be 1; or alternatively you can forget κ by comparing one value of θ, θ_1, with another, θ_2, in the ratio

$$(\theta_1/\theta_2)^{n-r} \times p(\theta_1)/p(\theta_2).$$

In this ratio form, take a case where θ_1 is larger than θ_2 and specifically where the former is twice the latter so that the ratio, occurring here, is 2. The result of each natural observation is to multiply $p(\theta_1)/p(\theta_2)$ by 2, so that

$$\frac{p(\theta_1 \mid N)}{p(\theta_2 \mid N)} = 2 \frac{p(\theta_1)}{p(\theta_2)},$$

where N means natural. Each further natural observation provides another doubling. Thus 10 naturals will multiply your initial opinion ratio by 1024, with the result that your revised probability for θ_1 is enormously greater than that for θ_2. Consequently, the large values of θ, near to 1, will have their probabilities increased substantially in comparison with the small values nearer to 0. Here is a numerical example of 10 values of θ with their initial probabilities, and their final values after 10 natural conclusions.

θ	0.1	0.2	0.3	0.4	0.5	0.6	0.7	0.8	0.9	0.99
$p(\theta)$	0.1	0.1	0.1	0.1	0.1	0.1	0.1	0.1	0.1	0.1
$p(\theta\|D)$							0.02	0.08	0.25	0.65

(10 equally spaced values of θ are taken, though 0.99 replaces the dogmatic 1.00; D denotes data of 10 natural observations and the probabilities unstated are all zero to two decimal places.) It will be seen that, as a result of the data, θ is almost surely not below 0.7 and the only really credible values are 0.9 and 0.99. This is with $n - r = 10$; the data quoted above had $n - r = 980$, with the result that θ must be very close to 1. More detailed analysis shows that θ, the chance of a natural explanation, then has probability about 0.95 of exceeding 0.997 and it may be concluded that very few, if any, aliens have arrived.

All the assumptions made above are reasonable, at least as good approximations, except for one, that the two errors that result in a doubtful classification are equal, $\alpha = \beta$. It is sensible to think that a situation that is truly UFO related is more likely to be classed as doubtful, chance β, than a natural phenomenon remaining doubtful, chance α. If this is so, it is necessary to return to (12.12), where the factorization of terms in θ from those in α and β, that was used above, does not obtain. Without the factorization, the number r of doubtful sightings becomes relevant and it is necessary to use methods of calculation that are more technical than can be contemplated here. The conclusion, using them, is that the same effect, of making your probability distribution of θ concentrate near 1, continues to hold but that the concentration is rather less dramatic. For example, with 980 cases classed as natural and 20 doubtful, the effect is similar to 40 doubtful under the assumption $\alpha = \beta$. Thus the conclusion that UFO visitations, if they occur at all, are extremely rare, persists.

12.9 CONGLOMERABILITY

In §5.4 it was explained that everything in the probability calculus follows from the three rules, except for the little matter of conglomerability, which we promised to discuss further. This we now do, the delay arising because the example of the envelopes in §12.6 is the first occasion where the notion is relevant. This section can be omitted, but it does provide some little insight into the difficulties mathematicians encounter when they introduce infinities, like the infinity of the integers 1, 2, 3, ... continuing forever. Infinity is such a useful concept that it cannot be jettisoned; nevertheless, it does require careful handling.

Suppose that you think the events $E_1, E_2, \ldots, E_n$, finite in number, are exclusive, only one of them can be true, and exhaustive, one of them must be true; then your probabilities for them, $p(E_i)$, must add to 1 and they are said to form a partition (§9.1). Consider another event F; your probability for it can be found by extending the conversation to include the E's and to do so will involve taking the products $p(F \mid E_i)p(E_i)$ for each E_i and adding all n, to obtain $p(F)$ as in §9.1. Now suppose that $p(F \mid E_i)$ is the same for each E_i and denote their common value by κ. The products will be $\kappa p(E_i)$ and their sum will be κ since the $p(E_i)$ add to 1. Consequently, for a finite partition and an event F, if your probability for F is the same conditional on each member of the partition, it has the same value unconditionally: $p(F \mid E_i) = \kappa$, for all E_i, implies $p(F) = \kappa$. This property is termed conglomerability and will be defined more precisely below. The question we now address is whether this need be true for an infinite partition. The surprising answer is "No". Here is an example of the property failing.

Suppose you are told that the amounts of money in the envelopes in §12.6 could be 1,2,3, ... without limit, in terms of a unit such as a penny, and that you think all values are equally probable. Consider the partition of these values into

$$(1\ 2\ 4) \quad (3\ 6\ 8) \quad (5\ 10\ 12) \quad \ldots, \text{and so on.}$$

Thus E_1 is the event of obtaining either 1, 2, or 4 pennies on opening the envelope. The rule here is clear, the odd values are each assigned one to a triplet, whereas the even values go in pairs to the triplets. Notice that such a partition is not available for a finite number of consecutive integers since you would run out of even numbers before all the odd ones had been introduced. Let G be the event of obtaining an even number of pennies. Then the assessment $p(G \mid E_i) = 2/3$ follows since E_i contains twice as many even values as odd and all values have the same probability for you. If conglomerability held for this infinite partition, it would follow that $p(G)$ is also 2/3.

Next consider another partition, $F_1, F_2, F_3, \ldots$

$$(1\ 3\ 2) \quad (5\ 7\ 4) \quad (9\ 11\ 6) \quad \ldots, \text{and so on,}$$

with the roles of odd and even reversed from the other partition. The same argument will give $p(G|F_i) = 1/3$, and, if conglomerable, $p(G) = 1/3$ in contradiction with the other partition. It is therefore impossible that the result for a finite partition holds for these particular infinite partitions. On the other hand, if $p(G|E_i) = \kappa$ for all E_i, it seems compelling that $p(G) = \kappa$. There is a natural connection with the sure thing principle of §10.5 in that if whatever happens (whatever event of the partition obtains) the result is the same, that result must be true overall. There is another important reason for thinking the result should hold. If it does, most difficulties in probability involving infinities are resolved. Care still needs to be exercised but no contradictions are known to arise. Notice that, in the example here, it was assumed that all amounts of money were equally probable and that the same assumption, Equation (12.10) of §12.6, led to anomalies with the two envelopes. It was seen that the envelope problem could be resolved by abandoning this assumption and the same feature applies here.

The upshot of the discussion is that it is usual to introduce a fourth rule of probability, which cannot, as the above example shows, be deduced from the three others. It cannot be derived from our standard without some assumption concerning an infinity of balls. The rule is therefore one of mathematical convenience that impinges on reality through examples like that of the envelopes. It remains only to state the rule precisely.

Conglomerable rule. If, on knowledge base K, the events $E_1, E_2, \ldots$ are, for you, exclusive and exhaustive; and F is another event that, for you, has $p(F|E_i) = \kappa$, the same for every E_i, then $p(F) = \kappa$.

12.10 EFRON'S DICE

We have often used, as an example of coherence, the assertion that if A is preferred to B, and B to C, then necessarily A is preferred to C. In

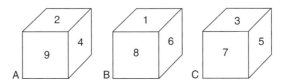

FIGURE 12.1 Efron's dice

§10.2 it was used for consequences and implied in §3.3 for events, when comparisons were made with urns. The preference is said to be transitive, the first two preferences being transferred to the third. Here we show, by a clever example, that the transitive property does not always obtain with some types of preferences.

Figure 12.1 shows three dice. These are the same as normal dice, solid cubes all of the same size, except that instead of having the numbers 1–6 on the six faces, they have the numbers shown in the figure. For faces not visible, the number thereon is the same as that on the opposite face. Thus for the left-hand die, labeled A, the bottom face bears the number 2, the left-facing one bears 4, and so on. The dice are supposed to be fair in the sense that when properly thrown, each face has the same probability, 1/6, of appearing uppermost.

Now suppose that you possess die A, while your opponent has die B, and that you both throw them once, randomly and independently with the winner being the one with the higher of the two scores on the upper, exposed faces. What is your probability that you will win? This will also be the probability for B that he will lose, since both have made a judgment of randomness for their own and their opponent's throws.

Table 12.1 is a table of the 9 possible outcomes, where the columns refer to A and the rows to B. The entries in the table are W for a win for you with die A, or L for your loss. Thus if you throw 2 and your opponent 1, then you win, the entry W in the top, left-hand corner.

TABLE 12.1 Competition between die A and die B

			A	
		2	4	9
B	1	W	W	W
	6	L	L	W
	8	L	L	W

Each entry has probability 1/9, since each outcome for A has probability 1/3, similarly for B, and they are independent so the multiplication rule applies. The table reveals 5 wins for you with A against only 4 losses. By the addition rule, your probability of winning is 5/9. In summary, die A is better than die B when they are in opposition. Similar calculations show that B is better than C, and that surprisingly C is better than A. This is the opposite of the notion with preference expressed in the first sentence of this section. The first two judgments of "better than" do not transfer across to the last judgment which, had it done so would have been that A is better than C. You would not use the magician's wand of §10.2 to replace B by A because your opponent might arrive with C.

This beautiful paradox of the dice does not impinge on the use of "transitive" employed in this book, essentially because A, B, and C are not consequences in the sense we have used the term. Suppose there was a decision tree (§10.10) with A at the end of a branch. Is A good or bad? It is good if your opponent has B but bad if they have C. How does B compare with A? It is good with C but bad with A. So it does not make sense in this situation to say that A is a good outcome. No die can be at the end of a decision tree, for there is still your uncertainty about what your opponent has. If another branch is added to the tree to include your opponent, the end of that branch will have, for example, A for you and B for them, with value 5/9, or A for you and C for them, value 4/9. In the enlarged tree every branch will end in either 4/9 or 5/9, when the fact that the latter exceeds the former is all that matters.

There are three lessons to be learned from Efron's dice. The first is that one should be very careful when making assumptions, especially when they have important consequences, as in decision making. The second teaches us to be aware of analogies. In debates, people often say one situation is analogous to another, and then go on to say that what is known to be true in one situation is true in the other. The dice show that preferences among consequences are not analogous to preferences among dice. Nothing beats the faculty of reason; analogy is no substitute.

The third lesson affects several criticisms that have been leveled against the use of numbers. It is often said, especially among those without experience of the scientific method, that one cannot put numbers to abstract concepts, like "happiness", so that our use of utility as a measure of concepts is flawed. The objection fails to recognize that utility applies to material consequences, to outcomes that you might experience, and not to abstract ideas. It may be that happiness is an ingredient of the outcome whose utility is being assessed, but we do not view it in isolation. Utility considers the whole package, the whole of the consequence. All the time we are in reality with consequences that you might experience. The outcome is concrete, not abstract.

CHAPTER **13**

Probability Assessment

13.1 NONREPEATABLE EVENTS

It has been shown in Chapter 7 how you may assess your probabilities in many cases using classical ideas of equiprobable outcomes or, more often, by employing frequency concepts. Historically, these have been the most important methods of assessment and have led to the most valuable applications. However, there remain circumstances where neither of these ideas is relevant and resort has to be made to other methods of assessment; to other methods of measuring your uncertainty. For example, if you live in a democracy, the event that the political party you support will win the next election is uncertain, yet no equiprobable cases or frequency data exist. It is clearly unsound to argue that because over the past century your party has been successful only 22% of the time, your probability of success now is around 0.22, for elections are not usually judged exchangeable. No really sound and tested methods exist for events such as elections and as a result, this chapter is perhaps the most unsatisfactory in the book. What is really needed is a cooperative attack on the problem by statisticians and psychologists. Unfortunately, the statisticians have been so entranced by results using frequency, and the psychologists have concentrated on

---

Understanding Uncertainty, Revised Edition. Dennis V. Lindley.
© 2014 John Wiley & Sons, Inc. Published 2014 by John Wiley & Sons, Inc.

328 PROBABILITY ASSESSMENT

valuable descriptive results, that a thorough treatment of the normative view has not been forthcoming. What follows is, hopefully, not without value but falls short of a sound analysis of the problem of assessing your probability for a nonrepeatable event.

The treatment makes extensive use of calculations using the three basic rules of probability. Readers who are apprehensive of their own mathematical abilities might like to be reminded that those rules only correspond to properties of proportions of different balls in an urn (§5.4) so that, if they wish, they can rephrase all the calculations that follow in terms of an urn with 100 balls, some of which, corresponding to the event A below, are red, the rest, A^c, white, while some are plain corresponding to B and others spotted for B^c. With a little practice, probabilities are easier to use, but the image of the urn is often found simpler for the inexperienced. An alternative strategy would be to write computer programs corresponding to the rules and use them. But initially it is better to experience the calculations for yourself rather than indulge in the mystique of a black box, however useful that may ultimately turn out to be.

Suppose you are contemplating a nonrepeatable, uncertain event, which we will refer to as A. You wish to assess your probability $p(A)$ for the event on some knowledge base that will be supposed fixed and omitted from both the discussion and the notation. Because readers are interested in different things and, even within a topic, have divergent views, it is difficult to produce an example that will appeal to all. The suggestion is that you take one of the examples encountered in Chapter 1 to help you think about the development that follows. Perhaps the simplest is Example 1, the event A of "rain tomorrow", with "rain on the day after tomorrow" as the event B introduced later. With one event being contemplated, the only logical constraints on your probability are convexity, that it lies between 0 and 1, both extremes being excluded by Cromwell, and that your probability for the complement A^c is $1 - p(A)$. In practice, you can almost always do better than that because some events are highly improbable, such as a nuclear accident at a named power plant next year, whence $p(A)$ is near 0. Others are almost certain and have $p(A)$ near to 1. In both these cases, the difficult question is how near the extremes are. Other events are highly balanced, as might be the election, and therefore $p(A)$ is nearer $\frac{1}{2}$. Generally, people who are prepared to cooperate and regard the assessment as worth thinking about are willing to provide an interval of values that seem reasonable for them. Suppose that you feel your probability $p(A)$ for the event A lies between 0.5 and 0.7 but are reluctant to be more precise than that. This is not to say that smaller or larger values are ruled out, but that you feel them rather unreasonable. It is this willingness to state an interval of values that has led to the concept of upper and lower probabilities in §3.5, an avenue not explored here, preferring the simplicity of the single value, if that can be assessed, for reasons already given.

13.2 TWO EVENTS

With a single event, this seems about as far as you can go and the attainment of a precise value, such as 0.6, let alone 0.6124, is beyond reach. You can think about your probability of the event being false, but this is so naturally $1 - p(A)$ that this scarcely helps. However, if a second related event, B, is introduced, the two other rules of probability, addition and multiplication, come into play and since, with convexity already used, you have all the basic rules upon which all others depend, there is a real opportunity for progress, essentially because coherence can be exploited in full. As has been seen in §4.2, with two events, A and B, there are three probabilities to be assessed, $p(A)$ already mentioned and two others that express your appreciation of the relationships between the events in the form of your probabilities for B, both when A is true and when it is false. These are $p(B|A)$ and $p(B|A^c)$ that, together with $p(A)$, completely express your uncertainty about the pair of events. Each of these probabilities can take any value between 0 and 1, irrespective of the values assumed by the other two. Again, in practice, people seem to be able to give intervals within which their probabilities lie. In the table, which will frequently be referred to in what follows, an example has been taken in which you feel $p(B|A)$ lies between 0.2 and 0.3, while $p(B|A^c)$ lies between 0.6 and 0.8. These values appear in the top, left-hand corner of the table and imply that the truth of A leads you to doubt the truth of B in comparison with your opinion when A is false, A^c is true. In the language of §4.4, you think the two events are negatively associated.

As an aside, it would be possible for you to procced differently and to contemplate four events derived from the two original ones, namely,

$$A \text{ and } B, \quad A \text{ and } B^c, \quad A^c \text{ and } B, \quad A^c \text{ and } B^c,$$

the last, for example, meaning that A and B are both false. This partition would lead to four assessments, which must necessarily add to 1, so to only three being free for you to assess, as with the method in the last paragraph. However, the partition is generally not as satisfactory as the method we go on to use because it only exploits the addition rule, in adding to 1, whereas ours uses the multiplication rule as well. Nevertheless, the choice is yours; you may be happier using the partition and be prepared to sacrifice numerical precision for psychological comfort, which is far from absurd. Moreover, from the partition values, you can calculate the conditional probabilities using the multiplication rule.

Returning to the position where you have made rough assessments for $p(A), p(B|A)$ and $p(B|A^c)$, we recall from §4.2 that it would be possible for you to contemplate the events and their probabilities in the reverse order, starting with $p(B)$ and then passing to the dependence of A on B through $p(A|B)$ and $p(A|B^c)$, these values being determined from the first three by the addition and multiplication rules, so that no new assessment is called for. To see how this works, take the midpoints of the three interval assessments already made and consider what these intermediate values imply for your probabilities when the events are taken in the reverse order. Recall, from the table, that the three intermediate values are

$$p(A) = 0.60, \quad p(B|A) = 0.25, \quad p(B|A^c) = 0.70$$

listed as (13.1) in the table.

$p(A)$	0.5 to 0.7	0.60	(13.1)	0.49	(13.4)	0.58	(13.6)	0.58 (13.7)
$p(B\|A)$	0.2 to 0.3	0.25		0.31		0.29	$\to$	0.26
$p(B\|A^c)$	0.6 to 0.8	0.70		0.55		0.61	$\to$	0.65
		$\downarrow$		$\uparrow$		$\uparrow$		$\downarrow$
$p(B)$		0.43	(13.2)	0.43	(13.3)	0.43	(13.5)	0.42 (13.8)
$p(A\|B)$		0.35		0.35	$\to$	0.40		0.36
$p(A\|B^c)$		0.79	$\to$	0.60	$\to$	0.72		0.74

The rule of the extension of the conversation in §5.6, here from B to include A, enables $p(B)$ to be found,

$$\begin{aligned}p(B) &= p(B|A)p(A) + p(B|A^c)p(A^c) \\ &= 0.25 \times 0.6 + 0.7 \times 0.4, \quad \text{using} \quad p(A^c) = 1 - p(A) \\ &= 0.15 + 0.28 = 0.43.\end{aligned}$$

Bayes rule (§6.3) enables your view of the dependence of A on B to be found.

$$\begin{aligned}p(A|B) &= p(B|A)p(A)/p(B) \\ &= 0.25 \times 0.6/0.43 = 0.15/0.43 = 0.35,\end{aligned}$$

$p(B)$ coming from the calculation just made. Similarly,

$$\begin{aligned}p(A|B^c) &= p(B^c|A)p(A)/p(B^c) \\ &= 0.75 \times 0.6/0.57 = 0.45/0.57 = 0.79,\end{aligned}$$

where the result, that your probability for the complement of an event is one minus your probability for the event, has been used twice. We repeat: if your probabilities had been

$$p(A) = 0.6, \quad p(B|A) = 0.25, \quad p(B|A^c) = 0.7, \quad (13.1)$$

then necessarily

$$p(B) = 0.43, \quad p(A|B) = 0.35, \quad p(A|B^c) = 0.79, \quad (13.2)$$

and you have no choice in the matter; this is coherence using the full force of the rules of probability. These implications, with the numbering of the equations, are shown in the table following the arrows.

You may legitimately protest that you did not state values originally but gave only ranges. True, and it would be possible to calculate intervals, using the rules of probability, for the new assessments, but this gets a little complicated and tedious, so let us just stay with the intermediate values (13.1) and their implications (13.2), not entirely forgetting the intervals. With these implications available, you can think whether they seem sensible to you. Alternatively, you could, before doing the calculations above that lead to (13.2), assess reasonable ranges for the probabilities in (13.2). Again, we will omit these complications and ask you to consider the values in (13.2) produced by straight calculations from (13.1).

In the hypothetical example, suppose that you consider the value for $p(A|B^c)$ at 0.79 to be excessively high, feeling that 0.60 is more sensible, but that the other two probabilities in (13.2) are reasonable. Then with

$$p(B) = 0.43, \quad p(A|B) = 0.35, \quad p(A|B^c) = 0.60, \qquad (13.3)$$

you may reverse the process used above, with Bayes rule and the extension of the conversation, to obtain the implication

$$p(A) = 0.49, \quad p(B|A) = 0.31, \quad p(B|A^c) = 0.55 \qquad (13.4)$$

in lieu of (13.1). The calculations are left to the reader and the results are displayed in the table following the arrow. Now these implications are disturbing, for each of the values in (13.4) lie outside your original intervals, the first two only slightly but the last more seriously. It therefore looks as though the shift of $p(A|B^c)$ from 0.79 in (13.2) to 0.60 in (13.3) is too extreme and requires amendment. Looking at (13.2) again, suppose you feel that the dependence of A on B that they express is too extreme, your probability of A changing from 0.35 to 0.79 according as B is true or false. Perhaps you were correct to lower the latter but that the same effect might be better achieved by raising the former and lowering the latter rather less, leading to

$$p(B) = 0.43, \quad p(A|B) = 0.40, \quad p(A|B^c) = 0.72, \qquad (13.5)$$

in place of (13.3).

Now you can apply Bayes rule and the extension to calculate the new implications for your original probabilities with the results

$$p(A) = 0.58, \quad p(B|A) = 0.29, \quad p(B|A^c) = 0.61, \qquad (13.6)$$

shown in the table. Again comparing these with your original intervals, you notice that all the values in (13.6) lie within them, which is an improvement on (13.5), but that both the conditional probabilities are at or near the ends of their respective intervals, which suggests bringing them in a little to

$$p(A) = 0.58, \quad p(B\,|\,A) = 0.26, \quad p(B\,|\,A^c) = 0.65, \qquad (13.7)$$

leaving $p(A)$ unaltered. Bayes and the extension imply

$$p(B) = 0.42, \quad p(A\,|\,B) = 0.36, \quad p(A\,|\,B^c) = 0.74, \qquad (13.8)$$

all of which are shown in the table.

13.3 COHERENCE

If we stand back from the numerical details and consider what has been done in the last section, it can be seen that, starting from a triplet of probabilities (13.1), each of which can freely assume any value in the unit interval, the implications for another triplet (13.2) have been calculated using coherence. This new triplet can be amended according to your views and the calculations reversed, with the events interchanged, leading back to new values for the original triplet. If that amendment does not work, another one can be tried and its implications tested. This process of going backward and forward between the two triplets of probabilities will hopefully lead to a complete sextet that adequately expresses your uncertainties about the two events, as we suppose (13.7) and (13.8) to do in the example. The key idea is to use coherence to the full by employing all three of the basic rules of probability, achieving this coherence by a series of adjustments to values that, although coherent, do not adequately express your uncertainties. Essentially, you look at the solution from two viewpoints, of A followed by B, and then B followed by A, until both views look sound to you. This section is concluded with a few miscellaneous remarks on the procedure.

The method just described uses two related events, A and B, but it can be improved by including a third event C. Contemplating them in the order A, B, and then C, the assessments with the first two proceed as above but the addition of C leads to four additional probabilities

$$p(C|AB), \quad p(C|AB^c), \quad p(C|A^cB), \quad p(C|A^cB^c),$$

each of which can freely assume any value in the unit interval. This requires seven assessments in all, three original and four new ones. There are six possible orders in which the three events can be contemplated, namely,

$$ABC, \quad ACB, \quad BAC, \quad BCA, \quad CAB, \quad CBA,$$

leading to passages backward and forward between them and vastly increased possibilities for exploiting coherence. This extension is naturally much more complicated but, with the help of computer programs that use Bayes rule and the extension of the conversation, is not unrealistic.

This method for probability assessment is analogous to that used for the measurement of distances, at least before the use of satellites, in that several measurements were made, surplus to the minimal requirements, and then fitted together by coherence. For distances, coherence is provided by the rules of Euclidean geometry, replacing the rules of probability that we used. With two events, six probabilities were used instead of the minimal three. Coherence, ordinarily expressed through rules described in the language of mathematics, is basic to any logical treatment of a topic, so that our use is in no way extraordinary.

There are situations where the procedure outlined above is difficult to pursue because some uncertainties are hard for you to think about. For example, suppose event A precedes event B in time, when $p(B|A)$ and $p(B|A^c)$ are both natural, expressing uncertainty about the present, B, given what happened with A in the past, whereas $p(A|B)$ and $p(A|B^c)$ are rather unnatural, requiring you to contemplate the past, given present possibilities. The method is still available but may be less powerful because the intervals you ascribe to the unnatural probabilities may be rather wide. Notice however that there are occasions when the unnatural values are the important ones, as when A is being guilty of a crime and B is evidence consequent upon the criminal act. The court is required to assess the probability of guilt, given the evidence, $p(A|B)$ or $p(G|E)$ in the notation of §6.6.

The coherent procedure can be simplified by the use of independence, though it is rather easy to misuse this elusive concept. For example, in considering three events, it might be reasonable to assume that A and C are, for you, independent, given B, so that $p(A|BC)$ reduces to $p(A|B)$ and others similarly, thereby reducing the number of probabilities to be assessed. The danger lies in confusing your independence of A and C, given B, with their independence, given only your knowledge base (see §8.8). There is one situation where independence has been used with great success in contemplating events that occur in time or space. Here we discuss only the temporal case. Let $A_1, A_2, \ldots$ be similar events that occur on successive days, thus A_i might be rain on day i. Then the natural, and ordinarily important, uncertainties concern rain today, given rainfall experience in the past, for example, $p(A_5 | A_4 A_3^c A_2^c A_1)$, your probability for rain on day 5, Thursday, given that it also rained on days 4 and 1, Wednesday and Sunday, but not on days 3 and 2, Tuesday and Monday. An extreme possibility is to assume that the past experience from Sunday to Wednesday does not affect your uncertainty about Thursday, when we have the familiar independence and the Bernoulli series of §7.4 if, in addition, $p(A_i)$ is the same for all values of i. A more reasonable assumption might be that today's rain depends only on yesterday's experience and not on earlier days, so that, in particular, the above probability becomes $p(A_5 | A_4)$. The general form of this assumption is to suppose that, given yesterday's experience, here A_4, today's A_5 is independent of all the past, A_3^c, A_2^c, A_1, and even further back. Such a sequence of events is said to have the Markov property. Independence is an important, simplifying assumption that should be used with care. The Markov form has been most successful, producing a vast literature. It is a popular generalization of exchangeability because, by using various tricks, so many phenomena can be judged to have the Markov property.

Mention of the scientists' use of small and large worlds was made in §11.7. Similar considerations apply here in the use of coherence to aid your assessment of your probabilities. Essentially, the thesis expounded in this chapter is that your small world can be too small and, by enlarging it, you can better determine your uncertainties. Confining your attention to a single event, and its complement, may be inadequate so that your world is far too small to take advantage of the power of coherence. By adding a second, related event, you can use the full force of coherence in the larger world in the manner described in §13.2. Even this may not be enough and a third event may need to be included before your uncertainties can be adequately described in the yet larger world with three events. A striking example of this was encountered with Simpson's paradox in §8.2 where the relationship between disease and treatment could only be understood by including a third factor, sex. There is an unfortunate tendency these days for discussion to take place in too small a world with a possible distortion of the situation. As these words are being written, there is a discussion being conducted about crime, its nature, its prevention, and its punishment. Yet there is one factor commonly omitted, namely, poverty and the role it plays in the types of crime under consideration. Another factor that is possibly relevant is drug taking. There comes a point where the enlargement of your small world has to stop because the analysis becomes impossibly complicated. Scientists have often been most successful in finding worlds that are sufficiently small to be understood, often using sophisticated mathematics, but are adequate to make useful predictions about future data. Economists have perhaps been less successful. The achievement of a balance between the simplicity of small worlds, the complexity of large ones, and the reality of our world is a delicate one. The essence of the approach here is that you should not make your world too small when discussing uncertainty.

13.4 PROBABILISTIC REASONING

In Chapter 2 it was emphasized that the approach adopted in this book would be based on reason. This is perhaps contrary to the practice in most writing where, to use the language of §2.5, the result is more descriptive than normative. Now that uncertainty has been studied and probability developed as the reasoned way to study the phenomenon, we can go back and look at the implications that the development has on the reasoning process itself. Though the earlier discussion may have deplored the lack of reasoning in everyday life, there are occasions where it is used with advantage. Here is a simple example.

Economists might reason that, were the government to increase taxation, people would have less money in their pockets and so would reduce their spending; traders would suffer and a recession would result. This is surely a reasoned argument, though some may claim that the reasoning is at fault, but there is one thing wrong with the reasoning process itself in that it does not allow for uncertainty. In other words, the methodology is defective irrespective of any flaws in the economic reasoning. It is simply not true that the increase in taxes *will* result in a recession; the most that could be said is that it is highly probable that increased taxation will result in a recession. In the style developed in this book, the probability of a recession, given increased taxes, is large. Notice incidentally, the condition here is a "do" operation, rather than "see" (§4.7). Our contention is that reasoning itself, with the emphasis on truth and implication, can be improved by incorporating uncertainty, in the form of probability, into the process. As has been mentioned in §5.4, logic deals with two states only, truth and falsity, often represented by the values 1 and 0, respectively, so that $A = 1$ means that the event A is true. On the other hand, probability incorporates the whole unit interval from 0 to 1, the two end points corresponding to the narrower demands of ordinary logic. Essentially, the calculus of probability is a significant generalization of logical reasoning. To support this claim, an example of probabilistic reasoning now follows but, in presenting it, it must be pointed out that the emphasis is on the probability aspect, not on the economics that it attempts to describe. The probabilities that appear are mine; I am the "you" of the treatment. A statistician's task is to help the expert, here an economist, articulate their uncertainties and really "you" should be an economist. The style of the analysis is sound; the numerical values may be inappropriate.

13.5 TRICKLE DOWN

A thesis, put forward in the years when Britain had a government led by Mrs. Thatcher, and more recently by other right-wing politicians, was that if the rich were to pay less tax, the top rate of tax being lowered from about 80% to around 40%, the consequent increase in their net salaries would encourage greater efficiency on the part of the rich, thereby increasing productivity, and ultimately the poor would share in the prosperity. In other words, more money for the rich would also mean more for the poor. It was termed the "trickle-down effect". Although said with some assurance by the politicians, there is clearly some uncertainty present so that a study using probability might be sensible.

We begin by contemplating two events:

L: there is *less* tax on the rich,
R: the poor get *richer*.

A more sophisticated approach would refer, not to events, but to uncertain quantities (§9.2) measuring the decrease in tax and the increase in wages for the poor but, to avoid technical problems, we here consider only events. In these terms, the trickle-down effect can be expressed by saying that the probability of the poor gaining is higher if the top rate of tax is reduced, than otherwise. In symbols,

$$p(R \mid L) \text{ is greater than } p(R \mid L^c)$$

for a "you" who believes in the effect. Since the effect must operate through the gross domestic product (GDP), the conversation is extended to include the event

G: the GDP increases by more than 2%,

during some period under consideration. Technological advances can account for a 2% increase whatever government is in power, so the best that the changes to taxation could achieve is an increase beyond 2%. With three events, L, R, and G, we are ready to introduce probabilities. The events arise in the natural order, L first, which affects G and then the poor share in the increase, R; so the events are taken in that order. L is an act, a "do", and has no uncertainty.

According to the reasoning used by the government, the event L of less tax will result in an increase in GDP, event G. Inserting the

13.5 TRICKLE DOWN

uncertain element, the firm assertion is replaced by saying G is, for you, more probable under L than under L^c. Suppose you think about this and come up with the values

$$p(G|L) = 0.8 \quad \text{and} \quad p(G|L^c) = 0.4. \tag{13.9}$$

The next stage is to include the poor through the event R. First consider the case where the GDP does increase beyond its natural value, event G, and contrast the two cases, L with the tax reduction, and L^c without. For a fixed increase in GDP, the rich will consume more of it with L than with L^c because in the former case they will have more money to spend, with the result that the poor will benefit less under L than with L^c. Essentially, the poor's share will diminish under L because the rich have the capacity to increase theirs, recalling that this is for a fixed increase in GDP. However, both groups will probably do well because of the increase in prosperity due to the higher GDP. Putting all these considerations together suggests that the values

$$p(R|GL) = 0.5 \quad \text{and} \quad p(R|GL^c) = 0.7 \tag{13.10}$$

reasonably reflect them, both probabilities being on the high side but that, given L, being the smaller.

Next pass to the case where the GDP does not increase beyond its natural value, event G^c. It will still probably remain true that, with the tax breaks, the rich will consume more of the GDP than if they had not had them, so that the poor will get less. On the contrary, neither group will do as well as with G because there is less to be shared. The values

$$p(R|G^cL) = 0.2 \quad \text{and} \quad p(R|G^cL^c) = 0.4 \tag{13.11}$$

might reasonably reflect these considerations.

It was seen in §8.8, that with three events, there are seven probabilities to be assessed in order to provide a complete structure. Here, one event, L, has no uncertainty, it is either done or not, so only six values have to be found and these are provided in (13.9), (13.10), and (13.11) above. The probability calculus can now be invoked and the conversation extended from the events of importance, R and L, to include G. First with the tax relief L

$$p(R|L) = p(R|GL)p(G|L) + p(R|G^cL)p(G^c|L)$$
$$= 0.5 \times 0.8 + 0.2 \times 0.2 = 0.4 + 0.04 = 0.44,$$

and then with L^c

$$p(R|L^c) = p(R|GL^c)p(G|L^c) + p(R|G^cL^c)p(G^c|L^c)$$
$$= 0.7 \times 0.4 + 0.4 \times 0.6 = 0.28 + 0.24 = 0.52.$$

As a result, you think that the poor will probably do better without the tax relief for the rich, 0.52, than with it, 0.44, and the probability development does not support the trickle-down effect.

The essence of the above argument is that if you include the GDP, then the poor are likely to have a smaller share of it if the rich get their tax breaks, whatever the size of the GDP. On averaging over values of the GDP, the reduction persists. Notice that what happens with both G and with G^c does not necessarily happen when the status of the GDP is omitted, as we saw with Simpson's paradox in §8.2, but here the values suggested in (13.10) and (13.11) do not lead to the paradox. (A reader interested in comparing the calculation here with that for Simpson may be helped by noting that R here corresponds to recovery, L to the treatment, and G to sex.)

Before leaving the discussion, let me emphasize the point made at the beginning, namely that the emphasis is on the methodology of the discussion and not on its economic soundness. It would be possible for the reader, acting as another "you", to replace some, if not all, of the probabilities used above, by other probabilities in order to produce an argument that supports the trickle-down effect. The discussion in this section, indeed throughout the book, is not intended to be partisan but only to demonstrate a form of reasoning, using uncertainty, intended to shed new light on a problem. One feature of the general approach is that it incorporates other considerations that may be relevant to the main issue. Here GDP has been included to relate tax on the rich to well-being of the poor. The tool here is coherence, fitting the uncertainties together in a logical manner. By being able to calculate other probabilities from those initially assessed, it is possible to look at different aspects of the problem. The inclusion of more features brings with it more opportunities to exploit coherence and more checks on the soundness of the uncertainties that have been assessed. More features involve more complexity, but the process only requires the three rules of probability. These are suitable for use on a computer. I envisage an analysis in which a decision maker, "you", thinks about some uncertainties, leaving the computer to calculate others. The presentation here has been in terms of beliefs but extends to action because utility is itself expressed in terms of probabilities as was seen in §10.2. The claim here is that we have a tool that enables you to both think and act, while a computer supplies checks on the integrity of your thoughts and actions.

13.6 SUMMARY

All the methods described in this chapter depend on the concept of coherence, of how your beliefs fit together. Indeed, it can be said that all the arguments used in this book revolve around coherence. With the single exception of Cromwell's rule, which excludes emphatic beliefs about events that are not logically proven, none of the material says what your beliefs should be; none of your probabilities are proscribed. There are many cases where it has been suggested that specific probabilities are rather natural, such as believing the tosses of a coin to be exchangeable; or based on good evidence, such as believing that a new-born child is equally likely to be of either sex. But there is no obligation on you to accept these beliefs, so that you can believe that, when you have tossed the coin 10 times with heads every time, the next toss will probably be tails, to make up the deficit; or a pregnant woman believes that the child she bears is male. Neither of these beliefs is wrong; the most that can be said is that they are unusual or incoherent.

At first sight, this extremely liberal view that you can believe what you like, looks set to lead to chaos in society, with all of us having different opinions and acting in contrary ways. However, coherence mitigates against this. We saw in the simple example of the red and white urns in §6.9 that, whatever your initial belief about the color of the urn, provided you updated this belief by Bayes rule, the continual withdrawal of more white balls than red would raise your probability that the urn was white to nearly 1, so that everyone would be in the same position, irrespective of any initial disagreements. Generally, if there are a number of theories, data will eventually convince everyone who has an open mind that the same theory is the correct one. It is our shared experiences that lead us to agreement. But notice that this agreement depends on your use of Bayes rule, or generally on coherence in putting all your beliefs together. Without coherence there is little prospect of agreement. I suggest that in coherence lies the best prospect of social unity on this planet.

In this chapter, we have not been nearly so ambitious, being content to argue that you should not contemplate beliefs and probabilities in isolation, but should always consider at least two beliefs so that the full force of the probability calculus may be used. Similarly, in decision making, it is important to fit all the parts of the tree together in a coherent way. The lesson of this book—

BE COHERENT.

CHAPTER 14

Statistics

14.1 BAYESIAN STATISTICS

There has been an extensive development of the ideas presented in this book within the field of statistics. Statistics (in the singular) is the art and science of studying statistics (in the plural), namely data, typically in numeric form, on a variety of topics. An early historical example was the collection of data on the production of wheat every year. Nowadays statistics covers any topic in which data are available, in particular within the scientific method. Once the data are available, statisticians have the task of presenting and analyzing them and, as a result, the activity of statistics (in the singular) has been developed. Within science, most of the effort has been devoted to models where, as we have seen, data, now denoted by x, are modeled in terms of a parameter ϕ, through a probability distribution of x for each value of ϕ, $p(x|\phi)$. A simple example is the measurement x of the strength ϕ of a drug, where there will be uncertainty because all people do not react in the same way to the drug. If x contains measurements on several people, the object of the investigation is to assess the strength ϕ; we say we want to make an inference about ϕ. A typical inference might be to say that ϕ is 0.56 ± 0.04 in suitable units.

Understanding Uncertainty, Revised Edition. Dennis V. Lindley.
© 2014 John Wiley & Sons, Inc. Published 2014 by John Wiley & Sons, Inc.

In the probability system developed in this book, the problem of inference about a parameter is easily solved, at least in principle, by introducing a prior distribution $p(\phi)$ for the parameter. This expresses your uncertainty about ϕ based solely on background knowledge that will remain fixed throughout the inference and therefore omitted from the notation. With the acquisition of data x, the posterior distribution $p(\phi \mid x)$ may be calculated using Bayes rule (§6.3):

$$p(\phi \mid x) = p(x \mid \phi) p(\phi) / p(x). \qquad (14.1)$$

In view of the central role played by the rule, this treatment of data is termed Bayesian statistics. The distinguishing feature is that the model is extended by the introduction of the prior, the justification being the general one that uncertainty, here of ϕ, is described by probability. Recall (§9.1) that the denominator in (14.1) is found by extending the conversation from x to include ϕ, which means adding the numerator of the rule over all values of ϕ under consideration. It is important to recognize that the posterior distribution $p(\phi \mid x)$ provides *all* the information you have about the parameter, given the data and background knowledge. The inference is complete and there is nothing more to be said. However, a distribution is a complicated concept and you may wish to extract features from it, if only for ease of comprehension. An obvious example is your expectation of ϕ, that is, the mean of the distribution of the parameter. Another popular choice is the spread of the distribution (§9.5), which is valuable in telling you how precise is your information. It is usual to refer to the expectation as an estimate of the parameter. There are other possible estimates; for example, with income distributions, which typically have some very high values, the median of the distribution, where your probability of exceeding it is $1/2$, may be preferred. Any feature of the distribution that helps you appreciate your uncertainty about the parameter can be employed. These are questions of comprehension, not logic.

One feature of Bayesian statistics is worth noting because many popular statistical procedures do not possess it. It is clear from the rule that once you have the data x, the only aspect of it that you need to make the complete inference is the function $p(x \mid \phi)$, for the fixed x seen, and all values of ϕ. If you doubt the truth of this statement because of the appearance of $p(x)$ in the denominator of the rule, recall that $p(x)$ is calculated by extending the conversation from x to include ϕ, so still only including the terms $p(x|\phi)$ for the observed data. This result is commonly stated as the following:

Likelihood principle. The only contribution that the model and the data make to your inference about the parameter is contained in the likelihood function of the parameter for the actual data observed.

Notice that this refers solely to the data's contribution to the inference; the other ingredient in Bayes rule is your prior for the parameter. What the likelihood principle requires you to do is to compare your probability of the data for one value of the parameter with its value for another, in essence, the likelihood ratio (§6.7). This is another example of the mantra that there are no absolutes, only comparisons. The principle was encountered in the two-daughter problem (§§12.4 and 12.5) where the omission of the likelihood in the formulation of the problem made it ambiguous.

In many scientific cases it is desirable to include more than one parameter in the model. In an example in the next section, it is necessary to include the spread of the data distribution, as well as the mean. With two parameters, ϕ and ψ (psi), the whole of the previous argument goes through with

$$p(\phi, \psi \,|\, x) = p(x \,|\, \phi, \psi) p(\phi, \psi) / p(x),$$

where $p(\phi, \psi)$ is your prior distribution for the two parameters. Again, in principle, this causes no difficulty because $p(\phi|x)$ is obtained by summing $p(\phi, \psi \,|\, x)$ over all values of ψ, essentially the marginal distribution of ϕ (§9.8). If ϕ is the parameter of interest, ψ is often called a nuisance parameter, and the nuisance can be eliminated by this summation. Extra nuisance parameters are often included because they make the specification of the model simpler, often by introducing exchangeability, a concept that is almost essential for an alternative approach to statistics studied in the next section. Sometimes it is useful to introduce many nuisance parameters and then have a distribution for them. The death-rate example of §9.10 provides an illustration, with ψ_i connected with region i. Then it is useful to have a probability distribution for these nuisance parameters. Models of this type are often termed hierarchical.

In principle, the Bayesian method is simple and straightforward. In practice there are mathematical difficulties mainly in the summation, over ϕ when calculating $p(x)$, and over ψ when eliminating a nuisance parameter. The development of Bayesian ideas was hindered by the lack of easy methods of summation, or what mathematicians term integration. However, with the arrival of modern computers, it has been found possible to do the integration and perform the necessary calculations to obtain your posterior distribution, and hence the complete inference. Aside from the mathematical difficulties, the usual objection to the Bayesian approach has been the construction of the probability distributions, especially the prior. Exchangeability is often available for the model distribution. It is the use of background knowledge to provide a prior that has led to most criticism. The rewards for having a prior are so great that the advantage usually outweighs the difficulty. Notice that the Bayesian method has two stages. In the first, you have to think about your probabilities. In the second, thinking is replaced by calculations that the computer can perform. In the next section, we present an example of a simple, but common use of the method. In the section on significance tests (§14.4), a further example of the use of Bayesian methods is provided.

14.2 A BAYESIAN EXAMPLE

In this section, a simple example of the Bayesian approach to statistics is examined that is often appropriate when you have a substantial amount of data, perhaps 100 observations. It is based on extensive use of the normal distribution (§9.9) and before reading further, the reader may wish to refresh their understanding of this distribution, so beloved by statisticians for its attractive properties, making it relatively easy to manipulate. Recall that it has two defining features, its mean and its spread. The maximum of the density occurs at the mean, about which the density is symmetric. The spread describes how far values can depart from the mean. The mean is alternatively termed the expectation of the uncertain quantity having that distribution. The spread is the standard deviation (s.d.). Figures 9.5 and 9.6 illustrate these features. The notation in common use is to say that an uncertain quantity has a normal distribution of mean ϕ and s.d. s, writing $N(\phi, s)$.

In the basic form of the example, the data consist of a single real number x, resulting from just one observation, which has a $N(\phi, s)$ distribution. Here ϕ is the parameter of interest and s is supposed to be known. The model might be appropriate if the scientist was measuring ϕ, obtaining the value x, using an apparatus with which she was familiar, so knew s from previous experiences with measuring other quantities using the apparatus. The method is often used where x has been extracted from many observations, since, due to a result in the calculus, such a quantity is often normally distributed. That describes the term $p(x \mid \phi)$ in Bayes rule. There is no need to refer to s in that rule because it is known and can therefore be regarded as part of your background knowledge. It remains to specify the prior, $p(\phi)$, which is also supposed to be normal, the mean denoted by μ(Greek 'mu'), and the spread by t, $N(\mu, t)$. In practice, μ will be your most reasonable value of ϕ, your expectation, and t will describe how close to μ you expect ϕ to be. If you know little about ϕ, t will be large, but if the data are being collected to check on a suspected value of ϕ, t will be small. With $p(x \mid \phi)$ and $p(\phi)$ both defined, all is ready to apply Bayes rule.

Theorem If x is $N(\phi, s)$ given ϕ and ϕ is $N(\mu, t)$, then the posterior distribution of ϕ, given x, is also normal with

$$\text{mean}: \quad \frac{x/s^2 + \mu/t^2}{1/s^2 + 1/t^2}, \qquad (14.2)$$

$$\text{spread}: \quad \frac{1}{[1/s^2 + 1/t^2]^{1/2}}. \qquad (14.3)$$

The most remarkable and convenient feature of the theorem is that normality of the data, and of the prior, results in the posterior of the parameter also being normal. This is part of the reason why normality is such an attractive property, for it is preserved under Bayes rule. That normality also frequently arises in practice makes the theorem useful as well as attractive. Unfortunately, it is not possible to prove that the theorem is correct with the limited mathematics used in this book (§2.9) because to do so would require an understanding of the exponential function. Throughout this book, I have tried to show you how the results, such as Bayes and the addition and multiplication rules, follow from some comparisons with the standard of balls from an urn. There you do not need to trust me, you can see it for yourself. Here I regretfully have to ask you to trust me.

The expression for the posterior mean (14.2) is best appreciated by comparison with Equation (7.2) where a posterior probability was a weighted average of data f, and prior assessment g, the weights being the number of observations n in the data and your prior confidence m in g. Here we again have a weighted sum; x is weighted by $1/s^2$, the prior μ by $1/t^2$. If the observation x is precise, s^2 is small and a lot of attention is paid to it. Similarly, if you are originally confident about μ, t^2 is small and it receives attention. In most practical cases, s^2 is smaller than t^2, so that most weight goes on x. Notice that it is the reciprocals of the squares of the spreads, $1/s^2$ and $1/t^2$, that provide the weights. There is no generally accepted term for them and we will refer to them as weights. With that definition, the formula (14.3) is readily seen to say that the posterior weight, the reciprocal of the square of the spread, is the sum of that for the data, $1/s^2$, and that for the prior $1/t^2$.

Suppose that a scientist was measuring an uncertain quantity and expressed her uncertainty about it as normal with expectation 8.5 but with a spread of 2.0. Recall from §9.9 that this means that she has about 95% probability that the quantity lies between $12.5 = 8.5 + 4$ and $4.5 = 8.5 - 4$, the values 4 being twice the s.d. or spread. Using the apparatus, assumed normal of spread 1.0, suppose she obtains the observation $x = 7.4$. Then the theorem says that she can now assess the quantity to have expectation:

$$\frac{7.4 + (8.5/4)}{1 + (1/4)} = 7.62.$$

The denominator here shows that the spread is $1/(1.25)^{1/2} = 0.89$. The expectation is a little larger than the direct measurement, which was lower than she had expected, but the increase is small since the apparatus spread was only half the spread of the prior view. As the prior view gets less reliable, the weight $1/t^2$ decreases, and the role of her prior opinion decreases, with the raw value of 7.4 ultimately accepted. Many statisticians do this as routine. According to Bayesian ideas, this is regrettable but here it is not a bad approximation unless you have strong prior knowledge. A possibility that can arise is that the observation x is well outside the prior limits for ϕ, here 4.5 and 12.5, even allowing for the spread s of x. If this happens, it may be desirable to rethink the whole scenario. Coherence cannot be achieved all the time. It is human to err; it is equally human to be incoherent.

It is common to take several measurements of an uncertain quantity, especially in medicine where biological variation is present. The above method easily extends to this case with interesting results. Suppose that n measurements are made, all of which are $N(\phi, s)$ and are independent, given ϕ. Then the likelihood function is dependent on the data only through their mean, the total divided by their number. In the language of §6.9, the mean, written $\bar{x}$, is sufficient. It is often called the sample mean to distinguish it from the true mean ϕ. (Another convenient property of normality.) We saw with the square-root rule in §9.5 that the mean has a smaller spread than any single observation, dividing the spread s by $\sqrt{n}$. It can then be proved that the theorem holds for n such observations with x above replaced by $\bar{x}$ and $1/s^2$ by n/s^2. Let us try it for the above numerical example with 10 measurements giving a mean of 7.4, the same as for a single measurement originally, so that increasing the number can be more easily appreciated. The posterior mean will be

$$\frac{(7.4 \times 10) + (8.5/4)}{10 + (1/4)} = 7.43,$$

and the spread $1/\sqrt{10.25} = 0.31$. The effect of increasing the number of measurements has been to bring the expectation down from 7.62 to 7.43, nearer to the observed mean of 7.4, because of the added weight attached to the mean, compared with that of a single measurement. Also, the posterior spread has gone down from 0.89 to 0.31. Generally, as n increases, the weight attached to the mean increases, whereas that attached to the prior value stays constant, so that ultimately the prior hardly matters and the posterior expectation is the mean $\bar{x}$ with spread $s/\sqrt{n}$ according to the square-root rule. This provides another example of the decreasing relevance of prior opinion when there is lot of data.

An unsatisfactory feature of this analysis with normal distributions is that the spread of the observations, s, is supposed known, since there is information about it through the spread observed in the data. The difficulty is easily surmounted within the Bayesian framework at some cost in the complexity of the math, which latter will be omitted here. It is overcome by thinking of the model, x as $N(\phi, s)$, being described, not in terms of one parameter ϕ, but two, ϕ and s. As a result of this, your prior has to be for both with a joint distribution $p(\phi, s)$. It is usually thought appropriate to suppose that ϕ and s are independent on background knowledge, so that the joint distribution may be written as $p(\phi)p(s)$. With $p(\phi)$ as before, it only remains to specify $p(s)$. Having done this, Bayes rule may be applied to provide the joint posterior distribution of ϕ and s, given the data. From this, it is only necessary to sum over the values of s to obtain the marginal distribution of ϕ, given the data. The details are omitted but a valuable result is that, for most practical cases, the result is essentially the same as that with known s, except for really small amounts of data. The previously known spread needs to be replaced by the spread of the data. As a result, save for small amounts of data, the posterior distribution of ϕ remains normal. When the normal approximation is unsatisfactory, the exact result is available. The treatment of this problem with s uncertain, though from a different viewpoint from that used here, was made a little over a century ago and was an early entry into modern statistical thinking. The exact distribution is named after its originator who wrote under the pseudonym "Student", and is well tabulated for all values of n.

Another example of the Bayesian method is given in §14.4 with a discussion of significance tests.

14.3 FREQUENCY STATISTICS

It is unfortunately true that many statisticians, especially the older ones, reject the Bayesian approach. They have two main, related reasons for the rejection: first that your prior $p(\phi)$ is unknown, second that it is your posterior, your inference, so that the procedure is subjective, with you as the subject whereas science is objective. These two objections are discussed in turn. It is certainly true that, prior to the data, ϕ is unknown, or as we would prefer to say, uncertain, but it would be exceptional not to know anything about it. In the relativity experiment (§11.8), the amount of bending could not be large, for if so, it would have been noticed, and the idea of it being negative, bending away from the sun, would be extraordinary. So that there was some information about the bending before the experiment. Indeed, what right has any scientist to perform an experiment to collect data, without understanding something of what is going on? Has the huge expenditure on the Hadron collider been made without some knowledge of the Higgs boson? In practice, scientists often use the information they have, including that about ϕ, to design the experiment to determine x; indeed it would be wrong not to use it, but then abandon that information when they make an inference about ϕ based on x. To my mind, this practice is an example of incoherence, two views in direct conflict. Some statisticians have tried to find a distribution of ϕ that logically expresses ignorance of ϕ but all attempts present difficulties.

The idea that nothing is known about the parameters before the data are observed, may be unsound, but those who support the idea do have a point: so often it is hard to determine your distribution for the parameters. We have seen examples of this difficulty, even in simple cases, in Chapter 13. Methods for the assessment of a single parameter are deficient; the assessment of several probabilities is more sensible, so that the full force of coherence may be exploited. The parameter ϕ in the model is part of the theory, so that there are opportunities to relate ϕ to other quantities in the theory. My view is that it would be more sensible to devote research into methods of assessment, rather than use, as many do, the incoherent techniques we investigate below. Another curious feature of this reluctance to assess a prior is the casual, almost careless, way in which the probability distribution of the data is selected; normality is often selected because of its simple properties. Sometimes there is a practical reason for choosing the data distribution using frequency considerations (§7.2). Many experiments incorporate some form of repetition, leading to exchangeability, and the concept of chance (§9.7). For example, with personal incomes, the subjects may be thought exchangeable with respect to income and their values plotted to reveal a distribution with a long tail to the right, corresponding to the few rich people.

The second objection to Bayesian ideas is that they are subjective. All the probabilities are your probabilities, whereas a great advantage of science is that it is objective. One counter to this is provided by the result that if two people have different views, expressed through different priors, then the use of Bayes rule, and the likelihood that they share, will ultimately lead them to agreement as the amount of data increases. We met a simple example of this in §11.6, when discussing the two urns, different initial odds converging to essentially equal odds with enough sampling from the urn. This is true rather generally in the Bayesian analysis, and scientists will ultimately agree, except for a few eccentrics; or else they accept that the model, or even the theory, is unsatisfactory. This is what happens in practice where a theory is initially a matter of dispute. At this stage, the science is subjective. Then the deniers collect data with a view to getting evidence against it, whereas the supporters experiment to see if the data support it. Eventually one side wins, with either the supporters admitting defeat, or the deniers accepting that they were wrong. In practice it is more complicated than these simple ideas suggest; for example, the theory may need substantial modification or be limited in its scope. Nevertheless, ultimately agreement will be reached and the subjectivity will be replaced by objectivity. Science is correctly viewed as objective when all but a few scientists agree. This can be seen happening now with climate change, though some of the skeptics are not scientists using Bayes, or other forms of inference, but groups who treat self-interest as more important than truth. A contrasting example is the geological theory of continental drift, which was for long not accepted but is now. Here, and in many other instances, data are not the only way to reach a definitive view of a theory; agreement can come through new ideas. In the case of continental drift, the new feature was an explanation of how the drift could happen.

Statisticians who reject the Bayesian approach, often for the reasons just discussed, still have probability as their principal tool but interpret it differently. The basic rules of convexity, addition and multiplication are used but thought of in a frequency sense (§7.2).

14.3 FREQUENCY STATISTICS

Thus, the use of $p(x|\phi)$ in a model with data x and parameter ϕ will be read as the frequency with which the value x will arise when the parameter has value ϕ. This view has substantial appeal when experiments are repeated in a laboratory, or when a sociologist records a quantity for each of a number of people. A consequence of this interpretation is that probability is only available for the data, never for the parameter, which is supposed to be fixed but unknown and cannot be repeated. We will refer to statistical methods based on this interpretation as frequency statistics. Most of the practical differences between the Bayesian and frequency views rest on the ubiquitous use of the likelihood principle by the former and its denial by the latter. There is even a frequency method called maximum likelihood that uses the likelihood function but denies the principle when it assesses how good is the result of using the method.

Nevertheless, the two attitudes do often, as with maximum likelihood, produce similar results, and practitioners of the frequency school have even been known to use a Bayesian interpretation of their results. For example, with a model of the type studied in the previous section with $p(x|\phi)$, a common frequency method is to provide a point estimate of ϕ, that is a function of the data x that is, in some sense, the most reasonable value for ϕ. In the method of maximum likelihood, the point estimate is that value of ϕ that maximizes the likelihood function $p(x|\phi)$ for the data x observed. Often the point estimate is the posterior mean, or very close to it. In the normal case of the last section, the point estimate is the mean $\bar{x}$ differing from the posterior mean only by the contribution from the prior, which will be negligible with a large amount of data. With many parameters, there can arise real differences between the results from the two approaches. An example was encountered in §9.10, where the apparently least-performing authority would have as its frequency point estimate the average over all relevant organizations within that authority, whereas the posterior expectation would be greater than that, and often substantially so, because of information about better-performing authorities.

In addition to point estimates, frequency methods use confidence intervals for ϕ, intervals in which they are confident the parameter really lies. The degree of confidence is expressed by a probability, namely the frequency with which the interval will include the true value on repetition of the procedure that produced the data. Again repetition replaces the Bayesian concept, repeats being with fixed ϕ. In interval estimation by maximum likelihood, the principle is denied. The violation of the likelihood principle is inevitable within the frequency viewpoint whenever uncertainty about a parameter is desired, since the only frequency probability concerns the data, the parameter being fixed but uncertain. Intervals of the posterior distribution can be used but they merely help interpret the inference, rather than being basic to it. Nevertheless, numerical agreement between the confidence and posterior intervals is often close, though gross discrepancies can arise.

One important way in which Bayesian and frequency views differ is in respect of "optional stopping". Suppose two drugs are being compared to see which is more effective. Matched pairs of patients are treated, one with drug A, the other with drug B, to see which is better. Finance is available for 100 pairs but after 50 have been compared it is clear that drug A is superior to drug B. Under these circumstances, it seems sensible to stop the trials, partly for financial reasons, but more importantly because it is wrong to continue to give the inferior drug to 50 further patients. The trial is therefore stopped and the limited data used to make an inference and perhaps reach a decision. Statisticians have then asked themselves if the data from optional stopping needs to be analyzed differently from the analysis that would have been used had the trial started with resources for only 50 trials and the same results been obtained.

To examine this question, consider a situation in which the trial data form a Bernoulli series (§7.4) with chance θ of "success", A being preferred to B, for each pair. In this context, consider two rules for terminating the series: in the first, the number n of trials is fixed; in the second, r is fixed, rather than n, and the series is halted when r successes have been observed. Suppose one experimenter fixes n, another fixes r, but that coincidentally the same results are obtained even down to the order of successes and failures. The relevant probability structures of the data are, for that with fixed n, $p(r \mid n, \theta)$, and with fixed r, $p(n \mid r, \theta)$.

The frequency method operates with the distribution of r in the first case, and n in the second, whereas the Bayesian needs only the observed values of r and n, using the likelihood $p(r, n \mid \theta)$ as a function of θ, which is common to both experimental procedures. We have seen

(§9.2) that in the first case, the distribution of r is binomial. The second distribution, whose details need not concern us here, is different. For example, with $n = 6$ and $r = 2$, we saw that the binomial extends from 0 to 6, whereas the other goes up from 2 without limit. It follows that the analyses within the frequency framework will be different for the two experiments, whereas the Bayesian, using only the likelihood, $\theta^r (1 - \theta)^{n-r}$, will employ identical analyses. Readers who have encountered these problems before do not need to be reminded that the likelihood may be multiplied by any constant without affecting the Bayesian inference. This is clear from the discussion at the end of §9.1.

14.4 SIGNIFICANCE TESTS

The most obvious differences of real importance between Bayesian and frequency methods arise with significance tests that are now examined in more detail. A reader not familiar with these tests may wish to consult §11.10 before proceeding. As explained there, in the framework of a model $p(x \mid \phi)$, one value of ϕ may be of special interest. For example, the theory may imply a value for ϕ, as in the relativity experiment in §11.8, or if ϕ corresponds to the strength of a drug being tested, then $\phi = 0$ implies that the drug is useless. It usually happens that the scientist thinks that the special value might obtain, in which case a Bayesian would assign positive probability to that event. We will suppose that special value to be zero. If, in the practical application, it was ϕ_0, redefine the parameter to be $\phi - \phi_0$. Our Bayesian interpretation of the situation, thought by frequentists to require a significance test, will therefore be described by saying $p(\phi = 0) = c$, where $c > 0$. Notice in the analysis of §14.2, when the parameter had a normal distribution, no value had positive probability since $p(\phi)$ is the density and $p(0)h$ is the probability that ϕ lies in a small interval of width h that includes $\phi = 0$ (§9.8). Within the Bayesian viewpoint, there are two classes of problem with any parameter that has a continuous range of values; in the first, the probability structure is defined by a density; in the second, there is a value of the parameter that has positive probability. It is the latter that gives rise to a significance test. The other case is sometimes referred to as a problem of estimation.

The particular significance test to be discussed has the same data structure as the estimation problem of §14.2, namely the data is a number x with model density $N(\phi, s)$ with s known. This will enable comparisons to be made between the posterior appreciations in the two cases, of tests and of estimates. As already mentioned, $p(\phi = 0) = c > 0$. The prior distribution of ϕ when it is not zero remains to be specified and again it will be supposed to be the same as in the estimation case, $N(0, t)$ centered on the value to be tested, with $\mu = 0$ in the earlier notation. Centering it on the value $\phi = 0$ to be tested means that, if not exactly 0, it is near to it, its nearness depending on t. It is usual to speak of the hypothesis H that $\phi = 0$. This leads to the complementary hypothesis H^c that ϕ is not zero. We refer to a significance test, or simply a test, of the null hypothesis H that $\phi = 0$ against the alternative H^c that $\phi \neq 0$. In this language, the prior distribution $N(0, t)$ is $p(\phi \mid H^c)$, the density of the parameter when the null hypothesis is false.

We are now in a position to use Bayes rule in its odds form (§6.5),

$$o(H \mid x) = \frac{p(x \mid H)}{p(x \mid H^c)} o(H), \tag{14.4}$$

replacing F there by H, evidence E by x and omitting reference to background knowledge. Two of the terms on the right-hand side are immediate, $o(H) = c/(1 - c)$ and $p(x \mid H)$ is $N(0, s)$, since under H, $\phi = 0$. It remains to calculate $p(x \mid H^c)$, which is done by extending the conversation from x to include ϕ, so that it consists of the sum, over ϕ, but excluding $\phi = 0$, of terms $p(x \mid \phi)p(\phi \mid H^c)$. The perceptive reader will query the omission of H^c from the first probability, explained by the fact that, given ϕ, H^c is irrelevant when $\phi \neq 0$. Unfortunately, this summation requires mathematics beyond the level of this book, so we must be content with stating the result, which is that x, with given H^c, is again normal with zero mean and spread $\sqrt{s^2 + t^2}$. This is another instance that makes normality so attractive. With both probabilities known, the likelihood ratio in (14.4) can be evaluated and the posterior odds found. These odds can be transformed into a posterior probability for H given x. It is this probability that we want to compare with the level of the significance test produced by frequency methods. Before doing this, we remark that, as in the estimation case of §14.2, we often have n data values, iid given ϕ. If so, the same result for the posterior odds persists with x replaced by the sample mean $\bar{x}$, which is still sufficient, and s by $s/\sqrt{n}$, using the square root rule (§9.5).

Recall that in a significance test, a small probability, usually denoted by α (alpha), is selected and if x departs significantly from the null value of the parameter, here zero, the result of the test is said to be significant at level α. In this sentence, "departs significantly" means that x falls in a region of values that, were H true, has probability α. The literature has extensive discussion as to how this region should be selected. Figure 11.1 illustrates the case where x is $N(0, s)$, where there is general agreement that the region is that shaded, where x differs from zero by more than a multiple of the spread s. We will take the case where $\alpha = 0.05$, where the multiple is 1.96, effectively 2. If this happens, the frequency view says that the data are significant at 5%, suggesting that an unusual result (falling in the region) has occurred were H true, so that H is probably not true. Here the word "probably" is used in a colloquial sense and not with the exact meaning used in this book. In what follows, it is supposed that x, or $\bar{x}$, is significant at 5%, in that it differs from zero by $2s/\sqrt{n}$, when we evaluate the posterior odds of H, given x or $\bar{x}$. One might hope that the frequency and Bayesian results would differ only a little, which would help to justify the near identification of "significant at 5%" with "posterior probability of 5%", an identification that is sometimes made even by experienced statisticians, and is often made by people using elementary methods. As this is being written, a scientist, writing in a newspaper, does just this. In fact we show that the identification hardly ever exists, even approximately; the two views are entirely different.

Table 14.1 gives the numerical value of the likelihood ratio in (14.4) when $t/s = 2$ and $\bar{x}$ is significant at 5%, for various values of n. It also gives your posterior probability $p(H\,|\,\bar{x})$ when your prior value was $\frac{1}{2}$, so that you originally thought that the null hypothesis had equal probabilities of being true or false. The most striking feature of Table 14.1 are the high values of your posterior probabilities; even when $n = 1$, it is a little over 33%. These are to be contrasted with the 5% significance, suggesting that H is not true. Your posterior values

TABLE 14.1 The likelihood ratio and posterior probability, when x is $N(\phi, s)$, H is $\phi = 0$, $p(H) = 1/2$, and $p(\phi \mid H^c)$ is $N(0, t)$ for n iid values giving mean $\bar{x}$ that yields a significance of 5%

n	$p(x \mid H)/p(x \mid H^c)$	$p(H \mid x)$
1	0.480	0.341
5	0.736	0.424
10	0.983	0.496
50	2097	0.677
100	2947	0.742

increase with n and even with $n = 100$, 74% is reached. It can be proved that with fixed significance level, here 5%, as n increases, your posterior probability will tend to one. In this situation, frequency ideas doubt the null hypothesis, whereas Bayes is nearly convinced that it is true. Here is a striking disagreement between the two schools of thought. The disagreement might be due to the particular prior used, though that is unlikely since other priors exhibit the same phenomenon. What is perhaps more serious is that, with a fixed significance level, here 5%, for different values of n, the posterior probabilities change. In this example, from 0.34 with $n = 1$, to any value up to the limit of one. This is an important illustration of incoherence, the phrase "significant at 5%" having different unacknowledged interpretations dependent on the number of observations that led to that level of significance. It is explained by the occurrence of the sample size, n, in the formula (14.5) below.

The significance test is one of the oldest statistical techniques and has repeatedly been used when the number of observations is modest, rarely more than 100. More recently it has been used with large numbers, for example in studies of extrasensory perception. Within the last decade, the test has been used with vast amounts of data that need to be analyzed by a computer. An example that has been widely reported in the world's press is the data collected in the search for the Higgs boson. Since the posterior probability for the null in Table 14.1 tends toward one, it could happen that the Bayesian analysis differed substantially from the frequency result that appeared in the press. The reporting of the physics in the press has necessarily been slight, so that the physics may be all right but the statistics is doubtful. The eclectic view that sometimes frequency methods can be used, yet in others Bayesian methods are appropriate, is hard to defend. My own view is that significance tests, violating the likelihood principle and using transpose conditionals (§6.1), are incoherent and not sensible.

These calculations apply only to the selected numerical values but similar conclusions will be reached for other values. The more serious criticism lies in the choice of the appropriate prior when a significance test is used. As far as I am aware, there is no prior that gives even approximate agreement with the test for all values of n. There is another aspect of these tests that needs to be addressed. They are often employed in circumstances in which a decision has to be taken, so that inference alone is inadequate. There Bayes has the edge because his methods immediately pass over to decision analysis, as we have seen in Chapter 10, whereas a significance test does not. As an example of the misuse of tests, take a situation where it is desired to test whether a new drug is equivalent to a standard drug, perhaps because the new drug is much cheaper. This is called a test of bioequivalence. The experiment is modeled in terms of ϕ, the difference in potency of the two drugs, with $\phi = 0$ as the null hypothesis H corresponding to bioequivalence. Often a significance test of H is used. This may be inadequate because ultimately a decision has to be made as to whether the new drug should be licensed. We have seen that this requires utility considerations (§10.2) leading to the maximization of expected utility (MEU). Omission of this aspect, as any inference does, would not reflect the operational situation adequately.

Finally, before we leave the critique of frequency statistics, and for the benefit of those who wish to extend the calculations given above, having at the same time, access to the exponential function, the formula for the likelihood ratio is provided. With the same notation as above, it is

$$(1 + n/\rho)^{1/2} \cdot \exp\left[-\tfrac{1}{2}\lambda^2/(1 + \rho/n)\right], \qquad (14.5)$$

where $\bar{x} = \lambda s/\sqrt{n}$ and $\rho = s^2/t^2$. Notice that the individual spreads are irrelevant, only their ratio s/t matters. In Table 14.1, $s/t = \tfrac{1}{2}$ and $\lambda = 1.9600$, corresponding to a significance level of 5%. Readers who would like to investigate the other two popular significance levels need $\lambda = 2.5758$ for 1% and $\lambda = 3.2905$ for 0.1%. Readers with a little familiarity with frequency statistics will recognize that two-tailed tests have been used throughout the discussion. The normal prior would be unsuitable for testing with a single tail, suggesting that $\phi \geq 0$. When n is large, the term ρ/n, within the square brackets of the exponential function, can be ignored in comparison with the 1 there. Similarly, the 1 before n/ρ in the square root can be ignored giving the good approximation

$$(n/\rho)^{1/2} \cdot \exp\left(-\tfrac{1}{2}\lambda^2\right).$$

Consequently the likelihood ratio increases slowly with the square root of n for fixed significance level, which determines λ, and fixed ρ. It is this dependence on n, the number of observations, that the tail area interpretation of the data ignores. 5% significance does not cast the same doubt on the null hypothesis for all n.

14.5 BETTING

In most circumstances, the presence of uncertainty is a nuisance; we prefer to know the truth and act without the impediment of doubt. Nevertheless, there are circumstances where uncertainty is enjoyed, adding to the pleasure of an experience, at least for some people. Two examples were mentioned in §1.2: card games (Example 7) and horse racing (Example 8), both of which involve gambling, in the popular sense of that term. In our development of uncertainty, we have used the term "gamble" to describe all cases of uncertainty (§3.1). We now look at the recreation of gambling and recognize that there are at least two types, distinguished by whether the probabilities are commonly agreed by all, or very much dependent on the gambler. Roulette is an example of the former where everyone accepts that if the wheel has 37 slots, the probability of the ball entering any one slot is 1/37, at least in an honest casino when the gambler has no lucky number. Horse racing provides an illustration of the second type because punters have different views on which animal will win the race, placing their bets accordingly. Roulette and similar games are referred to as games of chance because they concern independent repetitions of a stable situation that can have two results, success or failure, as discussed in §7.8. The rules of probabilities were historically first developed in connection with games of chance, where current advances therein have used mathematics to provide valuable results, somewhat outside the scope of this book. Here we investigate betting, where chances do not arise because the conditions are not stable, one race being different in several ways from another. Our emphasis here will be on bets in horse racing, though betting occurs in other contexts too, for example football. It has even been claimed that one can bet on any uncertain event.

Here betting will be discussed primarily in connection with horse racing as practiced in Britain where, for a given race, each horse is assigned odds before running and payment is made to the punter according to the odds displayed at the time the bet is placed. The odds are commonly presented in the form exemplified by 8 to 1, which means that if the person placing the bet, the bettor or punter, stakes 1 cent on a horse at these odds, they will receive 8 cents, and have their stake returned if the horse wins; otherwise the stake is lost. The set of odds for all the horses running in a race is called a book, and the person stating the odds is called a bookmaker, who accepts the stakes and pays the successful bettors. Odds are always of the form r to 1, where r is a positive number, 8 in the illustration. Readers may find it helpful to think of r as the leading letter of the reward to the bettor. In our discussion, we find it useful to work in terms of rewards, rather than odds. Often it is useful to refer explicitly to the event E, the horse winning, and write $r(E)$. Since the racing fraternity abhors fractions, a bet at odds of $2\frac{1}{2}$ to 1 will appear as 5 to 2. Odds such as 100 to 8, or 9 to 5 are not unknown. In writing, the word "to" is often replaced by a minus sign, thus 8−1 or 9−5, which can confuse. A mathematician would prefer to write simply 8, or $9/5 = 1.8$. At this point the reader might like to reread §3.8, where odds were introduced in connection with Bayes rule, the relationship with probability was examined, and the distinction between odds *against*, used by bookmakers, and odds *on* as used in this book was made.

In §3.8, the odds on were defined for an event E in terms of probability as

$$o(E) = p(E)/p(E^c)$$

in Equation (3.1). Had odds against been used, the ratio of probabilities would have been inverted to $p(E^c)/p(E)$. It is therefore tempting to write

$$r(E) = p(E^c)/p(E), \qquad (14.6)$$

for the reward if the event E is true, that the selected horse wins the race. The temptation arises because the racing fraternity, and bettors generally, speak of "odds against" where we have used "reward". It is often useful to do this and I personally prefer to turn the stated odds against into probabilities using the inverse of (14.6)

$$p(E) = 1/[1 + r(E)]. \qquad (14.7)$$

Thus, in this interpretation, odds against of 8 to 1 give a probability of 1/9 of the selected horse winning. However, while useful in some circumstances, the usage is dangerous because the left-hand side of (14.7) is *not* a probability; it does not satisfy the three basic rules, in particular the addition rule (§5.2). This result is now demonstrated for a race with only two runners but the argument is general and applies to realistic races with many runners, just as the addition rule extends to any number of events (§8.1).

Put yourself in the role of a potential bettor contemplating a two horse race in which the bookmaker has posted odds of r_1 to 1 and r_2 to 1, or in our terminology, rewards r_1 and r_2 for the two horses. The possibilities open to you are to place a stake of s_1 on the first horse and, simultaneously, stake s_2 on the second. Like the rewards, both stakes must be positive including zero, corresponding to not betting. Suppose you do this and the first horse wins, then you will be rewarded by $r_1 s_1$ but will lose your stake s_2 on the second horse, with an overall change in assets of $r_1 s_1 - s_2$. Similarly, if the second horse wins, the overall change is $r_2 s_2 - s_1$. At this point it occurs to you that perhaps, by judicious choice of the two stakes, it might be possible to make both changes positive; in other words, you win whatever horse wins. In the parlance of §10.5 you are on to a "sure thing". Clearly no bookmaker could allow this, for he would lose money for sure. All he can control are the odds, or rewards, so we might ask whether he can prevent you being on to a sure win.

For you to have a sure win, both changes in assets have to be positive,

$$r_1 s_1 > s_2 \text{ and } r_2 s_2 > s_1.$$

For this to happen, the ratio of your stakes must satisfy

$$1/r_1 < s_1/s_2 < r_2.$$

The left-hand inequality follows from the previous left-hand inequality, and similarly the right-hand one. It follows that you can only find stakes that will give you a sure win if $(1/r_1) < r_2$ or equivalently $r_1 r_2 > 1$; the product of the rewards (odds) must exceed one. The bookmaker would never allow this to happen since he could suffer sure loss against a wise punter if he did. He will always arrange for the product to be less than one. Recall that, from (14.6), $r_1 = (1 - p_1)/p_1$, where p_1 is the "probability" that the first horse wins, and similarly $r_2 = (1 - p_2)/p_2$, so that the bookmaker must ensure that

$$\frac{(1 - p_1)(1 - p_2)}{p_1 p_2} < 1.$$

Multiplying both sides of this inequality by $p_1 p_2$ and then subtracting this product from both sides reduces this to

$$p_1 + p_2 > 1. \tag{14.8}$$

We have two exclusive and exhaustive events—one and only one horse will win—so that if the p values were really probabilities, their sum would have to be one, in contradiction to (14.8). For a race with more than two horses, the extension of this result, which the readers can easily prove for themselves, is that the sum of the "probabilities" must exceed one to prevent a sure gain by a bettor. It follows that the left-hand sides of (14.7) cannot be termed probabilities. The reader may like to check this result for any book. As this is being written, a book is given for the result of a soccer match, with odds of 10 to 11 for a win by the home team, 3 to 1 for an away win, and 5 to 2 for a draw. The corresponding "probabilities" from (14.7) are 11/21, 1/4, and 2/7, adding to 1.06. The excess over one, here 0.06, determines the bookmaker's expected profit on the game.

An alternative expression for what has been established is that the bookmaker's odds cannot be a coherent expression of his beliefs about which horse will win the race. If not beliefs, what are they? They are numbers produced by the bookmaker that reflect the ideas of the people on the race course, or in betting shops, expressed in the stakes they have placed, and hopefully will give him the confidence to expect a profit.

We now turn to the behavior of the bettor when faced with a book for a race. The naïve approach is to look at what can be expected when a stake s is placed on a horse with reward r. If p is your, the bettor's, probability that the horse will win, then you have probability p of gaining rs and probability $1-p$ of losing s, so that your overall expected (§9.3) gain is $prs - (1-p)s$, which is equal to s times $pr - (1-p) = p(r+1) - 1$. You therefore expect to win if, and only if,

$$p > 1/(1+r),$$

that is whenever your probability exceeds the bookmaker's "probability", from (14.7). This can only happen with a few horses in the race since your probabilities coherently add to one, whereas the bookmaker's exceed one in total (14.8). Recognizing that the "probabilities" reflect the views of the crowd interested in the race, you should only bet, if at all, when your opinion of the horse is higher than that of the crowd.

The analysis in the last paragraph is not entirely satisfactory. One way to appreciate this is to study the expected gain, found above to be $s[p(r+1) - 1]$, which is positive if $p(r+1)$ exceeds one, but then increases without limit with s, the stake. In other words, the naïve analysis says that if the bet is to your advantage, you should stake all of your assets. This is hardly sensible since, with probability $(1-p)$, this would result in your losing your stake and hence everything. The analysis is deficient for the reason discussed in §10.4, where it is shown that the proper procedure for the decision over whether or not to bet is to maximize your expected utility. The utility for money was discussed in §10.13, and it is also mentioned in connection with finance in §14.6. It was also emphasized that it is your assets that enter the consequences, or outcomes, at the end of the tree needed to decide what stake, if any, to place; so it is your utility for your assets that is required to make a rational decision, not gains or losses.

The correct analysis of a betting situation of the type we have been discussing involves three different sums of money: your initial assets c before making the bet, your assets if you lose, $c - s$, and those if you win, $c + rs$. With p still your probability of winning, you should place a stake that maximizes over s your expected utility

$$pu(c + rs) + (1 - p)u(c - s), \tag{14.9}$$

provided the maximum exceeds $u(c)$. Here $u(x)$ is your utility for assets x. To discuss this further would involve detailed consideration of the utility function, which needs more mathematics than the rest of this book requires, and would also need to include problems that do not involve uncertainty, the book's main topic. Readers still interested may like to refer to Chapter 5 of my earlier book, *Making Decisions*, Wiley (1985). One complication that can be mentioned is the recognition that betting on a horse race, or other sporting fixture, is not entirely a question of money. Most of us experience more enjoyment in watching a race if we have a bet on a horse, than if we do not. We enjoy cheering on High Street even though we may often suffer disappointment. If so, perhaps $u(c + rs)$ in (14.9) should be increased to allow for the joy of winning, and $u(c - s)$ decreased. Recall that utility refers to consequences and the consequences of a race include joy and disappointment.

14.6 FINANCE

There has been a serious deterioration in economic prosperity in the United States and much of Europe since the first edition of this book in 2006, a change that has affected most of the world. Capitalism has wobbled, the wobble being mainly due to unwise investments by some bankers. An investment is a procedure in which a sum of money is placed in a situation where it will hopefully increase in value but may decrease. The outcome is uncertain, in the same sense as that used in this book, and the final yield of the investment is an uncertain quantity (§9.2). Uncertainty is therefore at the heart of the economic downturn, so it is relevant to ask whether the bankers responsible have used the methods outlined in this book. Unfortunately, the activities of bankers are shrouded in mystery, so that even the most inquisitive of investigators, such as government regulators, do not know the details of what goes on in banks. Consequently it may be difficult, if not impossible, to know whether bankers used probability in a sound manner. However, it is possible to shed some light on the mystery of what goes on in investment houses.

Business schools at universities throughout the world teach students, some of whom enter the banking profession and handle investments. One is therefore able to gain some insight into what these former students might do by reading the textbooks used in their instruction. These books have become more sophisticated over the last 50 years, reporting on research done in that period, much of which employs undergraduate mathematics. It has also been noticed that many graduates in math, where they would previously have gone into manufacturing, gain lucrative employment with investment houses. Put these facts together and there is a possibility that some light can be shed on the follies of investors by studying the textbooks. This view is supported by at least two recent essays by academics in which it is claimed that the use of erudite mathematics, based on unsatisfactory models, which are not understood by senior bankers, has contributed to the economic collapse. Let us then look at the methods described in the popular textbooks on finance.

When this is done, it soon becomes apparent that all the methods considered are based on an assumption that people, in particular bankers, are rational, whatever that means in an uncertain situation. Our thesis is based on coherence (§§3.5 and 13.3), which essentially means that all your activities fit together, like pieces in a well-made jigsaw, and that you will never experience two activities clashing, like pieces not fitting tightly, and where you would be embarrassed by having to admit an error. It came as a surprise to me when the books were examined, that there is rarely any reference to coherence, and the concept of probability is not adequately explored. A close examination shows that the methods, developed in detail and with powerful mathematics, are incoherent. In other words, in direct answer to our question above concerning the methods used by bankers, we can say that many of them are deficient in this respect. It is easy to demonstrate such incoherence; indeed it was first done in 1969 and the writing subsequently ignored. We now pass to a simple proof that one popular method is incoherent, using ideas developed in Chapter 9, especially §9.5.

An investment consists of an amount of money, placed in an activity now, in the anticipation that at some fixed time in the future, say a year, it will have increased in value, recognizing that it may alternatively decrease. The value of the money after a year will be termed the yield of the investment. We denote it by the Greek letter (§11.4) theta θ. At the time the investment is made, θ is an uncertain quantity for you, the investor. In the discussion that follows, the reader might find it useful to take the case of betting (§14.5), where a stake is invested in the outcome of an uncertain event such as a horse race. If the event occurs, the horse wins, then you have a reward and your stake is returned. If the contrary happens, you lose your stake. Here θ takes only two values, which are known to you at the time the bet is placed, at least when the odds are fixed. Thus, a 1dollar stake at 10−1 against will yield either 10 dollars or lose 1 dollar. This case, where θ takes only two values, will suffice to demonstrate incoherence of some, even most, textbook methods.

Returning to the general case, θ will have a distribution for you at the time the investment is made. As you will have appreciated from Chapter 9, a distribution is a complex idea involving many probabilities, so that it is natural to simplify the situation in some way. The obvious way is to consider the expectation of the investment or, what is usually called in this context, the mean yield (§9.3). Thus, in the example just mentioned, the mean is $10p - 1(1-p)$, where p is your probability, at the time you place the bet, that the chosen horse will win. This is only positive if p exceeds 1/11 and you might only place the bet if this held; that is, if your mean was positive and you expected to win. (Recall that p is *your* probability, not the bookmaker's (§14.5)). We have, at several places in this book, argued in favor of simplicity but the idea of replacing a whole distribution by its mean is carrying simplicity too far. There is another feature of a distribution that is usually relevant, namely the spread (§9.5) of θ. If the stake in the example is increased from 1 to 10 dollars, the mean is still positive and the bet worthwhile if p exceeds 1/11, but the spread is greater, the distribution ranging from -10 to $+100$. As a result, the bet may be thought sensible at a dollar stake but reckless at 10 dollars, especially if the latter is your total capital. Investors in the stock market can feel this, often preferring government stocks with a guaranteed yield, zero spread, to risky ventures that might do well but might lose them a lot of money, large spread. Utility considerations could be relevant here.

Considerations such as these lead to many methods in the financial textbooks replacing the distribution of θ, many numbers, by just two, the mean and the spread of θ. The spread is usually measured by the standard deviation (§9.5) but for our immediate purposes, it will not be necessary to discuss how it is calculated. With the problem of investment reduced from consideration of a whole distribution to just two numbers, the solution is much simpler and real progress can be made. We now argue that this is simplicity gone too far, with possible incoherence as a result. This incoherence is now demonstrated and only afterward will we return to an alternative specification that is not so simple but is coherent.

With both mean and spread to work with, the investor wishes to *increase* the mean but *decrease* the spread. Government stocks achieve the latter but have lower expectations than risky securities. A scenario like this has been met before (§10.13) where two different features were discussed, your health and your assets. It was found convenient to represent the situation by a Figure (10.6) with assets as the horizontal axis and health as the vertical one. Here in Figure 14.1, we do the same thing with the spread horizontal and the mean vertical. As in the earlier figure, curves are drawn such that all the values of the combination (assets, health) on a single curve are judged equivalent, any increase in health being balanced by a decrease in assets. There we wanted to increase both quantities, assets and health; here one desires an increase in mean but a decrease in spread, so the curves of spread and mean are differently placed, going from southwest to northeast in Figure 14.1. Consider points A and B on one of these curves. A corresponds to an

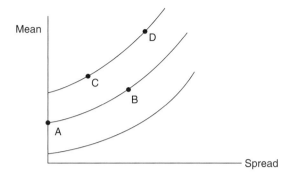

FIGURE 14.1 Curves of equal merit in an investment.

investment with zero spread, whereas B corresponds to an investment with larger mean, giving it an advantage, but is disliked for its larger spread. That A and B are on the same curve means that the two investments are of equal merit, the increase in spread in passing from A to B being balanced by the increase in mean. (Notice that the curves in Figure 10.6 were of constant utility. Here the curves are merely those where all points are judged equivalent. We will return to utility in connection with investments below.) We now show that the use of these curves to solve an investment problem leads to incoherence.

Suppose that you have settled on the curves in Figure (14.1) and, in particular, decided that an investment with mean 2 and spread 6 (C in the figure) is as good as one with mean 4 and spread 10 (D in the figure), the undesirable increase in spread in passing from C to D balanced by the increase in mean. Recall that it is being supposed that in evaluating investments, only mean and spread matter. We now produce two investments that correspond to C and D but in which D is obviously the better. It would not be an exaggeration to say that it is nonsense to judge the two investments, C and D, equal; the polite word is incoherent. The two investments are of the type described above.

C has $\theta = -1$ or $+5$ with equal probabilities $(\frac{1}{2}, \frac{1}{2})$

D has $\theta = -1$ or $+9$ again with equal probabilities $(\frac{1}{2}, \frac{1}{2})$

C corresponds to placing a stake of 1 on a bet at odds of 5−1 against, where you have probability $\frac{1}{2}$ of winning. D has odds of 9−1 against but otherwise is the same as C. It is immediate that

> C has mean 2 and spread 6, whereas
> D has mean 4 and spread 10.

The means are obvious. For C it is $-1 \cdot \frac{1}{2} + 5 \cdot \frac{1}{2} = 2$ (§9.3 where it is referred to as expectation); similarly, for D it is $-1 \cdot \frac{1}{2} + 9 \cdot \frac{1}{2} = 4$. The spread, in the familiar use of that term, is the difference between the higher and lower values that are possible, 4 for C, 6 for D. (Readers comfortable with the term standard deviation may like to be reminded that the measure of spread used here is twice the standard deviation, the factor 2 being irrelevant in this context.)

Now for the crunch: despite your rather sophisticated judgment in Figure (14.1) that C and D are equivalent, it is now obvious that D is better than C. In both bets, you stand to lose 1 with probability $\frac{1}{2}$, whereas your winnings would be 5 with C but a better 9 with D, again with the same probability of $\frac{1}{2}$. Everyone prefers D to C.

The approach used here with a numerical example can be generalized. If an investment with one pair of mean and spread is judged equivalent to a different pair, then two bets, each with the same means and spreads, can always be found in which one bet is blatantly superior to the other, so denying the sophisticated judgment of equivalence. The conclusion has to be that the comparison of investments solely through their means (expectations) and spreads (standard deviations) is unsound. It is surprising that so much effort has been put into methods based on an erroneous assumption. If these have been used in practice, then surprise turns to disquiet.

My own view is that this erroneous assumption could not, on its own, have led to the bankers acting as they did. It is true that many of the investments considered by them were difficult to understand but it is unlikely that this misunderstanding was enhanced by the use of mean–spread methods; rather the complexity of the methods may have disguised the nature of the investments. Another consideration is that, in practice, the erroneous methods may be good approximations to coherent, or even rational procedures. No, we need to look elsewhere for a convincing explanation for the credit crunch and recession.

A possible explanation can be found by going beyond the appreciation of the investment through the yield, θ, with your associated distribution of θ, to a recognition that to be effective, a decision has to be made about whether the investment should be taken and, if so, how much should be invested, the stake in the betting example. In other words, an investment is not just an opinion, it involves action. Bankers correctly refer to investment decisions and it is necessary to go beyond uncertainty, expressed through probability, to actions (§10.1). We have seen that doing this involves an additional ingredient, your utility (§10.2). Specifically, you need both a distribution (of probability) and a utility for the consequences of your actions in investing. Bankers should have had their utility function in mind when they acted, and should have then combined it with the distribution to enable them to decide by maximizing expected utility (MEU, §10.4). So a legitimate question that might be asked is, what utility function do bankers use? Or, since like the textbooks, little if any mention is made of utility in banking, perhaps the question needs modification to ask what utility should be used?

Bankers deal in money, so it is natural to think in terms of utility as a function of money. This was briefly discussed in §10.13 and a possible form of utility as a function of assets was provided in Figure 10.7. There the point was made that the money referred to had to be your, or the bank's, total assets. Thus, in investment C above, expressed as a bet, were your assets when contemplating the bet 100 dollars, you would need to consider assets of 99 dollars were you to lose, and 105 dollars were you to win. The increases of 100 dollars do not affect the argument used there but clearly, with the utility of Figure 10.7 in mind, investment C with assets of 1 dollar would be viewed differently with assets of 100 dollars, if only because in the former case, you might lose everything, whereas in the latter, a drop from 100 dollars to 99 dollars matters little.

Would it be sensible for bankers to consider their utility solely as a function of money? It might be provided it had the form of Figure 10.7 and the assets were the bank's, rather than those of individual bankers. For, if many investments failed together, the bank's assets would drop to near zero and the bank could fail. So perhaps they did not have a utility like that. Yet on the other hand, if the assets were those of an individual banker, the curve could account for the enormous bonuses they took. A glance at Figure 10.7 shows that a rich person needs a large increase in assets to achieve a small gain in utility, as can be seen by comparing the passage from Q to P with that from B to A. We just do not know if bankers had any utility in mind, and if they did, what form it took.

We might question whether the coherent banker should have a utility that depends solely on money, expressed either in terms of their own money or the combined assets of the bank. Consider the offer of a mortgage on a property. If the mortgagor fails, the bank will still have the property but the mortgagor may suffer what is to them a serious loss. Should the bank incorporate this serious loss into their utility? Should their utility involve more than their assets? My personal view is that bankers, like all of us, are social animals and should base their decisions on the effects they might have on society, on you and me. I strongly believe that bankers should have seriously discussed social issues that might have been affected by their actions. Only then should they have maximized the expected utility. What little evidence we have, through bankers being grilled by politicians, suggests that social utility did not enter into their calculations.

Of course, it is not only the bankers whose activities influence society. Pharmaceutical companies appear not to take adequate account of people either in choice of what drug to develop or in the prices they charge. Generally, commercial confidentiality, at the basis of capitalism, hides the motives behind management, so that we do not know whether big companies employ utility concepts, or if they do, what factors they take into consideration. Perhaps only the shareholders are considered.

It would appear from the textbooks on management and finance that the concept of utility is not used despite the demonstration that it, combined with probability, upon which it is based, giving MEU, provides coherent decisions. There are two reasons for this. First, the issue of confidentiality: decision makers would be embarrassed by a public demonstration of their utilities. The second reason is the hostility of so many people to the measurement of abstract concepts such as happiness. They either fail to appreciate the power of numbers in their ability to combine easily in two different ways (§3.1), or they fail to recognize what it is that we are trying to measure. Utility is a measure of the worth to you of a consequence or an outcome (§10.2).

No attempt is made to measure happiness, rather to measure your evaluation of some real outcome, which may involve an abstract idea, like happiness, in a material circumstance. All the measurements proposed here depend on the measurement of probability, which is discussed in Chapter 13 and the ideas developed there leave much to be desired, so that serious research on the topic needs to be done. Rather than award a Nobel Prize to someone who correctly shows how bad we are at decision making, we should support research into coherent decision making.

A wider issue here is the question of privacy. By suggesting that a company should provide us with its utility function, we are effectively invading its privacy. Indeed, whenever we ask for utilities in decisions that have social consequences, we are posing questions of privacy. My personal opinion is that the reasoning used in this book, leading to MEU, leads to the denial of privacy except in decisions that affect only the decision maker. Such questions are rare. For example, I think, as a result of MEU being the preferred method of decision making, all tax returns should be available online. It should significantly reduce tax avoidance and perhaps even tax evasion. Privacy is a quality endorsed by bad people so that they can do bad things.

Epilogue

It is convenient to look back over what has been accomplished in this book and to put the development into perspective, examining both its strengths and its weaknesses. Essentially what has been done is to establish the

logic of uncertainty.

Ordinary logic deals with truth and falsehood, whereas our subject has been uncertainty, the situation where you do not know whether a statement is true or false. Since most statements are, for you, uncertain, whereas knowledge, either of truth or falsity, is rare, the new logic has more relevance to you than the old. Furthermore, it embraces the old since truth and falsity are merely the extreme values, 0 and 1, on a probability scale. Cromwell's rule (§6.8) is relevant here.

We have also seen what this logic consists of, namely, the calculus of probability with its three basic rules of convexity, addition and multiplication. Many people think of probability merely as a number lying between 0 and 1, the convexity rule, describing your uncertainty of an event. This is only part of the story, and a rather unimportant part at that, because in reality you typically need to consider many uncertain events at the same time before combining them, often with other features, like utility, to produce an answer to your problem. The result established in this book is that these combinations *must* be effected by the addition and multiplication rules and not in any other way. It is this method of calculation, this calculus, that uniquely provides the logic of uncertainty.

Let us recall how this calculus was obtained. We began by making a number of assumptions, often called premises, which we hoped were obviously true and acceptable to all of us. From these premises, the ordinary logic of mathematics was used to derive the rules of probability as inevitable consequences. These results can most effectively be described as establishing

the inevitability of probability,

namely, if you want to handle uncertainty, then you *must* use probability to do it, there is no choice. This is one important reason why the material expounded here is so essential—there are no alternatives to probability, except simple transforms, like odds. If you have a situation in which uncertainty plays a role, then probability is the tool you have to use. To use anything else will lay you open to the possibility of your being shown up to be absurd, say in the sense of a Dutch book (§5.7) being made against you. Over the last quarter of a century, several procedures have been advocated to handle aspects of uncertainty in management and industry. Many of them flourish for a while, and make their advocates some money, only to disappear because their internal logic, often disguised under a torrent of words, is not that of probability. It is not a question of one method merely being preferred to another. Those not based on probability are wrong.

There is one serious objection to the line of reasoning in the last paragraph that merits consideration. The objection points out that the whole edifice of the probability calculus depends on the premises used to support it; if these fail, then the whole structure fails. This is surely correct since all our results are theorems that follow from the premises by the secure logic that is mathematics. I offer two responses to this legitimate rebuttal. The first one has been mentioned before in Chapter 5, namely, many different premises can be used that lead to probability, so lending strength to our claim of inevitability. Nevertheless, there are premises that do not lead to probability. Most of these center on the understandable objection that some people have to measuring uncertainty by a single number, perhaps feeling that this is an oversimplification of a reality that is more complex than can possibly be captured by anything as simple as 0.37. These doubts are reinforced by the very real difficulty all of us experience in determining whether our probability is 0.37; perhaps it should be 0.38 or 0.39. These ideas have led people to suggest replacing the single number by a pair of numbers called upper and lower probabilities, so that you would be able to say your uncertainty lay between 0.35 and 0.40, but that you cannot say precisely where, in this interval, it lay. This confuses the measurement problem with the argument in §3.4 that persuades you there is a unique value. Others have proposed using two numbers, rather than one, to describe uncertainty, but in a different form from lower and upper values. They point out that some uncertainties are more firm than others and that their "firmness" should be included as an additional measure. For example, your probability that the next toss of a coin will land heads may be $1/2$ and equally your probability that it will rain tomorrow may be $1/2$, but the first $1/2$ is firmer than the second in that you are sure the first is 0.50, whereas the second might be 0.51, or even 0.60. §7.7 touches on this. Both approaches lead to two numbers replacing our single value. My response to these ideas is to argue in favor of simplicity and not to venture into the complexity of two numbers until the single value has been seen to be inadequate in some way. Nothing that I have seen suggests that probability is inadequate. In particular, the difficulties raised both by upper and lower values and by firmness can, in my view, be handled within the probability calculus. For example, the value, denoted by the letter m in §7.7, expresses how firmly you hold to the initial probability g, all this strictly within the calculus of probability. Of course, there are those who will not admit measurement at all, preferring to use verbal descriptions of uncertainty like "often" or "sometimes". While these may be adequate for simple situations involving a single event, they are useless when two or more events are under discussion. If one event happens "often" and another "sometimes", how uncertain is it that they both will happen? "Seldom" perhaps. Language is inadequate on occasions like this and it is a pity that the inadequacy is sometimes not appreciated. Renouncing mathematical reasoning in favor of the verbal method can help a person to follow fallacious arguments to absurd conclusions without seeing that they are absurd. The problem of the two daughters (§§12.4 and 12.5) illustrates the danger.

As we have seen in §10.8, concern about the use of numbers is felt even more strongly when utility is used for consequences. There we pointed out that the numbers do not claim to cover every aspect of a consequence, but only those aspects relevant to the problem in hand, any more than the price of a theatre ticket completely describes "Hamlet". As with probability, even those who accept the need of numbers to describe outcomes, often feel that a single number is too extreme a description of complicated situations. Yet even in simple situations like those discussed in §10.13 with money, expressed in terms of assets and state of health, you have got to balance cost against improvement in well-being. Once this is recognized, the curves of fixed merit, a notion that does not employ measurement, but only comparisons of like with like, emerge naturally. We are then into the position where numerical comparison is sensible. And, as has been previously emphasized, the comparison we made was based on probability, because then it becomes possible to combine the two distinct notions of uncertainty and desirability of outcomes into a single measure appropriate for decision making. Always the need to combine ideas is an essential requirement. We may need to restrict ourselves to small worlds (§11.7), but they should not be too small.

Probability is important, but it is wrong to overestimate its importance as others, sometimes quite correctly, accuse me of doing. "Oh, Daddy" said one of my children, "you see Bayes in everything". So let us try to place probability into context and contemplate, not just its brilliance but also its limitations. An analogy may be helpful provided we recognize that analogies can be misleading. In my analogy, probability is a tool to help us understand and act in the real world, just as a spade is a valuable tool for gardeners, enabling them to prepare the soil for planting. A gardener needs several tools: fork, shears, hoe, and so on. Similarly we need concepts associated with probability: utility, expectation, and maximization of expected utility. The analogy may be pressed further by noting that a gardener does not just need the tools, he also needs to know how to use them and to judge when and how their use is appropriate. The analogous difficulty is even greater with probability where considerable experience is needed in structuring a problem so that the concept can be correctly and usefully employed. There is a distinction between decision making as art and as science. We have dealt almost entirely with the science, yet art is needed in relating the orderly mathematics to disorderly reality. Much emphasis has been placed on the differences between art and science, indeed in their conflict, in the concept of the two cultures. Yet they often complement one another, for example, the activities of scientists can never be entirely systematic. They must roam and explore before they can present the logic that we ultimately observe. Even mathematics, the most coldly logical of all subjects, has properly been described as the queen of the arts and, for those who can appreciate it, the proof that the square root of two is irrational is as beautiful as a piece of art by Rembrandt.

No decision problem is such that you can quickly write out the tree, fill in the utilities and probabilities, and leave the computer to maximize expected utility. It is much harder than that. All our analysis provides is a set of tools and you have to relate them to the circumstances you face, which is no easy task. What our analysis does is to provide a framework for your thoughts. For example, you are faced with an uncertain situation where one possibility is an event E. This immediately draws your attention to its complement E^c and what might arise if E did not occur. This in turn encourages you to think about many other possibilities so that you end up with several branches from the random node. You will never think of everything but the method encourages you to get near to the ideal where every possibility is foreseen. As we said in §10.7, an important element in good decision making is thinking of new possibilities. That is why decision making should be open because the openness will encourage others to criticize your ideas more effectively than you can by yourself. I have just read a book that is largely a catalog of disastrous decision making, where many of the disasters might have been avoided had the systematic analysis of expected utility been used and criticized. It would also have enabled the importance of small and large worlds, mentioned in §11.7, to be appreciated. Maximization of expected utility provides a framework for thinking, a tool that will improve your encounters with uncertainty. It should never be ignored, yet it is never entirely adequate.

What the analysis of this book provides is a framework for action in the face of uncertainty. It does not tell the whole story of decision making, but does provide a set of ideas that can be filled out to give a full and satisfactory analysis. The framework is science, the filling-out art, the whole being rounded off by science in the form of the computer calculating expectations and searching for maxima. Remember that you have no choice of framework, only probability will do. It is enormously helpful in any enterprise to have a framework on which to hang the ideas being put forward, especially one that will yield a solution, as does MEU. Another merit is that the ideas are firmly based on reason, so that rationality is forced into the process. Yet emotions have to be included as well and they find their proper place in utility, whose clear exposure reveals the features being incorporated into the decision.

Any sound appreciation of the thesis presented in this book must recognize a severe limitation of that thesis, a limitation that prevents it being used in many decision problems whose failure of us to resolve could result in the destruction of what civilization we possess. The limitation has been mentioned before (§5.11), but its importance justifies repetition.

The thesis is personal.

That is to say, it is a method for "you". "You" may be an individual, it may be an organization that has to make a decision, or a judgment of uncertainty, where the accountant, scientist, engineer, the personnel manager and the marketing expert can pool their resources to work within our framework. "You" could even be a government concerned with the welfare of its citizens. It could be a government wishing, as in the European Union, to operate in conjunction with others. But it could not be a government in conflict with another, nor a firm that is battling with another for the control of a market. Our theory admits only one probability and one utility. When cooperation is present, as in the jury room or in the board room, a "you" is not unreasonable, but not with conflict. The Prisoner's Dilemma of §5.11 illustrates.

My own experience was described in the Prologue. My hope is that today, somewhere in the world, there is a young person with sufficient skill and enthusiasm to be given at least five years to spend half their time thinking about decision making under conflict. They will need to be a person with considerable skill in mathematics, for only a mathematician has enough skill in reasoning and abstraction to capture what hopefully is out there, waiting to be discovered. Conflict is the most important mathematical and social problem of the present time.

People often accuse me of putting the Bayesian argument forward as if it was a religion. It is not. Religions are based on faith, though they do have some reasoned elements within them, for example, on moral and ethical questions. The ideas here are based entirely on reason. They do not require an injection of faith, but are the same throughout the world. They do not encompass all life, but merely provide a tool that should help you in handling this uncertain world. Christian and Muslim decision makings should have the same structure, but their probabilities and utilities may differ.

Subject Index

(Where there are several references, the principal ones are given in bold.)

Absolutes   293, 345
Abstraction   35, 326, 372
Acts   20, 106–7, 225–7, 269, 371
   in trees   246–9
Addition   38, 93
Addition rule   **87–9**, 95, 98,
   165–6, 213, 319, 325, 330,
   348, 352
Adversarial approach   13, 264
Advertising   26–7, 31
Agreement   135, 279, 342, 352
Agricultural experiments   **177–80**,
   204, 219
Almanac questions   3, 8, 13, 79
Alternatives   117
Alternative medicine   187, 242, 294
Ambiguity   53, 163, 308, 345
Analogies   325
Analysis of variance   205
Apprenticeship system   272
Area as probability   209–12
Art in the use of the calculus
   249–52, 283, 303
Assets   256–9, 365, 368, 371–2
Association   **77–9**, 83–4, 119, 172,
   181–3
Aunt Sally (see straw man)

Babbage   268
Bankers   366–7, 370–2
Bayes   XV, 32, 119, 378

Bayes Rule   XV, 37, 112, **118–23**,
   127, 129, 132, 157, 168, 190–1,
   212–3, 273, 276–80, 296,
   331–3, 344, 347, 352
   odds form   **123–5**, 126, 127, 136,
   278, 356
   examples   300, 304, 306, 311,
   313, 316, 319
Bayesian   XV, 278–81, 343–60
Belief   **18–20**, 49, 56–8, 63, 146,
   157–61, 225–7, 244–5
Bernoulli series   **151–2**, 161–3,
   191, 206, 306, 318–9, 335, 354
Betting   XII, 5, 15, 63, 67, **360–5**,
   367, 369–71
Binomial distribution   **191–5**, 203,
   208, 216, 355
Bioequivalence   359
Biology   189–90
Birthday problem   55, 96–7
Births   305
Bookmakers   63, 67, 361–4, 367
Borel field   36
Brackets   41, 62
Breeding   XIV
BSE   276, 278

Calculus of uncertainty   17–8,
   68, 94
Cancer test   99–100, 119–22,
   231–4

Understanding Uncertainty, Revised Edition. Dennis V. Lindley.
© 2014 John Wiley & Sons, Inc. Published 2014 by John Wiley & Sons, Inc.

Cardano's method 300, 309, **311–3**
Cards 5, 15, 52, 67, 80, 194, **300–1**
Causation 83, 180
Certain 8
Chance 124, **161–4**, 193, 198, 206–8, 277, 305, 318, 351
 distribution 206–8
Church of Scotland 15, 130, 220
Classical form of probability **143–5**, 327
Closure 112
Clusters 200
Coherence **54–6**, 67–8, 72, **74–5**, **93**, 131, 133, 160, 222, 226, 230, 255, 261, 282, 308, 318, 323, 329, 331–6, **341–2**, 349, 351, 358, 364, 366–70, 372
Combination of uncertainties 17, 46, 94, 243
Common sense 24, 200, 301, 306, 309
Comparison of consequences 227–31, 251–2
 with a standard 47, 107
Comparative 293
Compensation in a Bernoulli series 161
Complementary event **58–60**, 86, 96, 112, 190, 328, 331
Computers 106, 346
Computer packages 179
Conditional probability **72–5**, 90
Confidence intervals 353
Conflict 16, 27, **107–9**, 251, 380
Confounding **171–2**, 182
Conglomerability 94, **321–3**
Conjunction of events 85–6
Consequences 225, **227–31**, 246–9, 324–6, 365, 372
Consistent 56

Contingency table **70–2**, 101, 107, 122, 168, 177, 181, 228
Continuous quantity 209
Controlled experiment 173
Convexity rule **92**, 93, 104, **130**, 165, 328, 352
Conviction concerning a probability 159–60
Crime 79, 181–3, 184–5, 334–5
Cromwell's rule 15, **129–31**, 155, 220, 261, 269, 279, 308, 328, 341, 375
Cultures, two 28, 379

Darwin 295–6
Data 269–71, 277, 278, 296, 344, 346–54, 357, 358
 that did not arise 293
Daughters **305–13**, 345, 377
Davis score 46
Decimal place 43
Decision analysis 20–1, 106–7, 196, **225–64**, 305, 325, 359, 365, 373, 378–80
 trees **246–9**, 255, 325, 365
Defective 197
De Finetti 25
De Finetti's result XV, **152–4**, 206, 208
Denominator 41, 307, 344, 345, 348
Density 210–2, 355
 marginal 213
Dependent quantity 285
Descriptive **31–3**, 54, 67, 116, 221–2, 226, 315, 336
Desirability 57, 72, 226, 228–9
Diagnosis 119–22, 138–40
Diminishing returns 203
Disagreement amongst scientists 276
Discrete quantity 209

Disjunction of events 86–9
Distribution 193, 195, 315, 367
  joint 350
  marginal 345, 350
  Student's 350
Division 39
Doing and seeing **82–4**, 180, 286, 337
Doors 301–5
Double jeopardy 263
Dutch book **101–3**, 107, 376

Education 66, 104, 266
Efron's dice XII, 230, **323–6**
Einstein XV, 37, 119, 269, 279, 295
Elections 7, 251–2, 327
Ellsberg paradox 12, 32, 54, **219–24**
Elton John 30, 241, 270
Emotion 23, 29–31, 240–2
Envelopes 313–5
Epsilon 130–1
Equation 40, 42
Equality 38
Errors 25, 120, 121, 139
Estimation 353, 355–6
Event **18**, 45, 49–50, 58–60, 85–7, 245
  as uncertain quantity 193–6
Evidence 119, 133, 242, 261–3
Exchangeability **147–55**, 160, 161, 169, 176–80, 189, 194, 294, 318, 345, 346, 351–2
Exclusive events **88–9**, 91, 112, 165–6, 190, 311
Exhaustive events 112, **190**, 311
Expectation **195–7**, 199, 201, **220–4**, 235, 245–6, 285–7, 344, 347–9, 367, 378
Explanatory quantity 285

Exponential function 348, 359–60
Exposure to criticism 250
Extension of the conversation **98–101**, 120, 153, 156, 164, 168, **190–1**, 195, 221, 233, 245, 273, 306, 313, 331, 339, 344, 356

Factors **173**, 205
Facts 28–9, 81–2, 270
Faiths 8, 14–5, 241–2, 268, 296, 380–1
False positives (and negatives) 99–101
Finance XII, 257, 365–73
Forensic science 115, 125–7
Freedom of information 142
Frequency 53, 145–7, 155, **157–61**, 192, 198, 212, 277, 327
  statistics 350–60
Future data 156–8, 272, 278
Fuzzy logic 227

Gambles 15, 32, 48–9, 57, 67, 101–3, 230, 241, 360–5
Games of chance 361
Gaussian distribution 215
GDP 338–40
Genetics 145, 190, 194
Genotype 126–7
Geometry 19, 56, 334
Girls' names 310, 312
Given 65
Greek alphabet 286
Guilt 3–4, 18, 133, 244, 260–4

Half correction 216
Haphazard 176
History 4, 12, 18, 62, 67, 81
HIV 7, 12, 194
Horse race 5, 12, 15, 61, 67, 80, 103, 360–5

386 SUBJECT INDEX

How do you know? 187, 206, 280, 308, 312
Hypothesis 115, 135, 281–3, 287–91, 356, 358, 359

Ignorance 106, 144–5, 315, 351
Inadmissible evidence 263
Independence **75–7**, 91–2, 150, 157, 161–2, **183–6**, 253, 262, 273–5, 305, 319–20, 324, 335
Index of binomial 193, 208
Inequality 42
Inference 343, 345, 359
Information 16, **140–2**, 263
Interaction 173, 206
Investments 67, 366–71

Jain philosophy 130
Janus effect 114, 117, 139
Jury 244, 260–4

Knowledge base **63–5**, 119, 192, 195, 213, 300–1, 306, 344, 346, 350

Language 35, 124, 163, 377
Large numbers 154–7, 192
Large and small worlds 282–3, 335–6
Law 3–4, 12, 13, 27, 66, 79, 114, 115, 125–6, 244, **260–4**, 334–5
Learning 116–7, 277
Likelihood **124**, 163, 191, 238, 278, 308, 309, 352, 355
  function 319, 349, 354
  principle 125, 308, **345**, 353, 359
  ratio 112, **124**, **127–9**, 278–80, 289, 345, 356, 358–60
Linear 286
Literary argument 121
Literature 26

Logarithms 127
Logic XI, 95, 105–6, 297, 337, 375
Logical probability 65
Loss 240
Lottery 50, 63
Lower case 59
Lower and upper probability 54, 107, 159, 329, 377

Markov 335
Mathematics XI, 35–7, **38–43**, 124, 166, 297
Maxima and minima 93–4, 105
Maximum likelihood 353–4
Mean of a distribution 196, 199, 344–50, 356–7, 359, 367–70
Measurement **45–8**, 54
MEU 33, **234–6**, 239, 242, 243, 249, 283, 365, 371–3, 380
Median 344
Medical trial 15–6, 168–171, 176–80, 195, 248, 349, 354, 355, 359
Millennium bug 316
Minima (see maxima)
Mixture of probabilities 100
Models **284–287**, 343, 350, 352, 355
Monty Hall problem 301
Multiple choice 66
Multiplication 39, 93
Multiplication rule 74, 78, **89–92**, 98, 118, 149, **167**, 192, 283, 306, 309, 319, 325, 330, 348, 352
Music 62

Newton's laws 14, 25, 34, 37, 93, 269, 274, 279
Node **246–9**, 255
Nonrepeatable events 327–9

Norm   54
Normal distribution   **213–7**, 218, 346–51, 355–7, 360
Normative   **31–3**, 54, 116, 121, 219–23, 226, 270, 283, 315, 336
Nuisance parameter   **345**, 346
Numeracy   45–8, **242–5**
Numerator   41, 307, 344

Objectivity   163, 351–2
Observe, think, act   269
Odds   5–6, **60–3**, 94, 102–3, 105, 123–5, 278, 361–4
Opinion change   276
Opinion polls   7, 67–8
Optional stopping   195, **354–5**
Orthodox medicine   242
Outliers   281

Paradigm   268
Parameter   161, 193, 207, 284, 343–7, 350–1, 353–7
Parker score   45–6
Partition   112, **190–3**, 322–3, 330
Pattern   149
Perpetual money-making machine   230, 255
Personal (of probability)   56, 232, 251, 276, 290, 380
Physics   119, 173, 189, 358
Pictorial representation   208–12
Pins   146, 147, 155, 158
Placebo   168
Poisson distribution   **197–200**, 202, 203, 207–9, 216–18
Politics   12, 27, 81, 266
Posterior probability   344–58
Population   126
Poverty   336
Prediction   269, 271, 279
Preference   30, 244
Premises   23–5, 35, 51, 375

Prevision   196
Prior probability   191, 344–60
Privacy   373
Probability   50–4
  assessment   327–342, 351
  classical   143–5
  coherence   333–6
  distribution   193
  decisions   225–7, 253
  expectation   196–7
  frequency   145–7
  inevitability   376
  law   260
  notation   59, 65, 74
  odds   62, 123
  personal   18, 56
  reasoning   336–41
  rules   92–5, 106, 165–8
  utility   230
Product of events   91
Prosecutor's fallacy   127, 291, 293

Quadratic rule   104

Random   **49**, 174, 267, 303, 313, 324
  numbers   50, 175
  variable   193
Randomization   173–6
Ravens   135–8
Reason   23–5, 29, 240–2, 325, 336–7, 381
Reciprocal   41, 348
Reconciliation   295
Repetition   293
  words   38
Research   XIII–XV, 380
Reverse-Polish   41
Reverse time order   249

Sample   194, 358
Sample mean   349

Science   XV, 4, 14, 32, 81, 249, **265–98**, 343, 345, 351–2
Scoring rules   **103–5**, 111, 116
Screening test   121
Sensitivity   138
Shakespeare   XV, 26, 31, 111, 227
Significance test   32, 115, 128, **291–3, 355–60**
  level   115, 227, **292**, 356–60
Significant figures   43
Simplicity   33–4, 63, 159, 280, 286, 368, 377
Simpson's paradox   XII, 16, 29, 84, **168–70**, 174, **176–80**, 206, 270, 283, 286–7, 336, 340
Small (and large) worlds   112, 282–3, 335–6, 378, 379
Social utility   251
Sociology   173, 182
Specificity   138
Spin   27
Spread   **201–5**, 344–50, 356–7, 359, 368–70
Square-root rule   **202**, 205, 216–9, 349–50, 357
Standard   47
Standard deviation   201, 205, 215, 216, 218, 347, 368
Statistical mechanics   164
Statistics   XII, 191, 213, 308, **343–73**
Stigler's law   **XV**, 130, 170
Strategy   285
Straw man   115, 135, 289
Subjective   56, 351–2
Subjunctive   169
Subscripts   42
Subtraction   39
Sufficient   134, 148, 291, 349, 357
Sum of events   91
Supposition   81–2

Sure-thing principle   223, 238–9, 323, 363
Symbols   39
Symptoms   114–5, 116, 150, 156
Synonyms   38

Tactics   285
Tail area   292–3, 360
Tax   337–40
Technicalities   36
Television   27
Tennyson   268
Theories   187, **269–77**, 281–3, 296, 352
Thought experiment   254
Thumb tack (see pins)
Tomorrow's decision   **249**, 305
Transitive property   324–5
Transposed conditional   **113–7**, 124, 139, 300
Triangulation   47, 56
Trickle down   337–41
Truth table   86

UFO   317–21
Uncertain quantity   193, 195–6, 212, 284, 347, 349, 366, 367
Uncertainty   XI, **1–21**, 45–7, 105–6, 269, 276
Upper case   59
Upper probability (see lower)
Urn, standard   48–52, 83
  calculation   94, 100
  example   **131–5**, 152–3, 156–7, 272–5, 352
Utility   **230–60**, 371–2, 378
  assessment   **254**, 326

Variability   270
  as an experimental tool   204–6
Variance   205

Variation   46, 189–91, 217–9
Verdi   30, 241, 270
Volume as probability   212

Wand procedure   229, 233, 253–4, 325
Webster   48

Weights   159, 348
Wine tasting   45
Writing   37–8

Y2K   316–7
You   2, 18, 50, 105, 251, 264, 317, 341, 380

Index of Examples

(Material used only for purposes of illustration)

Aggression   138
Alcohol test   121
Allergy   187
Anastasia   10, 15
Armadillos   114
Astrology   56
Autism   188
Aviation   9

Bananas   187, 206
Barometer   180
Breast cancer   139
Buses   140

Casino   12, 144
Chemical engineering   84
Chess   305
Climate change   352
Clothing manufacture   217
Continental drift   352
Cookery   298

Dangerous driving   261
Diana, princess   30
Dice   51, 144, 223
Distance   60, 334
DNA   10, 126, 128

Eclipse   280, 284
Electricity supply   80
Ethnicity   181, 184

Evolution   269, 295
Extra-sensory perception   358

Fairies   280
Football   361, 364

Gardening   297
Garment choice   225
Glass   128
GM crops   9

Health   256
Health authorities   218, 353
Heights   217
Higg's boson   351, 358
Hip surgery   259
House purchase   238

Incomes   213
Inflation   6, 11, 69–79, 122, 145, 194
Influenza   189

Leukemia   199
Liberia   3, 13, 18, 28, 48, 66, 79

Milk   172
Mortgage   372
MS   259
Mugging   78, 181

Understanding Uncertainty, Revised Edition. Dennis V. Lindley.
© 2014 John Wiley & Sons, Inc. Published 2014 by John Wiley & Sons, Inc.

NHS   259
NRA   139
Nuclear accident   7, 21, 57, 63, 97, 226, 235

Opera   241

Painting   5
Pandora's box   13
Parking fine   263
Pharmaceutical companies   6, 372
Poetry   163
Portfolio   258
Princes in the tower   4, 18, 20, 67
Prussian cavalry   199

Rain   2, 11, 53, 58, 64, 79, 193, 335
Reincarnation   280
Relativity   280, 296, 351, 355
Roulette   51-2, 144, 194, 360-1

Saturated fat   8, 66, 80, 84
Selenium   4, 66, 80, 288
Skull   11
Smoking   XIV, 35, 84, 172

Solar system   34, 268
Stock market   6, 67, 219, 272
Sunrise   10

Tax returns   373
Teapots   189
Telephone operator   197, 207
Trivial Pursuit   3, 48, 66

Unemployment   69–79, 122, 181, 184
University fees   183

Vaccine   188, 267
Voting   81, 194, 227

Washing machine   238
Weather   147, 162, 335
   forecast   11, 110
Wedding anniversary   241
Whales   36
Wheelwright's shop   272
White Christmas   110

Yogurt   172

Index of Notations

$+$	38, 86	K	64
$-$	38	EF	85
$\times$	38	E or F	86
$\div$	38	$p(E)$	59
$=$	38	$p(E \mid K)$	65
ab	40	$p(M \mid L)$	74
a/b	41, 62	$p(E \mid F{:}GK)$	81
$\frac{a}{b}$	41	$o(E)$	61
$\sqrt{a}$	41	$o(F \mid K)$	123
a^2	40	$\ell(E \mid F)$	125
( )	41	α, β	286
[ ]	62	θ	151
$>$	42	μ	347
$<$	42	ϕ	284
$\propto$	191	iid	162
&	85	s.d	215
E^c	59	$N(\phi, s)$	347

WILEY SERIES IN PROBABILITY AND STATISTICS
ESTABLISHED BY WALTER A. SHEWHART AND SAMUEL S. WILKS

Editors: *David J. Balding, Noel A. C. Cressie, Garrett M. Fitzmaurice, Harvey Goldstein, Iain M. Johnstone, Geert Molenberghs, David W. Scott, Adrian F. M. Smith, Ruey S. Tsay, Sanford Weisberg*
Editors Emeriti: *Vic Barnett, J. Stuart Hunter, Joseph B. Kadane, Jozef L. Teugels*

The *Wiley Series in Probability and Statistics* is well established and authoritative. It covers many topics of current research interest in both pure and applied statistics and probability theory. Written by leading statisticians and institutions, the titles span both state-of-the-art developments in the field and classical methods.

Reflecting the wide range of current research in statistics, the series encompasses applied, methodological and theoretical statistics, ranging from applications and new techniques made possible by advances in computerized practice to rigorous treatment of theoretical approaches.

This series provides essential and invaluable reading for all statisticians, whether in academia, industry, government, or research.

† ABRAHAM and LEDOLTER · Statistical Methods for Forecasting
  AGRESTI · Analysis of Ordinal Categorical Data, *Second Edition*
  AGRESTI · An Introduction to Categorical Data Analysis, *Second Edition*
  AGRESTI · Categorical Data Analysis, *Third Edition*
  ALTMAN, GILL, and McDONALD · Numerical Issues in Statistical Computing for the Social Scientist
  AMARATUNGA and CABRERA · Exploration and Analysis of DNA Microarray and Protein Array Data
  AMARATUNGA, CABRERA, and SHKEDY . Exploration and Analysis of DNA Microarray and Other High-Dimensional Data, *Second Edition*
  ANDĚL · Mathematics of Chance
  ANDERSON · An Introduction to Multivariate Statistical Analysis, *Third Edition*
* ANDERSON · The Statistical Analysis of Time Series
  ANDERSON, AUQUIER, HAUCK, OAKES, VANDAELE, and WEISBERG · Statistical Methods for Comparative Studies
  ANDERSON and LOYNES · The Teaching of Practical Statistics
  ARMITAGE and DAVID (editors) · Advances in Biometry
  ARNOLD, BALAKRISHNAN, and NAGARAJA · Records
* ARTHANARI and DODGE · Mathematical Programming in Statistics
* BAILEY · The Elements of Stochastic Processes with Applications to the Natural Sciences
  BAJORSKI · Statistics for Imaging, Optics, and Photonics
  BALAKRISHNAN and KOUTRAS · Runs and Scans with Applications
  BALAKRISHNAN and NG · Precedence-Type Tests and Applications
  BARNETT · Comparative Statistical Inference, *Third Edition*
  BARNETT · Environmental Statistics
  BARNETT and LEWIS · Outliers in Statistical Data, *Third Edition*
  BARTHOLOMEW, KNOTT, and MOUSTAKI · Latent Variable Models and Factor Analysis: A Unified Approach, *Third Edition*
  BARTOSZYNSKI and NIEWIADOMSKA-BUGAJ · Probability and Statistical Inference, *Second Edition*
  BASILEVSKY · Statistical Factor Analysis and Related Methods: Theory and Applications

*Now available in a lower priced paperback edition in the Wiley Classics Library.
†Now available in a lower priced paperback edition in the Wiley–Interscience Paperback Series.

BATES and WATTS · Nonlinear Regression Analysis and Its Applications
BECHHOFER, SANTNER, and GOLDSMAN · Design and Analysis of Experiments for Statistical Selection, Screening, and Multiple Comparisons
BEIRLANT, GOEGEBEUR, SEGERS, TEUGELS, and DE WAAL · Statistics of Extremes: Theory and Applications
BELSLEY · Conditioning Diagnostics: Collinearity and Weak Data in Regression
† BELSLEY, KUH, and WELSCH · Regression Diagnostics: Identifying Influential Data and Sources of Collinearity
BENDAT and PIERSOL · Random Data: Analysis and Measurement Procedures, *Fourth Edition*
BERNARDO and SMITH · Bayesian Theory
BHAT and MILLER · Elements of Applied Stochastic Processes, *Third Edition*
BHATTACHARYA and WAYMIRE · Stochastic Processes with Applications
BIEMER, GROVES, LYBERG, MATHIOWETZ, and SUDMAN · Measurement Errors in Surveys
BILLINGSLEY · Convergence of Probability Measures, *Second Edition*
BILLINGSLEY · Probability and Measure, *Anniversary Edition*
BIRKES and DODGE · Alternative Methods of Regression
BISGAARD and KULAHCI · Time Series Analysis and Forecasting by Example
BISWAS, DATTA, FINE, and SEGAL · Statistical Advances in the Biomedical Sciences: Clinical Trials, Epidemiology, Survival Analysis, and Bioinformatics
BLISCHKE and MURTHY (editors) · Case Studies in Reliability and Maintenance
BLISCHKE and MURTHY · Reliability: Modeling, Prediction, and Optimization
BLOOMFIELD · Fourier Analysis of Time Series: An Introduction, *Second Edition*
BOLLEN · Structural Equations with Latent Variables
BOLLEN and CURRAN · Latent Curve Models: A Structural Equation Perspective
BOROVKOV · Ergodicity and Stability of Stochastic Processes
BOSQ and BLANKE · Inference and Prediction in Large Dimensions
BOULEAU · Numerical Methods for Stochastic Processes
* BOX and TIAO · Bayesian Inference in Statistical Analysis
BOX · Improving Almost Anything, *Revised Edition*
* BOX and DRAPER · Evolutionary Operation: A Statistical Method for Process Improvement
BOX and DRAPER · Response Surfaces, Mixtures, and Ridge Analyses, *Second Edition*
BOX, HUNTER, and HUNTER · Statistics for Experimenters: Design, Innovation, and Discovery, *Second Editon*
BOX, JENKINS, and REINSEL · Time Series Analysis: Forcasting and Control, *Fourth Edition*
BOX, LUCEÑO, and PANIAGUA-QUIÑONES · Statistical Control by Monitoring and Adjustment, *Second Edition*
* BROWN and HOLLANDER · Statistics: A Biomedical Introduction
CAIROLI and DALANG · Sequential Stochastic Optimization
CASTILLO, HADI, BALAKRISHNAN, and SARABIA · Extreme Value and Related Models with Applications in Engineering and Science
CHAN · Time Series: Applications to Finance with R and S-Plus®, *Second Edition*
CHARALAMBIDES · Combinatorial Methods in Discrete Distributions
CHATTERJEE and HADI · Regression Analysis by Example, *Fourth Edition*
CHATTERJEE and HADI · Sensitivity Analysis in Linear Regression
CHERNICK · Bootstrap Methods: A Guide for Practitioners and Researchers, *Second Edition*
CHERNICK and FRIIS · Introductory Biostatistics for the Health Sciences
CHILÈS and DELFINER · Geostatistics: Modeling Spatial Uncertainty, *Second Edition*
CHOW and LIU · Design and Analysis of Clinical Trials: Concepts and Methodologies, *Third Edition*

---

*Now available in a lower priced paperback edition in the Wiley Classics Library.
†Now available in a lower priced paperback edition in the Wiley–Interscience Paperback Series.

CLARKE · Linear Models: The Theory and Application of Analysis of Variance
CLARKE and DISNEY · Probability and Random Processes: A First Course with Applications, *Second Edition*
COCHRAN and COX · Experimental Designs, *Second Edition*
COLLINS and LANZA · Latent Class and Latent Transition Analysis: With Applications in the Social, Behavioral, and Health Sciences
CONGDON · Applied Bayesian Modelling
CONGDON · Bayesian Models for Categorical Data
CONGDON · Bayesian Statistical Modelling, *Second Edition*
CONOVER · Practical Nonparametric Statistics, *Third Edition*
COOK · Regression Graphics
COOK and WEISBERG · An Introduction to Regression Graphics
COOK and WEISBERG · Applied Regression Including Computing and Graphics
CORNELL · A Primer on Experiments with Mixtures
CORNELL · Experiments with Mixtures, Designs, Models, and the Analysis of Mixture Data, *Third Edition*
COX · A Handbook of Introductory Statistical Methods
CRESSIE · Statistics for Spatial Data, *Revised Edition*
CRESSIE and WIKLE · Statistics for Spatio-Temporal Data
CSÖRGŐ and HORVÁTH · Limit Theorems in Change Point Analysis
DAGPUNAR · Simulation and Monte Carlo: With Applications in Finance and MCMC
DANIEL · Applications of Statistics to Industrial Experimentation
DANIEL · Biostatistics: A Foundation for Analysis in the Health Sciences, *Eighth Edition*
DANIEL · Fitting Equations to Data: Computer Analysis of Multifactor Data, *Second Edition*
DASU and JOHNSON · Exploratory Data Mining and Data Cleaning
DAVID and NAGARAJA · Order Statistics, *Third Edition*
DEGROOT, FIENBERG, and KADANE · Statistics and the Law
DEL CASTILLO · Statistical Process Adjustment for Quality Control
DeMARIS · Regression with Social Data: Modeling Continuous and Limited Response Variables
DEMIDENKO · Mixed Models: Theory and Applications with R, *Second Edition*
DENISON, HOLMES, MALLICK and SMITH · Bayesian Methods for Nonlinear Classification and Regression
DETTE and STUDDEN · The Theory of Canonical Moments with Applications in Statistics, Probability, and Analysis
DEY and MUKERJEE · Fractional Factorial Plans
DILLON and GOLDSTEIN · Multivariate Analysis: Methods and Applications
DODGE and ROMIG · Sampling Inspection Tables, *Second Edition*
DOOB · Stochastic Processes
DOWDY, WEARDEN, and CHILKO · Statistics for Research, *Third Edition*
DRAPER and SMITH · Applied Regression Analysis, *Third Edition*
DRYDEN and MARDIA · Statistical Shape Analysis
DUDEWICZ and MISHRA · Modern Mathematical Statistics
DUNN and CLARK · Basic Statistics: A Primer for the Biomedical Sciences, *Fourth Edition*
DUPUIS and ELLIS · A Weak Convergence Approach to the Theory of Large Deviations
EDLER and KITSOS · Recent Advances in Quantitative Methods in Cancer and Human Health Risk Assessment
ELANDT-JOHNSON and JOHNSON · Survival Models and Data Analysis
ENDERS · Applied Econometric Time Series, *Third Edition*
ETHIER and KURTZ · Markov Processes: Characterization and Convergence

---

* Now available in a lower priced paperback edition in the Wiley Classics Library.
† Now available in a lower priced paperback edition in the Wiley–Interscience Paperback Series.

EVANS, HASTINGS, and PEACOCK · Statistical Distributions, *Third Edition*
EVERITT, LANDAU, LEESE, and STAHL · Cluster Analysis, *Fifth Edition*
FEDERER and KING · Variations on Split Plot and Split Block Experiment Designs
FELLER · An Introduction to Probability Theory and Its Applications, Volume I, *Third Edition,* Revised; Volume II, *Second Edition*
FITZMAURICE, LAIRD, and WARE · Applied Longitudinal Analysis, *Second Edition*
* FLEISS · The Design and Analysis of Clinical Experiments
FLEISS · Statistical Methods for Rates and Proportions, *Third Edition*
† FLEMING and HARRINGTON · Counting Processes and Survival Analysis
FUJIKOSHI, ULYANOV, and SHIMIZU · Multivariate Statistics: High-Dimensional and Large-Sample Approximations
FULLER · Introduction to Statistical Time Series, *Second Edition*
† FULLER · Measurement Error Models
GALLANT · Nonlinear Statistical Models
GEISSER · Modes of Parametric Statistical Inference
GELMAN and MENG · Applied Bayesian Modeling and Causal Inference from ncomplete-Data Perspectives
GEWEKE · Contemporary Bayesian Econometrics and Statistics
GHOSH, MUKHOPADHYAY, and SEN · Sequential Estimation
GIESBRECHT and GUMPERTZ · Planning, Construction, and Statistical Analysis of Comparative Experiments
GIFI · Nonlinear Multivariate Analysis
GIVENS and HOETING · Computational Statistics
GLASSERMAN and YAO · Monotone Structure in Discrete-Event Systems
GNANADESIKAN · Methods for Statistical Data Analysis of Multivariate Observations, *Second Edition*
GOLDSTEIN · Multilevel Statistical Models, *Fourth Edition*
GOLDSTEIN and LEWIS · Assessment: Problems, Development, and Statistical Issues
GOLDSTEIN and WOOFF · Bayes Linear Statistics
GREENWOOD and NIKULIN · A Guide to Chi-Squared Testing
GROSS, SHORTLE, THOMPSON, and HARRIS · Fundamentals of Queueing Theory, *Fourth Edition*
GROSS, SHORTLE, THOMPSON, and HARRIS · Solutions Manual to Accompany Fundamentals of Queueing Theory, *Fourth Edition*
* HAHN and SHAPIRO · Statistical Models in Engineering
HAHN and MEEKER · Statistical Intervals: A Guide for Practitioners
HALD · A History of Probability and Statistics and their Applications Before 1750
† HAMPEL · Robust Statistics: The Approach Based on Influence Functions
HARTUNG, KNAPP, and SINHA · Statistical Meta-Analysis with Applications
HEIBERGER · Computation for the Analysis of Designed Experiments
HEDAYAT and SINHA · Design and Inference in Finite Population Sampling
HEDEKER and GIBBONS · Longitudinal Data Analysis
HELLER · MACSYMA for Statisticians
HERITIER, CANTONI, COPT, and VICTORIA-FESER · Robust Methods in Biostatistics
HINKELMANN and KEMPTHORNE · Design and Analysis of Experiments, Volume 1: Introduction to Experimental Design, *Second Edition*
HINKELMANN and KEMPTHORNE · Design and Analysis of Experiments, Volume 2: Advanced Experimental Design
HINKELMANN (editor) · Design and Analysis of Experiments, Volume 3: Special Designs and Applications
HOAGLIN, MOSTELLER, and TUKEY · Fundamentals of Exploratory Analysis of Variance

*Now available in a lower priced paperback edition in the Wiley Classics Library.
†Now available in a lower priced paperback edition in the Wiley–Interscience Paperback Series.

* HOAGLIN, MOSTELLER, and TUKEY · Exploring Data Tables, Trends and Shapes
* HOAGLIN, MOSTELLER, and TUKEY · Understanding Robust and Exploratory Data Analysis
HOCHBERG and TAMHANE · Multiple Comparison Procedures
HOCKING · Methods and Applications of Linear Models: Regression and the Analysis of Variance, *Third Edition*
HOEL · Introduction to Mathematical Statistics, *Fifth Edition*
HOGG and KLUGMAN · Loss Distributions
HOLLANDER, WOLFE, and CHICKEN · Nonparametric Statistical Methods, *Third Edition*
HOSMER and LEMESHOW · Applied Logistic Regression, *Second Edition*
HOSMER, LEMESHOW, and MAY · Applied Survival Analysis: Regression Modeling of Time-to-Event Data, *Second Edition*
HUBER · Data Analysis: What Can Be Learned From the Past 50 Years
HUBER · Robust Statistics
† HUBER and RONCHETTI · Robust Statistics, *Second Edition*
HUBERTY · Applied Discriminant Analysis, *Second Edition*
HUBERTY and OLEJNIK · Applied MANOVA and Discriminant Analysis, *Second Edition*
HUITEMA · The Analysis of Covariance and Alternatives: Statistical Methods for Experiments, Quasi-Experiments, and Single-Case Studies, *Second Edition*
HUNT and KENNEDY · Financial Derivatives in Theory and Practice, *Revised Edition*
HURD and MIAMEE · Periodically Correlated Random Sequences: Spectral Theory and Practice
HUSKOVA, BERAN, and DUPAC · Collected Works of Jaroslav Hajek— with Commentary
HUZURBAZAR · Flowgraph Models for Multistate Time-to-Event Data
JACKMAN · Bayesian Analysis for the Social Sciences
† JACKSON · A User's Guide to Principle Components
JOHN · Statistical Methods in Engineering and Quality Assurance
JOHNSON · Multivariate Statistical Simulation
JOHNSON and BALAKRISHNAN · Advances in the Theory and Practice of Statistics: A Volume in Honor of Samuel Kotz
JOHNSON, KEMP, and KOTZ · Univariate Discrete Distributions, *Third Edition*
JOHNSON and KOTZ (editors) · Leading Personalities in Statistical Sciences: From the Seventeenth Century to the Present
JOHNSON, KOTZ, and BALAKRISHNAN · Continuous Univariate Distributions, Volume 1, *Second Edition*
JOHNSON, KOTZ, and BALAKRISHNAN · Continuous Univariate Distributions, Volume 2, *Second Edition*
JOHNSON, KOTZ, and BALAKRISHNAN · Discrete Multivariate Distributions
JUDGE, GRIFFITHS, HILL, LÜTKEPOHL, and LEE · The Theory and Practice of Econometrics, *Second Edition*
JUREK and MASON · Operator-Limit Distributions in Probability Theory
KADANE · Bayesian Methods and Ethics in a Clinical Trial Design
KADANE AND SCHUM · A Probabilistic Analysis of the Sacco and Vanzetti Evidence
KALBFLEISCH and PRENTICE · The Statistical Analysis of Failure Time Data, *Second Edition*
KARIYA and KURATA · Generalized Least Squares
KASS and VOS · Geometrical Foundations of Asymptotic Inference
† KAUFMAN and ROUSSEEUW · Finding Groups in Data: An Introduction to Cluster Analysis
KEDEM and FOKIANOS · Regression Models for Time Series Analysis
KENDALL, BARDEN, CARNE, and LE · Shape and Shape Theory

*Now available in a lower priced paperback edition in the Wiley Classics Library.
†Now available in a lower priced paperback edition in the Wiley–Interscience Paperback Series.

KHURI · Advanced Calculus with Applications in Statistics, *Second Edition*
KHURI, MATHEW, and SINHA · Statistical Tests for Mixed Linear Models
* KISH · Statistical Design for Research
KLEIBER and KOTZ · Statistical Size Distributions in Economics and Actuarial Sciences
KLEMELÄ · Smoothing of Multivariate Data: Density Estimation and Visualization
KLUGMAN, PANJER, and WILLMOT · Loss Models: From Data to Decisions, *Third Edition*
KLUGMAN, PANJER, and WILLMOT · Loss Models: Further Topics
KLUGMAN, PANJER, and WILLMOT · Solutions Manual to Accompany Loss Models: From Data to Decisions, *Third Edition*
KOSKI and NOBLE · Bayesian Networks: An Introduction
KOTZ, BALAKRISHNAN, and JOHNSON · Continuous Multivariate Distributions, Volume 1, *Second Edition*
KOTZ and JOHNSON (editors) · Encyclopedia of Statistical Sciences: Volumes 1 to 9 with Index
KOTZ and JOHNSON (editors) · Encyclopedia of Statistical Sciences: Supplement Volume
KOTZ, READ, and BANKS (editors) · Encyclopedia of Statistical Sciences: Update Volume 1
KOTZ, READ, and BANKS (editors) · Encyclopedia of Statistical Sciences: Update Volume 2
KOWALSKI and TU · Modern Applied U-Statistics
KRISHNAMOORTHY and MATHEW · Statistical Tolerance Regions: Theory, Applications, and Computation
KROESE, TAIMRE, and BOTEV · Handbook of Monte Carlo Methods
KROONENBERG · Applied Multiway Data Analysis
KULINSKAYA, MORGENTHALER, and STAUDTE · Meta Analysis: A Guide to Calibrating and Combining Statistical Evidence
KULKARNI and HARMAN · An Elementary Introduction to Statistical Learning Theory
KUROWICKA and COOKE · Uncertainty Analysis with High Dimensional Dependence Modelling
KVAM and VIDAKOVIC · Nonparametric Statistics with Applications to Science and Engineering
LACHIN · Biostatistical Methods: The Assessment of Relative Risks, *Second Edition*
LAD · Operational Subjective Statistical Methods: A Mathematical, Philosophical, and Historical Introduction
LAMPERTI · Probability: A Survey of the Mathematical Theory, *Second Edition*
LAWLESS · Statistical Models and Methods for Lifetime Data, *Second Edition*
LAWSON · Statistical Methods in Spatial Epidemiology, *Second Edition*
LE · Applied Categorical Data Analysis, *Second Edition*
LE · Applied Survival Analysis
LEE · Structural Equation Modeling: A Bayesian Approach
LEE and WANG · Statistical Methods for Survival Data Analysis, *Fourth Edition*
LePAGE and BILLARD · Exploring the Limits of Bootstrap
LESSLER and KALSBEEK · Nonsampling Errors in Surveys
LEYLAND and GOLDSTEIN (editors) · Multilevel Modelling of Health Statistics
LIAO · Statistical Group Comparison
LIN · Introductory Stochastic Analysis for Finance and Insurance
LINDLEY · Understanding Uncertainty, *Revised Edition*
LITTLE and RUBIN · Statistical Analysis with Missing Data, *Second Edition*
LLOYD · The Statistical Analysis of Categorical Data
LOWEN and TEICH · Fractal-Based Point Processes
MAGNUS and NEUDECKER · Matrix Differential Calculus with Applications in Statistics and Econometrics, *Revised Edition*

*Now available in a lower priced paperback edition in the Wiley Classics Library.
†Now available in a lower priced paperback edition in the Wiley–Interscience Paperback Series.

MALLER and ZHOU · Survival Analysis with Long Term Survivors
MARCHETTE · Random Graphs for Statistical Pattern Recognition
MARDIA and JUPP · Directional Statistics
MARKOVICH · Nonparametric Analysis of Univariate Heavy-Tailed Data: Research and Practice
MARONNA, MARTIN and YOHAI · Robust Statistics: Theory and Methods
MASON, GUNST, and HESS · Statistical Design and Analysis of Experiments with Applications to Engineering and Science, *Second Edition*
McCULLOCH, SEARLE, and NEUHAUS · Generalized, Linear, and Mixed Models, *Second Edition*
McFADDEN · Management of Data in Clinical Trials, *Second Edition*
* McLACHLAN · Discriminant Analysis and Statistical Pattern Recognition
McLACHLAN, DO, and AMBROISE · Analyzing Microarray Gene Expression Data
McLACHLAN and KRISHNAN · The EM Algorithm and Extensions, *Second Edition*
McLACHLAN and PEEL · Finite Mixture Models
McNEIL · Epidemiological Research Methods
MEEKER and ESCOBAR · Statistical Methods for Reliability Data
MEERSCHAERT and SCHEFFLER · Limit Distributions for Sums of Independent Random Vectors: Heavy Tails in Theory and Practice
MENGERSEN, ROBERT, and TITTERINGTON · Mixtures: Estimation and Applications
MICKEY, DUNN, and CLARK · Applied Statistics: Analysis of Variance and Regression, *Third Edition*
* MILLER · Survival Analysis, *Second Edition*
MONTGOMERY, JENNINGS, and KULAHCI · Introduction to Time Series Analysis and Forecasting
MONTGOMERY, PECK, and VINING · Introduction to Linear Regression Analysis, *Fifth Edition*
MORGENTHALER and TUKEY · Configural Polysampling: A Route to Practical Robustness
MUIRHEAD · Aspects of Multivariate Statistical Theory
MULLER and STOYAN · Comparison Methods for Stochastic Models and Risks
MURTHY, XIE, and JIANG · Weibull Models
MYERS, MONTGOMERY, and ANDERSON-COOK · Response Surface Methodology: Process and Product Optimization Using Designed Experiments, *Third Edition*
MYERS, MONTGOMERY, VINING, and ROBINSON · Generalized Linear Models. With Applications in Engineering and the Sciences, *Second Edition*
NATVIG · Multistate Systems Reliability Theory With Applications
† NELSON · Accelerated Testing, Statistical Models, Test Plans, and Data Analyses
† NELSON · Applied Life Data Analysis
NEWMAN · Biostatistical Methods in Epidemiology
NG, TAIN, and TANG · Dirichlet Theory: Theory, Methods and Applications
OKABE, BOOTS, SUGIHARA, and CHIU · Spatial Tesselations: Concepts and Applications of Voronoi Diagrams, *Second Edition*
OLIVER and SMITH · Influence Diagrams, Belief Nets and Decision Analysis
PALTA · Quantitative Methods in Population Health: Extensions of Ordinary Regressions
PANJER · Operational Risk: Modeling and Analytics
PANKRATZ · Forecasting with Dynamic Regression Models
PANKRATZ · Forecasting with Univariate Box-Jenkins Models: Concepts and Cases
PARDOUX · Markov Processes and Applications: Algorithms, Networks, Genome and Finance
PARMIGIANI and INOUE · Decision Theory: Principles and Approaches
* PARZEN · Modern Probability Theory and Its Applications
PEÑA, TIAO, and TSAY · A Course in Time Series Analysis

*Now available in a lower priced paperback edition in the Wiley Classics Library.
†Now available in a lower priced paperback edition in the Wiley–Interscience Paperback Series.

PESARIN and SALMASO · Permutation Tests for Complex Data: Applications and Software
PIANTADOSI · Clinical Trials: A Methodologic Perspective, *Second Edition*
POURAHMADI · Foundations of Time Series Analysis and Prediction Theory
POURAHMADI · High-Dimensional Covariance Estimation
POWELL · Approximate Dynamic Programming: Solving the Curses of Dimensionality, *Second Edition*
POWELL and RYZHOV · Optimal Learning
PRESS · Subjective and Objective Bayesian Statistics, *Second Edition*
PRESS and TANUR · The Subjectivity of Scientists and the Bayesian Approach
PURI, VILAPLANA, and WERTZ · New Perspectives in Theoretical and Applied Statistics
† PUTERMAN · Markov Decision Processes: Discrete Stochastic Dynamic Programming
QIU · Image Processing and Jump Regression Analysis
* RAO · Linear Statistical Inference and Its Applications, *Second Edition*
RAO · Statistical Inference for Fractional Diffusion Processes
RAUSAND and HØYLAND · System Reliability Theory: Models, Statistical Methods, and Applications, *Second Edition*
RAYNER, THAS, and BEST · Smooth Tests of Goodnes of Fit: Using R, *Second Edition*
RENCHER and SCHAALJE · Linear Models in Statistics, *Second Edition*
RENCHER and CHRISTENSEN · Methods of Multivariate Analysis, *Third Edition*
RENCHER · Multivariate Statistical Inference with Applications
RIGDON and BASU · Statistical Methods for the Reliability of Repairable Systems
* RIPLEY · Spatial Statistics
* RIPLEY · Stochastic Simulation
ROHATGI and SALEH · An Introduction to Probability and Statistics, *Second Edition*
ROLSKI, SCHMIDLI, SCHMIDT, and TEUGELS · Stochastic Processes for Insurance and Finance
ROSENBERGER and LACHIN · Randomization in Clinical Trials: Theory and Practice
ROSSI, ALLENBY, and McCULLOCH · Bayesian Statistics and Marketing
† ROUSSEEUW and LEROY · Robust Regression and Outlier Detection
ROYSTON and SAUERBREI · Multivariate Model Building: A Pragmatic Approach to Regression Analysis Based on Fractional Polynomials for Modeling Continuous Variables
* RUBIN · Multiple Imputation for Nonresponse in Surveys
RUBINSTEIN and KROESE · Simulation and the Monte Carlo Method, *Second Edition*
RUBINSTEIN and MELAMED · Modern Simulation and Modeling
RUBINSTEIN, RIDDER, and VAISMAN · Fast Sequential Monte Carlo Methods for Counting and Optimization
RYAN · Modern Engineering Statistics
RYAN · Modern Experimental Design
RYAN · Modern Regression Methods, *Second Edition*
RYAN · Sample Size Determination and Power
RYAN · Statistical Methods for Quality Improvement, *Third Edition*
SALEH · Theory of Preliminary Test and Stein-Type Estimation with Applications
SALTELLI, CHAN, and SCOTT (editors) · Sensitivity Analysis
SCHERER · Batch Effects and Noise in Microarray Experiments: Sources and Solutions
* SCHEFFE · The Analysis of Variance
SCHIMEK · Smoothing and Regression: Approaches, Computation, and Application
SCHOTT · Matrix Analysis for Statistics, *Second Edition*
SCHOUTENS · Levy Processes in Finance: Pricing Financial Derivatives
SCOTT · Multivariate Density Estimation: Theory, Practice, and Visualization
* SEARLE · Linear Models
† SEARLE · Linear Models for Unbalanced Data

---

*Now available in a lower priced paperback edition in the Wiley Classics Library.
†Now available in a lower priced paperback edition in the Wiley–Interscience Paperback Series.

† SEARLE · Matrix Algebra Useful for Statistics
† SEARLE, CASELLA, and McCULLOCH · Variance Components
SEARLE and WILLETT · Matrix Algebra for Applied Economics
SEBER · A Matrix Handbook For Statisticians
† SEBER · Multivariate Observations
SEBER and LEE · Linear Regression Analysis, *Second Edition*
† SEBER and WILD · Nonlinear Regression
SENNOTT · Stochastic Dynamic Programming and the Control of Queueing Systems
* SERFLING · Approximation Theorems of Mathematical Statistics
SHAFER and VOVK · Probability and Finance: It's Only a Game!
SHERMAN · Spatial Statistics and Spatio-Temporal Data: Covariance Functions and Directional Properties
SILVAPULLE and SEN · Constrained Statistical Inference: Inequality, Order, and Shape Restrictions
SINGPURWALLA · Reliability and Risk: A Bayesian Perspective
SMALL and McLEISH · Hilbert Space Methods in Probability and Statistical Inference
SRIVASTAVA · Methods of Multivariate Statistics
STAPLETON · Linear Statistical Models, *Second Edition*
STAPLETON · Models for Probability and Statistical Inference: Theory and Applications
STAUDTE and SHEATHER · Robust Estimation and Testing
STOYAN · Counterexamples in Probability, *Second Edition*
STOYAN, KENDALL, and MECKE · Stochastic Geometry and Its Applications, *Second Edition*
STOYAN and STOYAN · Fractals, Random Shapes and Point Fields: Methods of Geometrical Statistics
STREET and BURGESS · The Construction of Optimal Stated Choice Experiments: Theory and Methods
STYAN · The Collected Papers of T. W. Anderson: 1943–1985
SUTTON, ABRAMS, JONES, SHELDON, and SONG · Methods for Meta-Analysis in Medical Research
TAKEZAWA · Introduction to Nonparametric Regression
TAMHANE · Statistical Analysis of Designed Experiments: Theory and Applications
TANAKA · Time Series Analysis: Nonstationary and Noninvertible Distribution Theory
THOMPSON · Empirical Model Building: Data, Models, and Reality, *Second Edition*
THOMPSON · Sampling, *Third Edition*
THOMPSON · Simulation: A Modeler's Approach
THOMPSON and SEBER · Adaptive Sampling
THOMPSON, WILLIAMS, and FINDLAY · Models for Investors in Real World Markets
TIERNEY · LISP-STAT: An Object-Oriented Environment for Statistical Computing and Dynamic Graphics
TSAY · Analysis of Financial Time Series, *Third Edition*
TSAY · An Introduction to Analysis of Financial Data with R
TSAY · Multivariate Time Series Analysis: With R and Financial Applications
UPTON and FINGLETON · Spatial Data Analysis by Example, Volume II: Categorical and Directional Data
† VAN BELLE · Statistical Rules of Thumb, *Second Edition*
VAN BELLE, FISHER, HEAGERTY, and LUMLEY · Biostatistics: A Methodology for the Health Sciences, *Second Edition*
VESTRUP · The Theory of Measures and Integration
VIDAKOVIC · Statistical Modeling by Wavelets
VIERTL · Statistical Methods for Fuzzy Data
VINOD and REAGLE · Preparing for the Worst: Incorporating Downside Risk in Stock Market Investments

---

*Now available in a lower priced paperback edition in the Wiley Classics Library.
†Now available in a lower priced paperback edition in the Wiley–Interscience Paperback Series.

WALLER and GOTWAY · Applied Spatial Statistics for Public Health Data
WEISBERG · Applied Linear Regression, *Third Edition*
WEISBERG · Bias and Causation: Models and Judgment for Valid Comparisons
WELSH · Aspects of Statistical Inference
WESTFALL and YOUNG · Resampling-Based Multiple Testing: Examples and Methods for *p*-Value Adjustment
* WHITTAKER · Graphical Models in Applied Multivariate Statistics
WINKER · Optimization Heuristics in Economics: Applications of Threshold Accepting
WOODWORTH · Biostatistics: A Bayesian Introduction
WOOLSON and CLARKE · Statistical Methods for the Analysis of Biomedical Data, *Second Edition*
WU and HAMADA · Experiments: Planning, Analysis, and Parameter Design Optimization, *Second Edition*
WU and ZHANG · Nonparametric Regression Methods for Longitudinal Data Analysis
YIN · Clinical Trial Design: Bayesian and Frequentist Adaptive Methods
YOUNG, VALERO-MORA, and FRIENDLY · Visual Statistics: Seeing Data with Dynamic Interactive Graphics
ZACKS · Examples and Problems in Mathematical Statistics
ZACKS · Stage-Wise Adaptive Designs
* ZELLNER · An Introduction to Bayesian Inference in Econometrics
ZELTERMAN · Discrete Distributions—Applications in the Health Sciences
ZHOU, OBUCHOWSKI, and McCLISH · Statistical Methods in Diagnostic Medicine, *Second Edition*

*Now available in a lower priced paperback edition in the Wiley Classics Library.
†Now available in a lower priced paperback edition in the Wiley–Interscience Paperback Series.